Engineering Chemistry with Laboratory Experiments

M.S. Kaurav

Professor and Head
Department of Chemistry
Alwar Institute of Engineering and Technology
Alwar (Rajasthan)

Former Head (Department of Chemistry)
and
Dean–Administration
BRCM College of Engineering and Technology
Bahal, Bhiwani
Haryana

PHI Learning Private Limited

New Delhi-110001
2011

₹275.00

ENGINEERING CHEMISTRY WITH LABORATORY EXPERIMENTS
M.S. Kaurav

© 2011 by PHI Learning Private Limited, New Delhi. All rights reserved. No part of this book may be reproduced in any form, by mimeograph or any other means, without permission in writing from the publisher.

ISBN-978-81-203-4174-6

The export rights of this book are vested solely with the publisher.

Published by Asoke K. Ghosh, PHI Learning Private Limited, M-97, Connaught Circus, New Delhi-110001 and Printed by V.K. Batra at Pearl Offset Press Private Limited, New Delhi-110015.

*Dedicated
to*
"My Parents"

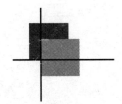

CONTENTS

Preface xv
Acknowledgements xvii

1. **THERMODYNAMICS** 1–42
   1.1 Introduction *1*
   1.2 Objectives and Limitations of Thermodynamics *1*
       1.2.1 Objectives of Thermodynamics *1*
       1.2.2 Limitations of Thermodynamics *1*
   1.3 The Basic Concepts *1*
       1.3.1 System *2*
       1.3.2 Surroundings *2*
       1.3.3 Types of Phase Systems *2*
       1.3.4 Types of Thermodynamic Systems *2*
       1.3.5 State Functions *2*
       1.3.6 State of the System *2*
       1.3.7 Properties of System *2*
   1.4 Thermodynamic Equilibrium *3*
       1.4.1 Conditions of the System *3*
       1.4.2 Thermodynamics Processes *3*
   1.5 Work and Heat *3*
       1.5.1 Work *3*
       1.5.2 Heat *5*
   1.6 Internal Energy *5*
   1.7 First Law of Thermodynamics *6*
       1.7.1 Mathematical Form of the First Law *6*
       1.7.2 Some Useful Conclusions Drawn from the First Law *6*
       1.7.3 Limitations of the First Law of Thermodynamics *7*
   1.8 Enthalpy or Heat Content *7*
       1.8.1 Enthalpy Change *7*

v

1.9 Second Law of Thermodynamics 7
   1.9.1 Conclusions Drawn from the Second Law of Thermodynamics 8
1.10 Spontaneous (Irreversible) and Non-spontaneous Processes 8
1.11 Entropy 8
   1.11.1 Carnot Cycle 8
   1.11.2 Entropy—Definition 10
1.12 Entropy Change for Ideal Gas 10
1.13 Entropy Change in Reversible and Irreversible Processes 11
   1.13.1 Entropy Change in Reversible Process 11
   1.13.2 Entropy Change in Irreversible Isothermal Process 12
   1.13.3 Entropy Change for Ideal Gas with Temperature and Pressure 13
   1.13.4 Entropy Change During Phase Transitions 14
1.14 Entropy of Mixture of Ideal Gases 15
1.15 Entropy of Mixing 16
1.16 Physical Significance of Entropy 16
   1.16.1 Entropy is the Measure of Randomness or Disorderness of the System 16
   1.16.2 Entropy is the Measure of Probability or Feasibility of the Process 17
   1.16.3 Entropy and Unavailable Energy 17
1.17 Criteria for Determining the Reversibility and Spontaneity 17
1.18 Free Energy and Work Function 18
   1.18.1 Free Energy Function is Superior Over Entropy Function 19
1.19 Variation of Free Energy with Temperature and Pressure 20
1.20 Partial Molar Properties 21
   1.20.1 Concept of Partial Molar Quantities (Properties) 22
   1.20.2 Partial Molar Free Energy (Chemical Potential) 23
   1.20.3 Physical Significance of Chemical Potential 24
   1.20.4 Variation of Chemical Potential with Temperature 24
   1.20.5 Variation of Chemical Potential with Pressure 26
1.21 Gibbs–Helmholtz Equation 26
   1.21.1 Integrated Form of Gibbs–Helmholtz Equation 27
   1.21.2 Applications of Gibbs–Helmholtz Equation 28
   1.21.3 Limitations of Gibbs–Helmholtz Equation 28
1.22 Clausius–Clapeyron Equation 29
   1.22.1 Integrated Form of Clausius–Clapeyron Equation 30
   1.22.2 Applications of Clausius–Clapeyron Equation 31
1.23 Third Law of Thermodynamics 31
1.24 Zeroth Law of Thermodynamics 31

Solved Numerical Problems 31
Summary 36
Exercises 38

2. PHASE RULE 43–81

2.1 Introduction 43
2.2 The Phase Rule 43
   2.2.1 Definition of Terms Used 44

      2.2.2   Derivation of Phase Rule   46
      2.2.3   Some Special Cases of Phase Rule   48
      2.2.4   Applications of Phase Rule   48
      2.2.5   Limitations of Phase Rule   49
2.3 Phase Diagram   49
2.4 One-component System   50
      2.4.1   Water System   50
      2.4.2   Carbon Dioxide System   53
2.5 Polymorphism   55
      2.5.1   Polymorphism of Water at High Pressure   56
2.6 Two-component Systems   57
2.7 Eutectic System   57
      2.7.1   Lead–Silver (Pb–Ag) System   57
2.8 Congruent System   60
      2.8.1   Zinc–Magnesium (Zn–Mg) System   60
2.9 Incongruent Systems   64
      2.9.1   Sodium–Potassium (Na–K) System   64
      2.9.2   Sodium Sulphate–Water System   67
2.10 Thermal Analysis–Cooling Curves   72

Solved Numerical Problems   73
Summary   77
Exercises   78

3. CATALYSIS   82–107
3.1 Introduction   82
3.2 Catalyst and Catalysis   82
      3.2.1   Role of the Catalyst   82
      3.2.2   Activity of Catalyst   83
      3.2.3   Catalyst and Equilibrium Constant   83
3.3 General Characteristics of Catalyst/Catalysis   84
3.4 Classification of Catalytic Reactions (Catalysis)   85
      3.4.1   Homogeneous Catalysis   85
      3.4.2   Heterogeneous Catalysis   86
3.5 Types of Catalyst   87
      3.5.1   Positive Catalyst   87
      3.5.2   Negative Catalyst   87
      3.5.3   Catalytic Promoter or Activator   87
      3.5.4   Catalytic Poisons   88
      3.5.5   Auto Catalyst   88
      3.5.6   Induced Catalyst   88
3.6 Theories of Catalysis   89
      3.6.1   Intermediate Compound Formation Theory   89
      3.6.2   Adsorption Theory   90
      3.6.3   Facts of Heterogeneous Catalysis on the Basis of Adsorption Theory   91
      3.6.4   Role of Heterogeneous Catalysts on the Basis of Adsorption Theory   92

viii ♦ Contents

- 3.7 Solid Catalyst 92
  - 3.7.1 Shape-selective Catalysts (Zeolites) 93
- 3.8 Acid–Base Catalysis 93
  - 3.8.1 Mechanism of Acid–Base Catalysis 94
  - 3.8.2 Kinetics of Acid–Base Catalyzed Reactions 94
- 3.9 Enzyme Catalysis 96
- 3.10 Kinetics of Enzyme-Catalyzed Reaction 97
  - 3.10.1 Michaelis–Menten Equation 98
- 3.11 Characteristics of Enzyme Catalyst 100
- 3.12 Effect of Temperature on Enzyme Catalysis 101
- 3.13 pH Dependence of Rate Constant of Catalyzed Reaction 102
  - 3.13.1 Acid-Catalyzed Reaction 102
  - 3.13.2 Base-Catalyzed Reaction 103

Summary 104
Exercises 106

4. WATER TREATMENT—PART I 108–132

- 4.1 Introduction 108
- 4.2 Sources of Water 108
  - 4.2.1 Surface Water 108
  - 4.2.2 Underground Water 108
- 4.3 Impurities in Water 109
- 4.4 Properties Imparted by Impurities Present in Water 109
  - 4.4.1 Colour 109
  - 4.4.2 Taste 109
  - 4.4.3 Odour 109
  - 4.4.4 Turbidity and Sediments 110
  - 4.4.5 Acidity 110
- 4.5 Hardness of Water 110
  - 4.5.1 Types of Hardness 110
  - 4.5.2 Alkaline and Non-alkaline Hardness 110
- 4.6 Calcium Carbonate Equivalent (Degree of Hardness) 111
  - 4.6.1 Degree of Hardness 111
  - 4.6.2 Units of Hardness 112
- 4.7 Determination of Hardness 112
  - 4.7.1 EDTA Titration Method 112
- 4.8 Disadvantages of Hard Water 114
  - 4.8.1 Domestic Use 114
  - 4.8.2 Industrial Use 115
- 4.9 Alkalinity of Water 115
  - 4.9.1 Types of Alkalinity 116
  - 4.9.2 Experimental Determination of Alkalinity 116
  - 4.9.3 Calculation of Types of Alkalinity 117
- 4.10 Sludge and Scale 118
  - 4.10.1 Sludge 118
  - 4.10.2 Scales 119
  - 4.10.3 Internal Treatment Methods 121

4.11 Boiler Corrosion *122*
    4.11.1 Causes of Boiler Corrosion *122*
    4.11.2 Control of Boiler Corrosion *122*
4.12 Caustic Embrittlement *122*
4.13 Priming and Foaming *123*
    4.13.1 Priming *123*
    4.13.2 Foaming *123*
4.14 Carry Over *124*
    4.14.1 Causes *124*
    4.14.2 Minimization of Carry Over *124*

Solved Numerical Problems *124*
Summary *129*
Exercises *130*

5. WATER TREATMENT—PART II   133–159

5.1 Introduction *133*
5.2 Characteristics of Potable Water *133*
5.3 Treatment of Water for Domestic Use *133*
    5.3.1 Removal of Floating and Suspended Impurities *133*
    5.3.2 Removal of Micro-organism (Disinfection or Sterilization of Water) *135*
5.4 Removel of Hardness (Softening) of Water *138*
    5.4.1 Lime-Soda Method *138*
    5.4.2 Zeolite or Permutit Method *141*
    5.4.3 Ion Exchange or Deionization or Demineralization Method *142*
    5.4.4 Mixed Bed Deionization *144*
5.5 Intrinsic or Polished Water *145*
5.6 Boiler Feed Water *145*
5.7 Desalination of Brackish Water *145*
    5.7.1 Electrodialysis *146*
    5.7.2 Reverse Osmosis *147*

Solved Numerical Problems *149*
Summary *155*
Exercises *156*
Numerical Problems *158*

6. CORROSION AND ITS PREVENTION   160–190

6.1 Introduction *160*
6.2 Causes of Corrosion *160*
6.3 Consequences of Corrosion *160*
6.4 Electrochemical Cell/Galvanic Cell *161*
    6.4.1 Chemical or Formation Cell *161*
    6.4.2 Electromotive Force (Emf) or Cell Potential of a Cell *162*
    6.4.3 Concentration Cell *163*
6.5 Electrochemical Series (Activity Series) *164*

- 6.6 Types of Corrosion 165
  - 6.6.1 Dry or Chemical Corrosion 165
  - 6.6.2 Electrochemical or Wet Corrosion 167
  - 6.6.3 Electrochemical Theory of Corrosion 167
- 6.7 Other Types of Electrochemical Corrosion 169
  - 6.7.1 Galvanic Corrosion 169
  - 6.7.2 Pitting Corrosion 170
  - 6.7.3 Stress Corrosion or Stress Cracking 171
  - 6.7.4 Concentration Cell Corrosion 172
  - 6.7.5 Soil Corrosion 174
  - 6.7.6 Microbiological Corrosion 175
- 6.8 Passivity or Passivation 175
- 6.9 Factors Affecting Rate of Corrosion 175
  - 6.9.1 Nature of Metal 176
  - 6.9.2 Nature of Environment 178
- 6.10 Corrosion Control–Protection from Corrosion 179
  - 6.10.1 Material Selection 179
  - 6.10.2 Proper Designing 179
  - 6.10.3 Barrier Protection 180
  - 6.10.4 Metallic Coatings 181
  - 6.10.5 Metal Cladding 182
  - 6.10.6 Galvanization (Sacrificial Protection) 182
  - 6.10.7 Tinning 182
  - 6.10.8 Cathodic Protection or Electrical Protection or Sacrificial Protection 183
  - 6.10.9 Anodic Protection 184
- 6.11 Inhibiters 185
- *Summary* 185
- *Exercises* 187

7. LUBRICATION AND LUBRICANTS   191–217

- 7.1 Introduction 191
- 7.2 Friction 191
- 7.3 Lubricants 191
  - 7.3.1 Functions of Lubricant 191
- 7.4 Mechanism of Lubrication 192
  - 7.4.1 Fluid Film or Thick Film or Hydrodynamic Lubrication 192
  - 7.4.2 Thin Film or Boundary Line Lubrication 192
  - 7.4.3 Extreme-Pressure Lubrication 193
- 7.5 Classification of Lubricants 194
  - 7.5.1 Liquid Lubricants or Lubricating Oils 194
  - 7.5.2 Semisolid Lubricants or Greases 195
  - 7.5.3 Solid Lubricants 197
- 7.6 Emulsions 198
  - 7.6.1 Types of Emulsions 198
- 7.7 Synthetic Lubricants 199
  - 7.7.1 Characteristics of Synthetic Lubricants 199

      7.7.2   Important Synthetic Lubricants   *199*
      7.7.3   Uses of Synthetic Lubricants   *200*
  7.8   Biodegradable Lubricants   *200*
      7.8.1   Conditions of Biodegradation   *200*
      7.8.2   Mechanism of Biodegradation of Lubricants   *201*
      7.8.3   Types of Biodegradable Lubricants   *201*
      7.8.4   Advantages of Biodegradable Lubricants   *202*
      7.8.5   Disadvantages of Biodegradable Lubricants   *202*
  7.9   Properties of Lubricating Oils   *202*
      7.9.1   Viscosity and Viscosity Index (VI)   *202*
      7.9.2   Flash and Fire Point   *205*
      7.9.3   Oiliness   *206*
      7.9.4   Cloud and Pour Points   *206*
      7.9.5   Aniline Point   *207*
      7.9.6   Saponification Value or Koettsdoerfer Number   *208*
      7.9.7   Acid Value or Neutralization Number   *208*
      7.9.8   Iodine Value or Iodine Number   *209*
      7.9.9   Specific Gravity   *209*
      7.9.10   Carbon Residue   *210*
      7.9.11   Volatility   *211*
      7.9.12   Mechanical Stability   *211*
  7.10   Properties of Greases   *212*
      7.10.1   Consistency or Yield Value   *212*
      7.10.2   Drop Point   *213*

Solved Numerical Problems   *214*
Summary   *214*
Exercises   *216*

8. POLYMERS AND POLYMERIZATION   218–251
  8.1   Introduction   *218*
  8.2   Polymers   *218*
  8.3   Degree of Polymerization   *218*
  8.4   Functionality   *218*
  8.5   Types of Polymers   *219*
      8.5.1   Classification on the Basis of Origin   *219*
      8.5.2   Classification on the Basis of Structure   *219*
      8.5.3   Classification on the Basis of Nature of Monomer Units   *219*
      8.5.4   Classification on the Basis of Mode of Synthesis   *220*
      8.5.5   Classification on the Basis of Molecular Forces   *220*
      8.5.6   Classification on the Basis of Mechanism   *221*
  8.6   Types of Polymerization   *221*
      8.6.1   Addition or Chain Polymerization   *222*
      8.6.2   Condensation Polymerization   *222*
      8.6.3   Coordination Polymerization   *222*
      8.6.4   Chain Growth Polymerization   *224*

- 8.7 Stereo-specific Polymerization 226
- 8.8 Effect of Structure on Properties of Polymers 226
  - 8.8.1 Strength or Toughness 226
  - 8.8.2 Physical State 228
  - 8.8.3 Crystallinity 229
  - 8.8.4 Plastic Deformation 229
  - 8.8.5 Elasticity 229
  - 8.8.6 Chemical Reactivity 229
  - 8.8.7 Electrical Properties 230
- 8.9 Some Commercially Important Polymers: Thermoplastics 230
  - 8.9.1 Polyvinyl Chloride (PVC) 230
  - 8.9.2 Polyvinyl Acetate (PVA) 231
  - 8.9.3 Polytetrafluoroethylene (PTFE) or Teflon or Fluon 232
- 8.10 Thermosetting Polymers (Thermosets) 232
  - 8.10.1 Phenol Formaldehyde (PF) Resin or Bakelite 232
  - 8.10.2 Urea Formaldehyde (UF) Resin 234
- 8.11 Elastomers 235
- 8.12 Vulcanization of Rubber 235
- 8.13 Synthetic Rubber 235
  - 8.13.1 Styrene Butadiene Rubber (SBR) or Buna-S 235
  - 8.13.2 Nitrile Rubber (NBR or GR-N or Buna-N) 236
- 8.14 Biopolymers 237
  - 8.14.1 Characteristics of Biopolymers 237
  - 8.14.2 Types of Biopolymers 237
  - 8.14.3 Benefits of Biopolymers 239
- 8.15 Biodegradable Polymers 239
  - 8.15.1 Classification of Biodegradable Polymers 240
  - 8.15.2 Conditions of Biodegradation 240
  - 8.15.3 Mechanism of Biodegradation of Polymers 240
  - 8.15.4 Applications and Limitations of Biodegradable Polymers 241
  - 8.15.5 Benefits of Biodegradable Polymers 242
- 8.16 Silicones or Polysiloxanes (Inorganic Polymers) 242
  - 8.16.1 Preparations of Silicones 242
  - 8.16.2 Properties of Silicones 243
  - 8.16.3 Uses of Silicones 243
  - 8.16.4 Types of Silicones 244
- 8.17 Determination of Molecular Mass of Polymer 246
  - 8.17.1 Number Average Molecular Mass ($\bar{M}_n$) 246
  - 8.17.2 Weight Average Molecular Mass $\bar{M}_w$ 246
  - 8.17.3 Viscosity Average Molecular Mass $\bar{M}_v$ 247
- 8.18 Polymer Composites 247
  - 8.18.1 Fibre Reinforced Plastics (FRP) 247
  - 8.18.2 Polymer Blends 248
  - 8.18.3 Polymer Alloys 248

Summary 248
Exercises 250

9. ANALYTICAL METHODS　　　　　　　　　　　　　　　　　　　　　　　　252–297

- 9.1 Introduction　252
- 9.2 Types of Chemical Analysis　252
  - 9.2.1 Gravimetric Methods　252
  - 9.2.2 Volumetric Methods　253
  - 9.2.3 Optical Methods　253
  - 9.2.4 Electrical Methods　253
  - 9.2.5 Separation Methods　253
- 9.3 Thermogravimetric Analysis (TGA)　253
  - 9.3.1 Types of TGA　254
- 9.4 Differential Thermal Analysis (DTA)　256
- 9.5 Differential Scanning Calorimetry (DSC)　258
- 9.6 Spectroscopy　260
  - 9.6.1 Properties of Electromagnetic Radiations (EMR)　260
  - 9.6.2 Origin of Spectra　262
  - 9.6.3 Selection Rule　263
  - 9.6.4 Spectral Lines　264
  - 9.6.5 Broadening of Spectral Lines　264
  - 9.6.6 Chromophores　265
  - 9.6.7 Types of Chromopores　265
  - 9.6.8 Auxochrome　265
  - 9.6.9 Types of Molecular Spectra　266
  - 9.6.10 Types of Spectroscopy on the Basis of Nature of Sample　267
  - 9.6.11 Types of Spectroscopy on the Basis of Absorption and Emission of Radiations　267
  - 9.6.12 Some Common Terms Used in Laws of Absorption　268
  - 9.6.13 Laws of Absorption: Beer–Lambert Law　268
- 9.7 Spectroscopy as an Analytical Tool　272
- 9.8 Types of Absorption Spectroscopy　272
  - 9.8.1 UV–Visible Spectroscopy　272
  - 9.8.2 Vibrational–Rotational Spectra–Infrared Spectroscopy　276
  - 9.8.3 Electronic Spectroscopy　279
  - 9.8.4 Spectrophotometry　281
  - 9.8.5 Flame Photometry (Flame Emission Spectroscopy)　283
- 9.9 Conductometeric Analysis　285
  - 9.9.1 Some Common Terms Used in Conductometric Analysis　285
  - 9.9.2 Conductometric Titrations　287
  - 9.9.3 Types of Conductometric Titration　288
  - 9.9.4 Advantages of Conductometric Titration　289

Solved Numerical Problems　289
Summary　292
Exercises　294

CHEMISTRY LABORATORY EXPERIMENTS (List of Experiments) 299–337

1. Determination of Ca^{2+}, Mg^{2+} Hardness of Water Using EDTA Solution *301*
2. Determination of Alkalinity of Water Sample *305*
3. Determination of Dissolved Oxygen (DO) in the Given Water Sample *308*
4. To Find the Melting and Eutectic Point for a Two-component System by Using a Method of Cooling Curve *311*
5. Determination of Viscosity of Lubricant by Redwood Viscometer (No. 1 and No. 2) *313*
6. To Determine Flash Point and Fire Point of an Oil by Pensky–Marten's Flash Point Apparatus *315*
7. To Prepare Phenol Formaldehyde and Urea Formaldehyde Resin *317*
8. To Find Out the Saponification Number of a Given Oil *320*
9. To Determine the TDS of Water Samples of Different Sources *322*
10. Determination of Concentration of $KMnO_4$ Solution Spectrophotometerically *323*
11. Determination of Strength of HCl Solution by Titrating it against NaOH Solution Conductometrically *327*
12. To Determine Amount of Sodium and Potassium in a Given Water Sample by Flamephotometer *331*
13. Estimation of Total Iron in an Iron Ore *334*
14. Estimation of Calcium in Limestone and Dolomite *336*

SAMPLE PAPERS *339–344*
LOG TABLES *345–348*
INDEX *349–354*

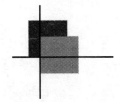

PREFACE

This book has been written exclusively for the students of various engineering courses (B.E./B.Tech. 1st year, MD University, Rohtak and Kurukshetra University, Kurukshetra, Haryana) taking their professional requirements into consideration, after a long practical experience with the students in classes and the problems raised and discussed.

Chemistry is a very interesting subject which touches almost all the aspects of our lives, our culture and our environment. It has changed our civilization to a great extent. The present-day chemistry has provided man with more comfort for a healthier and happier life. The synthetic fibres produced in chemical factories are most widely used. Modern chemistry has given man new plastics, fuels, metals, alloys, fertilizers, building materials, drugs, energy sources, etc.

In the field of engineering, many of the new products of the chemical industries are found to be applicable. The role of chemistry and chemical products in every branch of engineering expanded greatly with each successive year. The chemical composition of the substances, the strength of materials, their behaviour when subjected to different treatments and environments, and the law of heat and dynamic energy enter in almost every activity of modern life. The emergence of material science (chemistry) as a discipline is an acknowledgement of the need to interpret the behaviour of materials in terms of their structure and properties. The knowledge of chemical analysis and the study of properties and behaviour of different materials using various instrumental techniques are parts of engineering. Therefore, the engineers and technologists dealing with materials must have some training in chemistry, a certain minimum to prepare them for their profession. The study contents given in this book of engineering chemistry provide sound and solid foundation needed for the understanding of the principles of chemistry and for the applications of chemistry in engineering and technology. The laboratory experiments given at the end of the book will help the students in enhancing their practical knowledge and expertize.

M.S. Kaurav

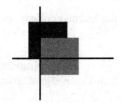

ACKNOWLEDGEMENTS

With the deepest gratitude I wish to thank the college management: Dr. V.K. Agarwal, Chairman, Dr. Manju Agarwal, Executive Director, Sh. Sanjay Prakash Garg, Director, IET group of institution, Alwar for their moral support in publishing the manuscript.

I feel greatly indebted to Sh. H.K. Chaudhary, Chairman, BRCM CET Bahal, Bhiwani and Prof. R.C.D. Kaushik, Director General TERII, Kurukshetra and Dr. B.K. Khan, Director GIT Sonepat for being main source of inspiration and motivation behind this book.

I express my gratitude to Dr. S. Ray Chaudhary and Dr. A.K. Srivastava (Former Principal, BRCM CET Bahal) and Dr. D.P. Gupta, Principal, BRCM CET, Bahal for their whole hearted support and cooperation for providing me the facilities for the preparation of manuscript.

I express my sincere thanks to my Ph.D. guide Prof. K. Dwivedi (Head, Department of Chemistry, Jiwaji University, Gwalior) for his continuous inspiration and guidance. I take this opportunity to express my gratitude to professors of various renowned institutions: Dr. K.C. Singh (Head, Department of Chemistry, M.D. University, Rohtak), Dr. Ishwar Singh (former Head, Department of Chemistry, M.D. University, Rohtak), Dr. M.R. Maurya (Department of Chemistry, IIT Roorkee), Dr. B.P. Singh (Department of Chemistry, DCRUST, Murthal), Dr. D.P. Singh (Department of Chemistry, NIT, Kurukshetra), Dr. Surendra Kaushik (Department of Chemistry, TITS, Bhiwani), Dr. Ashok Yadav (Director, DAV Engineering College, Kanina), Dr. R.B. Bajpai (Department of Chemistry, ITM, Gurgaon), Amit Vashishtha (Department of Chemistry, HCTM, Kaithal) who have helped me in enhancing my knowledge to prepare the book.

I am also very thankful to my colleagues Dr. Y. Tiwari (Principal—AIET, Alwar), Mr. Rajendra Gaur (Sr. Manager, LGC, Teddington, UK), Dr. H.S. Gupta (Sr. Manager, Cadila Pharma, Ahmedabad), Dr. (Mrs.) Deepa Gautam, Mr. Dinesh Sharma (BRCM CET, Bahal), Prof. (Dr. Sanjay Sharma, IET, Alwar), Dr. Pankaj Gupta (AIET—Alwar) for their moral support.

I am very thankful to my father, Sh. Ram Swaroop Singh Kaurav, mother, Smt. Kishori Devi Kaurav for their inspiration and moral support. I cannot forget my wife, Mrs. Asha Kaurav, daughter, Jyoti and son, Vinayak without whose cooperation it could have not been possible to complete this book.

I am also thankful to Mr. Pawan Ahuja, Mr. M.K. Arora and Mr. Rituraj for typing the manuscript, and also to the publisher, PHI Learning and its editorial and production staff for publishing this book.

Suggestions for the improvement are highly invited and appreciated from the learned colleagues and dear students.

M.S. Kaurav
Email: dr.mskaurav@gmail.com

CHAPTER 1

THERMODYNAMICS

1.1 INTRODUCTION

The term *thermodynamics* means the flow of heat (Thermo = heat, dynamics = flow). "Thermodynamics is the branch of science which deals with the study of transformation of energy" or "It is the branch of science which involves the study of interrelation of various kinds of energy accompanying physical or chemical changes".

1.2 OBJECTIVES AND LIMITATIONS OF THERMODYNAMICS

1.2.1 Objectives of Thermodynamics

1. To interrelate the different kinds of energy changes during physical or chemical transformation.
2. To predict the spontaneity (feasibility) of given changes.
3. To deduce various laws, e.g. distribution law, phase rule, law of mass action, etc. thermodynamically.
4. To arrange the experimental data on the basis of heat change.
5. To derive the condition at which the equilibrium is attained by a change.

1.2.2 Limitations of Thermodynamics

1. Laws of thermodynamics are valid only for bulk of matter and do not supply information about individual atom.
2. They fail to suggest the rate of reactions.
3. They fail to explain the systems away from equilibrium.

1.3 THE BASIC CONCEPTS

For the purpose of study in physical chemistry, the universe is divided into two parts, i.e. system and its surroundings.

1.3.1 System

A system is the specified part of the universe which is selected for the purpose of study of effect of certain variables such as temperature, pressure, etc. For example, chemical reactions taking place in a beaker.

1.3.2 Surroundings

Rest part of universe other than the system separated from the system by real or imaginary boundaries is called *surroundings*.

1.3.3 Types of Phase Systems

On the basis of phases of the components of the system it can be classified as:
  (a) *Homogeneous system:* The system which is uniform throughout and all the substances present in it have only one phase is called *homogeneous system*.
  (b) *Heterogeneous system:* The system which is not uniform throughout, i.e. it has more than one phases is called *heterogeneous system*.

1.3.4 Types of Thermodynamic Systems

On the basis of exchange of mass and energy the system can be classified as:
  (a) *Open system:* A system which can exchange matter as well as energy with surroundings is called *open system*. Example: tea contained in a cup is an open system.
  (b) *Closed system:* A system which can exchange only energy but not mass with surroundings is called *closed system*. Example: liquid contained in a closed vessel is a closed system.
  (c) *Isolated system:* A system which can neither exchange matter nor energy with surroundings is called *isolated system*. Example: tea contained in thermos flask is an isolated system.

1.3.5 State Functions

The measurable properties of a system which depend only upon the state of the system (i.e. initial and final state of the system) are known as *state functions*. For example—pressure, temperature, volume, composition, energy, etc. are state functions.

1.3.6 State of the System

When the fundamental properties such as pressure, temperature, volume, number of moles and composition have definite values, the system is said to be in a definite state. When there is any change in one of these properties the system is said to be under a change of state.

1.3.7 Properties of System

The thermodynamic properties of a system are divided into two categories:
  (a) *Extensive properties:* The properties which depend upon amount of the system are called *extensive properties*. Example: mass, volume, energy, heat, entropy, etc. are extensive properties.
  (b) *Intensive properties:* The properties which do not depend upon amount of the system are called *intensive properties*. Examples: temperature, pressure, viscosity, density, refractive index, surface tension, chemical potential, etc. are intensive properties.

1.4 THERMODYNAMIC EQUILIBRIUM

Thermodynamic equilibrium of any system is that condition in which the fundamental properties of the system do not undergo any change with time.

1.4.1 Conditions of the System

There are three different thermodynamic equilibrium conditions. These are:
  (a) *Thermal equilibrium:* In this equilibrium there is no flow of heat in any part of the system including surroundings. In such an equilibrium the temperature remains constant throughout the system and surroundings.
  (b) *Mechanical equilibrium:* In this equilibrium no mechanical work is done between different parts of the system or between the system and surroundings.
  (c) *Chemical equilibrium:* In this equilibrium the compositions of the system remains definite and fixed.

1.4.2 Thermodynamics Processes

The operations through which the system changes from one state to another is called *thermodynamic process*. The processes can be carried out in the following ways:
  (a) *Isothermal process:* The process in which the temperature of the system remains constant throughout the studies is called *isothermal process*. The process is carried out in thermostatic bath. During this process, difference in temperature $\Delta T = 0$ and also difference in internal energy $\Delta E = 0$.
  (b) *Adiabatic process:* A process during which there is no exchange of heat between the system and surroundings is called *adiabatic process*. This process is achieved by insulating the system boundaries. During this process, heat $q = 0$.
  (c) *Cyclic process:* A process in which initial state of system is regained after a series of operations is called *cyclic process*. During cyclic process, $\Delta E = 0$ and change in enthalpy $\Delta H = 0$.
  (d) *Isobaric process:* A process in which pressure of the system remains constant throughout the studies (i.e. $\Delta P = 0$) is called *isobaric process*.
  (e) *Isochoric process:* A process in which volume of the system remains constant throughout the studies (i.e. $\Delta V = 0$) is called *isochoric process*.
  (f) *Reversible process:* A process in which all changes occurring at any part of the process are exactly reversed, i.e. when it is carried out in opposite direction is called *reversible process*. The process may take place in either direction and maximum work is obtained in this process.
  (g) *Irreversible process:* A process where direction cannot be reversed by small change in variables is called *irreversible process*. The process is unidirectional.

1.5 WORK AND HEAT

1.5.1 Work

In thermodynamics, the fundamental physical property is work. Work is done when an object moves against an opposing force. Work is expressed as the product of two factors, i.e. intensity factor and capacity factor. Intensity factor is the measure of force against which work is done and capacity factor is the extent to which work is done.

Mechanical work = force (F) × displacement (d)
Gravitational work = gravitational force (mg) × height (h)
Electrical work = electrical potential (E) × charge (q)
Work of expansion = pressure (P) × Change in volume (ΔV)

A negative sign is given for the work done by the system and a positive sign is given for the work done on the system.

Unit of work: In CGS system, the unit of work is erg. Since the erg is very small unit so it is convenient to use the bigger unit, i.e. the joule (J) or kilojoule (kJ). 1 joule = 10^7 erg, 1 kJ = 1000 J.

(a) Work done in isothermal reversible expansion

Consider a system of an ideal gas enclosed in a cylinder fitted with a weightless and frictionless piston as shown in Figure 1.1. The cylinder is not insulated and external pressure P_{ext} is equal to pressure of the gas P_{gas} which is equal to P, i.e.

$$P_{ext} = P_{gas} = P \tag{1.1}$$

If the external pressure (P_{ext}) is decreased by an infinitesimal amount dP, i.e. it falls from P to $P - dP$, the gas will expand by an infinitesimal volume dV. Due to this expansion the pressure of the gas in the cylinder falls to $P_{gas} - dP$, i.e. it becomes equal to the external pressure and thus piston comes to rest. In each step the gas expands by volume dV.

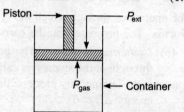

Figure 1.1 Reversible isothermal expansion.

Since the system is in thermal equilibrium the temperature remains constant. The work done by the system in each step of expansion can be given as

$$dw = -(P_{ext} - dP)\, dV$$

or
$$dw = -P_{ext} \cdot dV + dP \cdot dV \tag{1.2}$$

Since the product $dP \cdot dV$ is very small, hence it can be neglected.

∴
$$dw = -P \cdot dV \tag{1.3}$$

The total work done in isothermal reversible expansion of ideal gas from volume V_1 to V_2 can be given as

$$w = -nRT \int_{V_1}^{V_2} \frac{dV}{V} \tag{1.4}$$

∴ For an ideal gas,
$$P = \frac{nRT}{V}$$

∴
$$w = -nRT \ln \frac{V_2}{V_1} \tag{1.5}$$

or
$$w = -2.303\, nRT \log \frac{V_2}{V_1} \tag{1.6}$$

According to Boyle's law, $\quad P_1V_1 = P_2V_2 \quad$ or $\quad \dfrac{V_2}{V_1} = \dfrac{P_1}{P_2}$

Therefore, from Eq. (1.6), we get $\quad w = -2.303\, nRT \log \dfrac{P_1}{P_2}$ (1.7)

(b) Work done in isothermal irreversible expansion

There may be two types of irreversible isothermal expansion, i.e. free expansion and intermediate expansion.

In *free expansion*, the external pressure, P_{ext} is zero and hence work done will be zero when gas expands in vacuum. During *intermediate expansion*, when there is some external pressure on the gas which is less than the gas pressure, P_{gas}, the work done will be due to volume change from V_1 to V_2. Thus,

$$w = -P_{ext}\, dV \quad (1.8)$$

or

$$w = -P_{ext} \int_{V_1}^{V_2} dV$$

or

$$w = -P_{ext}\, \Delta V \quad (1.9)$$

1.5.2 Heat

It may be defined as the quantity of energy which flows between the system and surroundings due to the temperature difference. It always flows from a high temperature point to a low temperature point.

Mathematically, Heat $\quad q = m \times s \times t \quad$ (1.10)

where m is the mass of the substance, s is the specific heat and t is the temperature difference.

Units of Heat: The unit of heat is calorie (cal). A calorie is defined as the quantity of heat required to raise the temperature of one gram of water by 1°C. The heat and work are interrelated, the SI unit of heat is the joule (J), 1 J = 0.2390 cal, 1 cal = 4.184 J, 1 kcal = 4.184 kJ, 1 l atm = 101.3 J = 1.013×10^9 erg = 24.206 cal.

1.6 INTERNAL ENERGY

Any thermodynamic system which has some quantity of matter is associated with a definite amount of energy. This energy is known as *internal energy*. This is denoted by E.

"The internal energy may be the sum of different kinds of energies such as translational energy (E_t), rotational energy (E_r), vibrational energy (E_v), kinetic energy (E_k), potential energy (E_p), electronic energy (E_e), and bonding energy (E_b)."

Hence, $\quad E = E_t + E_r + E_v + E_p + E_e \quad$ (1.11)

Since, the internal energy includes various types of energies, therefore, its accurate measurement for a given substance in a given state is not possible. The internal energy of a particular system is a definite quantity irrespective of the manner by which it is obtained. Hence it is a state function like temperature, pressure, volume, etc. Thus, "the total of all possible kinds of energies of a system is called its *internal energy*."

Internal energy is an extensive property as its value depends on the mass of the mater contained in a system. In the thermodynamics it is concerned only with the energy changes when a system undergoes change from one state to another (e.g. from state A to state B):

$$\underset{E_i}{A} \longrightarrow \underset{E_f}{B}$$

where E_i and E_f are internal energies of the system in initial and final state, respectively.

∴ Change in internal energy of the system,

$$\Delta E = E_f - E_i \tag{1.12}$$

if $\qquad E_f > E_i,\ \Delta E = +\text{ve}\quad \text{and}\quad E_i > E_f,\ \Delta E = -\text{ve}$

Unit of internal energy: SI unit of internal energy is joule (J) and other unit is calorie.

1.7 FIRST LAW OF THERMODYNAMICS

This law states that "energy can neither be created nor destroyed, although it can be transformed from one form to another". This is also known as *law of conservation of energy*. Other statements of first law are:

"The total energy of the universe is constant."

or "Whenever a quantity of one kind of energy disappears, an exactly equivalent quantity of energy in some other form must appear."

or "It is impossible to construct a perpetual motion machine which could produce work without consuming energy."

It is now known that energy can be produced by the destruction of mass according to the Einstein's mass energy equation $E = mc^2$ (where m = mass of the substance and c = velocity of light). Hence, the modified law states that "the total mass and energy of an isolated system remains unchanged".

1.7.1 Mathematical Form of the First Law

Consider a system whose internal energy is E_1. Now if q amount of heat is supplied to the system then its internal energy becomes $E_1 + q$. If work w is done on the system then its internal energy further increases and becomes E_2. Therefore,

$$E_2 = E_1 + q + w \tag{1.13}$$

or $\qquad E_2 - E_1 = q + w \qquad (\because\ E_2 - E_1 = \Delta E) \tag{1.14}$

or $\qquad \boxed{\Delta E = q + w} \tag{1.15}$

If the work is done by the system, the negative sign will be used and the above equation can be written as

$$\boxed{\Delta E = q - w} \tag{1.16}$$

1.7.2 Some Useful Conclusions Drawn from the First Law

1. When $\Delta E = 0$, i.e. there is no change in internal energy of the system then,

$$q + w = 0 \quad \text{or} \quad q = -w \quad \text{or} \quad w = -q.$$

2. If no work is done, i.e. $w = 0$ then $\Delta E = q$, i.e. increase in internal energy of the system is equal to heat absorbed by the system and vice versa.

3. If there is no heat exchange between the system and surroundings, i.e. $q = 0$ then $\Delta E = w$ (in adiabatic process).
4. If a gas expands against the constant external pressure P_{ext}, the mechanical work done by the gas is equal to $P\Delta V$. Hence, $\Delta E = q + w$ or $\Delta E = q - P\Delta V$.
5. If a gas expands against constant volume, i.e. $\Delta V = 0$ and therefore, $\Delta E = q$ or q_v (heat change at constant volume).

1.7.3 Limitations of the First Law of Thermodynamics
1. This does not explain: Why chemical reactions proceed to completion?
2. This law does not tell about source of heat and direction of flow of heat.
3. This law is not able to explain: Why natural (spontaneous) processes are unidirectional?
4. This law does not tell about: How much quantity of energy can be converted into another form and up to which extent?
5. It does not tell about the attainment of thermodynamic equilibrium.
6. It does not tell about the spontaneity of the process.

1.8 ENTHALPY OR HEAT CONTENT

Generally, the chemical reactions are carried out at constant pressure (atmospheric pressure). At constant volume the heat content of the system are equal to the internal energy of system. To express the energy changes at constant pressure a new term called *enthalpy* is used.

Enthalpy may be defined as "the sum of internal energy and product of pressure and volume of the system". It is the total heat content of the system at constant pressure and denoted by H.

Mathematically, $$H = E + PV \quad (1.17)$$

Enthalpy is a state function like E, P and V, i.e. its value depends on initial and final state of the system and independent of the path.

1.8.1 Enthalpy Change
For a system $A \rightarrow B$, The change in enthalpy for this system may be given as:
$$H_B - H_A = \Delta H = \Delta E + P\Delta V \quad (1.18)$$

From the first law of thermodymies, $\Delta E = q + w$ ($\because w = -P\Delta V$) $\therefore \Delta E = q - P\Delta V$ (1.19)

From Eq. (1.18)
$$\Delta H = q - P\Delta V + P\Delta V$$
$$\Delta H = q_p \text{ (at constant pressure)} \quad (1.20)$$

Hence, the enthalpy change is a measure of heat exchange (evolved or absorbed) taking place during a process at constant pressure. Thus, if q amount of heat is absorbed by the system at constant pressure, then $q = \Delta H$ and at constant volume $q = \Delta E$.

For an isothermal process, ΔT and ΔE are zero, respectively, hence $\Delta H = 0$.

1.9 SECOND LAW OF THERMODYNAMICS

The second law of thermodynamics has been stated in the following various forms:

Clausius statement: "It is not possible for a self-acting machine to transfer heat by itself from a colder body to hotter body without the intervention of external energy."

Kelvin statement: "It is impossible to use a cyclic process to extract heat from a reservoir and to convert it into work without transferring at the same time a certain amount of heat from a hotter to colder part of the body," i.e. it is impossible to construct perpetual motion machine.

Carnot statement: "It is impossible to convert heat into work without compensation."

Generalized concept: "All the natural processes occur spontaneously, i.e. irreversibly it follows that the available energy is decreasing and entropy of universe is increasing continuously."

1.9.1 Conclusions Drawn from the Second Law of Thermodynamics

1. From this law we can determine the direction of flow of heat or energy.
2. This law has no mathematical derivation and is based on experiences about engines and refrigerators.
3. All the natural processes are spontaneous (irreversible).
4. Increase in randomness favours a spontaneous change.
5. The entropy of the universe tends towards a maximum.

1.10 SPONTANEOUS (IRREVERSIBLE) AND NON-SPONTANEOUS PROCESSES

Spontaneous processes are those processes which proceed by themselves and bring the system closer to the equilibrium. These are also called *positive* or *natural processes*.

All natural processes are spontaneous and follow a non-equilibrium path and are thermodynamically irreversible in nature. Although the natural process proceeds to attain the state of equilibrium but stop at equilibrium in the direction so as to bring the state of equilibrium without any external force. The system tries to acquire the equilibrium state when it comes in lower energy state by the dispersal of energy.

Examples:
1. Heat flows spontaneously from hotter to colder part of the body. This flow is continued till the temperature of both bodies equalize and we say that system has attained the equilibrium state.
2. Water falls downwards spontaneously and the direction cannot be reversed without applying external force.
3. Electricity flows spontaneously from a point of higher potential to a point of lower potential.
4. Air flows from a point of higher pressure to a point of lower pressure.

1.11 ENTROPY

Like internal energy and enthalpy the entropy is a state function, i.e. it is a definite quantity and depends only on initial and final state of the system. The entropy does not depend on the path or the manner by which the change is made. The value of entropy depends on the mass of the system hence it is an extensive property.

1.11.1 Carnot Cycle

The concept of entropy can be explained by taking the cycle of reversible isothermal changes (Figure 1.2). This cycle is known as Carnot cycle. Consider a cycle working between temperature T_1 and T_2 ($T_2 > T_1$). Between these two temperatures the efficiency of heat engine can be given as:

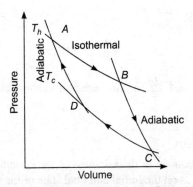

Figure 1.2 The basic structure of a Carnot cycle.

Efficiency of heat engine,
$$\eta = \frac{q_2 - q_1}{q_2} = \frac{T_2 - T_1}{T_2} \tag{1.21}$$

where
 T_2 = temperature of host source with corresponding heat quantity q_2
 T_1 = temperature of host source with corresponding heat quantity q_1

or
$$1 - \frac{q_1}{q_2} = 1 - \frac{T_1}{T_2} \tag{1.22}$$

or
$$\frac{q_1}{q_2} = \frac{T_1}{T_2}$$

or
$$\frac{q_1}{T_1} = \frac{q_2}{T_2} \tag{1.23}$$

or
$$\frac{q_2}{T_2} - \frac{q_1}{T_1} = 0$$

or
$$\frac{q_2}{T_2} + \frac{-q_1}{T_1} = 0 \tag{1.24}$$

The ratio of q/T in Eq. (1.24) is called the *reduced heat*. It shows that the algebraic sum of reduced heats for a reversible Carnot cycle is zero. On the other hand, the algebraic sum of reduced heats for a irreversible Carnot cycle is less than zero.

For an infinitesimally small reversible Carnot cycle, Eq. (1.24) can be written as

$$\frac{dq_2}{T_2} + \frac{-dq_1}{T_1} = 0 \tag{1.25}$$

where dq is the elementary reduced heat.

The reversible cycle is made of number of infinitesimally small Carnot cycles. Hence by integrating Eq. (1.25),

$$\int \frac{dq_2}{T_2} + \int \frac{-dq_1}{T_1} = 0 \tag{1.26}$$

The Eq. (1.26) can be written as

$$\oint \frac{dq}{T} = 0 \qquad (1.27)$$

where $\oint$ is the cyclic integral and $\frac{dq}{T}$ is the variation in the state function of the system and is called *entropy*.

1.11.2 Entropy—Definition

"Entropy is the measure of randomness or disorderness of the system". The absolute value of entropy cannot be determined as it depends on the initial and final state of the system. To determine the change in entropy the change between initial and final state is considered. Thus, change in entropy

$$\Delta S = S_f - S_i \qquad (1.28)$$

Thermodynamic definition of entropy: The thermodynamic definition of entropy is based on the change in entropy (dS) that occurs during physical or chemical changes, which is expressed as:

$$dS = \frac{dq_{\text{rev}}}{T} \qquad (1.29)$$

For a measurable change between the initial and final state, this expression integrated to

$$\Delta S = S_f - S_i = \int_{S_1}^{S_2} \frac{dq_{\text{rev}}}{T} \qquad (1.30)$$

or

$$\boxed{\Delta S = \frac{q_{\text{rev}}}{T}} \qquad (1.31)$$

This equation is applicable only for reversible processes, but not for irreversible processes because for irreversible processes the heat exchanged is indefinite and uncertain quantity. Thus, entropy change of a system during a change of state may be defined as "the summation of all the terms each involving heat exchange (q) reversibly during each infinitesimally small change in the process divided by the absolute temperature (T)".

Units of entropy: From Eq. (1.29) the entropy change,

$$\Delta S = \frac{\text{Heat absorbed}}{\text{Absolute temperature}}$$

i.e. calories per degree or cal deg^{-1}. In SI system, its unit is joules per degree kelvin or J K^{-1}

The molar entropy is expressed in joules per kelvin per mole (J K^{-1} mol^{-1}).

1.12 ENTROPY CHANGE FOR IDEAL GAS

From the definition of entropy, we know that $\Delta S = \frac{q_{\text{rev}}}{T}$ \qquad (1.32)

The heat absorbed during reversible isothermal expansion of an ideal gas can be calculated from the first law of thermodynamics,

for an isothermal process,
$$\Delta E = q_{rev} + w$$
$$\Delta E = 0$$
Therefore,
$$q_{rev} = -w \quad (1.33)$$

The work done in the expansion of an ideal gas from a volume of V_1 to V_2 at constant temperature T is given as:
$$w = -nRT \ln \frac{V_2}{V_1} \quad (1.34)$$
or
$$-w = nRT \ln \frac{V_2}{V_1} \quad (1.35)$$

From Eq. (1.33) and (1.34),
$$q_{rev} = -w = nRT \ln \frac{V_2}{V_1} \quad (1.36)$$

Hence,
$$\Delta S = \frac{nRT}{T} \ln \frac{V_2}{V_1} \quad (1.37)$$

or
$$\Delta S = nR \ln \frac{V_2}{V_1} \quad (1.38)$$

or
$$\boxed{\Delta S = 2.303 \, n \, R \log \frac{V_2}{V_1}} \quad (1.39)$$

1.13 ENTROPY CHANGE IN REVERSIBLE AND IRREVERSIBLE PROCESSES

The change in entropy of the system as well as surroundings during the changes in the system is called entropy change of the process.

1.13.1 Entropy Change in Reversible Process

(a) *Reversible isothermal process*: Consider an isothermal reversible expansion of an ideal gas at temperature T. During this process the system absorbs q amount of heat from the surroundings at temperature T and the entropy of the system (ΔS_{sys}) increases which is expressed as,

$$\Delta S_{sys} = +\frac{q}{T} \quad (1.40)$$

At the same time, surroundings loose the same quantity of heat q at same temperature. This results in the decrease of entropy of the surroundings (ΔS_{surr})

i.e.
$$\Delta S_{surr} = -\frac{q_{rev}}{T} \quad (1.41)$$

Hence the total change in entropy of the process

$$\Delta S = \frac{q_{rev}}{T} + \left(-\frac{q_{rev}}{T}\right) = 0 \quad \text{(i.e. } \Delta S \text{ process} = 0\text{)} \quad (1.42)$$

Thus, in reversible isothermal process the total entropy change is zero while the system and surroundings taken together.

(b) *Reversible adiabatic process:* During such process no heat is exchanged between the system and surroundings, i.e. $q = 0$,

Hence,
$$\Delta S_{total} = +\frac{q}{T} = \frac{0}{T} = 0 \tag{1.43}$$

$$\boxed{\Delta S_{sys} + \Delta S_{surr} = 0} \tag{1.44}$$

Thus, entropy change in adiabatic process is zero.

1.13.2 Entropy Change in Irreversible Isothermal Process

Suppose an ideal gas expands isothermally into vacuum at constant temperature T. This expansion of gas into vacuum is spontaneous and irreversible process. In the absence of opposing force the work done by the system will be zero and for isothermal process change in internal energy,

$$\Delta E = 0 \quad \text{and} \quad w = 0.$$

Therefore, from the law, $\Delta E = q + w$ or $0 = q + 0 = 0$ or $q = 0$, i.e. no heat is supplied to or removed from the surroundings. Hence entropy of the surroundings remains unchanged.

As we know the entropy of the system is the state function, i.e. it depends upon the initial and final state of the system consider a process in which gas expands from a volume V_1 to a volume V_2 at constant temperature. The entropy change for this expansion is given by

$$\Delta S_{sys} = R \ln\left(\frac{V_2}{V_1}\right) \tag{1.45}$$

Thus, total entropy of the system and surroundings is:

$$\Delta S_{total} = R \ln\left(\frac{V_2}{V_1}\right) > 0 \tag{1.46}$$

Alternatively: Suppose the system is at higher temperature (T_1), then its surroundings temperature (T_2) and the quantity of heat q passes irreversible from the system to surroundings. In such condition the entropy of the system decreases while the entropy of the surroundings increases. Thus,

$$\Delta S_{sys} = -\frac{q}{T_1} \tag{1.47}$$

and

$$\Delta S_{surr} = +\frac{q}{T_2} \tag{1.48}$$

Total entropy change of the process

$$\Delta S_{process} = -\frac{q}{T_1} + \frac{q}{T_2} = q\left(\frac{T_2 - T_1}{T_1 \cdot T_2}\right) \quad (\because T_1 > T_2) \tag{1.49}$$

$$T_1 - T_2 = +\text{ve}$$

Hence, $\Delta S_{process} = +\text{ve}$, i.e. entropy increases in irreversible process or $[\Delta S_{process} > 0]$

From Eqs. (1.45) and (1.46) we can conclude that:
1. All the natural processes in universe are irreversible (spontaneous) and the entropy of the universe is increasing, i.e. $\Delta S_{univ} > 0$.
2. We can predict whether a given process can take place spontaneously or not.

1.13.3 Entropy Change for Ideal Gas with Temperature and Pressure

As the entropy is a state function, hence its value for a pure gaseous substance depends upon any two of the three variables, i.e. temperature, pressure and volume. Generally, T is taken as one variable and other may be either V or P.

(a) *Entropy change (ΔS) with T and V variables:* For infinitesimal heat exchanged (dq) by the system (gas) with the surrounding at temperature T, the entropy change may be given as:

$$dS = \frac{dq_{rev}}{T} \quad (1.50)$$

From the first law of thermodynamics

$$dE = dq_{rev} + dw \quad \text{or} \quad dq_{rev} = dE - dw \quad (1.51)$$

or
$$dq_{rev} = dE + PdV \qquad [\because -dw = PdV] \quad (1.52)$$

From an ideal gas equation for n mole

$$PV = nRT \quad \text{or} \quad P = \frac{nRT}{V} \quad (1.53)$$

And also $\left(\dfrac{dE}{dT}\right)_V = C_V$ (heat capacity at constant volume)

or
$$dE = nC_V \cdot dT \quad (1.54)$$

$\therefore$ From Eqs. (1.52), (1.53) and (1.54), we get

$$dq_{rev} = nC_V\, dT + nRT\, \frac{dV}{V} \quad (1.55)$$

Now, from Eqs. (1.50) and (1.55), we get

$$dS = nC_V\, \frac{dT}{T} + nR\, \frac{dV}{V} \quad (1.56)$$

On integrating Eq. (1.56) with temperature limit $T_1 \to T_2$ and volume $V_1 \to V_2$

$$\Delta S = nC_V \int_{T_1}^{T_2} \frac{dT}{T} + nR \int_{V_1}^{V_2} \frac{dV}{V}$$

or
$$\Delta S = nC_V \ln \frac{T_2}{T_1} + nR \ln \frac{V_2}{V_1} \quad (1.57)$$

or
$$\boxed{\Delta S = 2.303\, n \left(C_V \log \frac{T_2}{T_1} + R \log \frac{V_2}{V_1} \right)} \quad (1.58)$$

i.e., entropy change depends upon initial and final temperatures as well as initial and final volumes.

(b) *Entropy change (ΔS) with T and P variables:* From the ideal gas equation

$$P_1V_1 = RT_1 \quad \text{and} \quad P_2V_2 = RT_2$$

or

$$\frac{P_1V_1}{P_2V_2} = \frac{RT_1}{RT_2} \quad \text{or} \quad \frac{V_2}{V_1} = \frac{P_1T_2}{P_2T_1} \tag{1.59}$$

From Eqs. (1.57) and (1.59)

$$\Delta S = n\left[C_V \ln \frac{T_2}{T_1} + R \ln \frac{P_1T_2}{P_2T_1}\right] \tag{1.60}$$

or

$$\Delta S = n\left[C_V \ln \frac{T_2}{T_1} + R \ln \frac{T_2}{T_1} + R \ln \frac{P_1}{P_2}\right] \tag{1.61}$$

$$\Delta S = n\left[(C_V + R) \ln \frac{T_2}{T_1} + R \ln \frac{P_1}{P_2}\right] \tag{1.62}$$

or

$$\Delta S = n\left[C_P \ln \frac{T_2}{T_1} - R \ln \frac{P_2}{P_1}\right] \tag{1.63}$$

$$[\because C_P - C_V = R \quad \text{or} \quad C_P = C_V + R]$$

$\therefore$

$$\Delta S = 2.303\, n\left[C_P \log \frac{T_2}{T_1} - R \log \frac{P_2}{P_1}\right] \tag{1.64}$$

Thus, the entropy change of the system depends upon initial and final temperatures as well as initial and final pressures.

1.13.4 Entropy Change During Phase Transitions

The change of matter from one state to another is called *phase transition*. These changes accompanied by the absorption or evolution of heat. The entropy change during phase transition can be given as

$$\Delta S = \frac{q_{rev}}{T} = \frac{\Delta H_{rev}}{T} \tag{1.65}$$

$(\because q_{rev} = \Delta H_{rev}$ at constant pressure$)$

Different types of entropies of phase transitions are discussed as follows:

(a) *Entropy of fusion:* It is defined as the entropy change when one mole of the solid substance changes into liquid form at its melting point, e.g. melting of ice

$$\text{Water (s)} \rightleftharpoons \text{Water (l),}$$

Hence, entropy of fusion, $S_{water} - S_{ice} = \Delta S_{fus} = \dfrac{\Delta H_{fus}}{T_{fus}}$ \hfill (1.66)

where ΔH_{fus} is the entropy of fusion and T_{fus} is the fusion temperature.

(b) *Entropy of vaporization:* It is the entropy change when one mole of liquid changes into vapours at its boiling point, e.g. vaporization of water

$$\text{Water (l)} \rightleftharpoons \text{Water (g)}$$

Hence, entropy of vaporization, $\Delta S_{vap} = \dfrac{\Delta H_{vap}}{T_b}$ \hfill (1.67)

where ΔH_{vap} is enthalpy of vaporization and T_b is the boiling temperature.

(c) *Entropy of sublimation:* It is the entropy change when one mole of a solid substance changes into vapours at is sublimation temperature

$$\text{Water (s)} \rightleftharpoons \text{Water (g)}$$

Hence, entropy of sublimation

$$\Delta S_{sub} = \dfrac{\Delta H_{sub}}{T_{sub}} \quad (1.68)$$

where ΔH_{sub} is enthalpy of sublimation, T_{sub} is the sublimation temperature.

(d) *Entropy of allotropic transformation:* It is the entropy change during transformation of one mole of substance from one state to another at its transitions temperature

$$S_r \rightleftharpoons S_m$$

Hence, entropy of transition

$$\Delta S_{trans} = \dfrac{\Delta H_{trans}}{T_{trans}} \quad (1.69)$$

where ΔH_{trans} is enthalpy of transformation and T_{trans} transformation temperature.

1.14 ENTROPY OF MIXTURE OF IDEAL GASES

We know that entropy change

$$dS = C_V \dfrac{dT}{T} + R \dfrac{dV}{V} \quad (1.70)$$

Integrating this equation without limit, we get

$$S = C_V \ln T + R \ln V + S_0 \quad (1.71)$$

where S_0 is the integration constant $\left(\because V = \dfrac{RT}{P} \right)$

$$S = C_V \ln T + R \ln \dfrac{RT}{P} + S_0 \quad (1.72)$$

$$S = C_V \ln T + R \ln T - R \ln P + R \ln R + S_0 \quad (1.73)$$

$$S = C_V \ln T + R \ln T - R \ln P + R \ln R + S_0 \quad (1.74)$$

$$S = C_V \ln T + R \ln T - R \ln P + S'_0 \quad (1.75)$$

$$\because \quad R \ln R + S_0 = S'_0 \quad (1.76)$$

$$S = (C_V + R) \ln T - R \ln P + S'_0 \quad (1.77)$$

where S'_0 is another constant.

Entropy of a mixture of gases can be given as the sum of entropies of individual constituents at existing pressure or composition of the mixture. For a mixture, consists of various gases with number of moles n_1, n_2, n_3, n_4, etc., and their partial pressures p_1, p_2, p_3, p_4, etc respectively, the entropy may be given as:

$$S = n_1(C_p \ln T - R \ln p_1 + S_0') + n_2(C_p \ln T - R \ln p_2 + S_0') + n_3(C_p \ln T - R \ln p_3 + S_0') \quad (1.78)$$

$$S = \sum n (C_p \ln T - R \ln p + S_0') \quad (1.79)$$

The partial pressure of an ideal gas, $p = xP$ (where x is the mole fraction of particular gas in the mixture and P is the total pressure of the mixture)

∴ From Eq. (1.79), we get $\quad S = \sum n (C_p \ln T - R \ln xP + S_0') \quad (1.80)$

$$\boxed{S = \sum n (C_p \ln T - R \ln x - R \ln P + S_0')} \quad (1.81)$$

This equation represents the change in entropy of a mixture of ideal gases.

1.15 ENTROPY OF MIXING

It is defined as the difference between the entropies of the mixture of ideal gases and sum of entropies of the individual gases in the mixture at a pressure P.

Hence, $\quad \Delta S_{mix} = \sum n (C_p \ln T - R \ln x - R \ln P + S_0') - \sum n(C_p \ln T - R \ln xP + S_0') \quad (1.82)$

$$\Delta S_{mix} = -R \sum n \ln x \quad (1.83)$$

$$\Delta S_{mix} = -R(n_1 \ln x_1 + n_2 \ln x_2 + n_3 \ln x_3 + \cdots) \quad (1.84)$$

$$\Delta S_{mix} = -R \sum n_i \ln x_i \quad (1.85)$$

(n_i = number of moles, x_i = mole fraction of each constituent, respectively).

Total number of moles $\quad n = n_1 + n_2 + n_3 + \cdots$

or $\quad 1 = \dfrac{n_1}{n} + \dfrac{n_2}{n} + \dfrac{n_3}{n} + \cdots \quad$ or $\quad 1 = x_1 + x_2 + x_3 + \cdots \quad$ or $\quad 1 = \sum x_i \quad (1.86)$

Hence, entropy of mixing for one mole of gaseous mixture may be given as

$$\Delta S_{mix} = -R \sum x_i \ln x_i \quad (1.87)$$

As the fraction x_i for a mixture is less than 1, hence the entropy of mixing is always positive.

1.16 PHYSICAL SIGNIFICANCE OF ENTROPY

1.16.1 Entropy is the Measure of Randomness or Disorderness of the System

As we know all the natural processes are spontaneous (irreversible). For example, flow of water from a high point to a low point, flow of heat from a hot end to a cold end, expansion of gas in vacuum, falling of an apple from tree, etc. All these processes are accompanied by increase in the disorderness (randomness) of the system. Consider the example of flow of heat from hotter to colder part of the conductor. At the colder end of the conductor there is certain amount of orderness. But as soon as the heat starts flowing from hotter end to colder end, due to energy distraction at the colder end there is movement of electrons and which results in the increase of disorderness (randomness).

From the equation of entropy change, $\Delta S = q_{rev}/T$ it shows that entropy increases on increasing the heat contents q as the heat flows from hotter to colder end of the conductor. Thus, we can say entropy is the measure of randomness of the system.

1.16.2 Entropy is the Measure of Probability or Feasibility of the Process

A spontaneous (irreversible) process tends to increase in entropy and hence disorderness, i.e. it tends to proceed from less probable state to more probable state. As the entropy and thermodynamic probability both increases simultaneously in a process hence the state of equilibrium is the state of maximum probability.

Boltzmann proposed the following mathematical relation between the entropy and thermodynamic probability P

$$S = k \ln P + C \qquad (1.88)$$

where k is Boltzmann constant and C is another constant. This equation is called *Boltzmann equation*. Planck suggested the constant $C = 0$.

Hence,
$$S = k \ln P \qquad (1.89)$$

This equation is called *Boltzmann–Planck equation*.

1.16.3 Entropy and Unavailable Energy

When heat is supplied to the system, some portion of this heat is utilized in doing work and called available or workable energy and rest part of heat remains unutilized. This unutilized portion is called unavailable energy. From the equation of efficiency of heat engine, derived from Carnot cycle working between two temperatures

$$\eta = \frac{q_2 - q_1}{q_2} = \frac{T_2 - T_1}{T_2} \qquad (1.90)$$

$w = q_2 - q_1 = $ Energy available for doing useful work

$q_1 = $ unavailable portion of heat (unavailable energy)

or from Eq. (1.90)
$$q_1 = \left(\frac{q_2}{T_2}\right) T_1 \qquad (1.91)$$

but q_2/T_2 is the increase in entropy during isothermal process at temperature T_1. Therefore, from Eq. (1.91) the unavailable energy (q_1) is determined by q_2/T_2. Hence, we can say that entropy is the measure of unavailable energy of the system. The entropy of the system may be given as:

$$\text{Entropy} = \frac{\text{Unavailable Energy}}{\text{Temperature (°C)}}.$$

1.17 CRITERIA FOR DETERMINING THE REVERSIBILITY AND SPONTANEITY

Spontaneity of the process can be determined with help of following thermodynamic conditions:
1. Spontaneous changes are irreversible, i.e. unidirectional and take place by its own and during such changes no work is done. While reversible changes takes place in both direction and some work is done in reversible changes.

2. Spontaneous changes are real changes and can take place rapidly or very slowly.
3. Entropy change may determine the thermodynamic conditions of the system.

From the second law of thermodynamics,
We know that entropy change,

$$dS = \frac{dq}{T} \tag{1.92}$$

and from the first law of thermodynamics,

$$dq = dE + PdV \tag{1.93}$$

$$\therefore \quad Tds = dE + PdV \tag{1.94}$$

If E is constant, $dE = 0$,

and V is constant, then, $dV = 0$, $\therefore$ $(dS) \geq 0$ (1.95)

from Eq. (1.95) we can conclude that the criteria for determining spontaneity of the process are summarized in Table 1.1.

Table 1.1 Criteria for determining spontaneity of the Process

S. No.	Spontaneity	Equilibrium	Reversibility
1	$(\partial S)_{E,V} > 0$	S = Max	$(\partial S)_{E,V} = 0$
2	$(\partial E)_{E,V} < 0$	E = Min	$(\partial E)_{E,V} = 0$
3	$(\partial H)_{E,V} > 0$	H = Min	$(\partial H)_{E,V} = 0$
4	$(\partial A)_{E,V} > 0$	A = Min	$(\partial A)_{E,V} = 0$
5	$(\partial G)_{E,V} > 0$	G = Min	$(\partial G)_{E,V} = 0$

1.18 FREE ENERGY AND WORK FUNCTION

In thermodynamics, the quantities E, H and S are dependent only on the state of the system and many of the processes cannot be explained with the help of these functions. It is convenient to introduce further two more functions to explain various physical and chemical processes taking place at constant temperature and pressure with the same characteristic properties, i.e.

1. Free energy function (Gibbs free energy, free enthalpy or Gibbs potential)

$$G = H - TS \tag{1.96}$$

2. Work function (Helmholtz free energy or Helmholtz function)

$$A = E - TS \tag{1.97}$$

Like E, H and S, the functions A and G are also extensive properties.

For an isothermal change, this Eq. (1.96) may be written in the form

$$\Delta G = \Delta H - T\Delta S \tag{1.98}$$

$$\therefore \quad \Delta H = \Delta E + P\Delta V \quad \text{and} \quad \Delta E = q - w$$

$$\therefore \quad \Delta G = q - w + P\Delta V - T\Delta S \tag{1.99}$$

Entropy change,

$$\Delta S = q/T \quad \text{or} \quad q = T\Delta S$$

Hence
$$\Delta G = T\Delta S - w + P\Delta V - T\Delta S$$

or
$$\Delta G = -w + P\Delta V$$

or
$$-\Delta G = w - P\Delta V \quad (\because w - P\Delta V = w_{net})$$

$\therefore$
$$\boxed{-\Delta G = w_{net}} \tag{1.100}$$

Thus, the decrease in free energy in a given isothermal process at constant pressure is equal to the net work, i.e. the work other than that due to volume change.

For the change in work function, from Eq. (1.97), we get

$$\Delta A = \Delta E - T\Delta S \tag{1.101}$$

and
$$\Delta E = q - w \quad \text{and} \quad q = T\Delta S$$

$\therefore$
$$\Delta A = T\Delta S - w - T\Delta S \tag{1.102}$$

or
$$\Delta A = -w$$

or
$$\boxed{-\Delta A = w_{max}} \tag{1.103}$$

Thus, the decrease in work function is the measure of the maximum (reversible) work which can be made available in the given isothermal process. It should be noted here that work, w will include all forms of work, i.e. mechanical work against the external pressure, electrical work, etc. The differences between the free energy function and the work function are shown in Table 1.2.

Table 1.2 Differences between free energy function and work function

Free energy function	Work function
It is the amount of energy available to the system which can be converted into useful work at the time of process.	It is the amount of energy available to the system which can be converted into maximum useful work.
It does not includes pressure-volume work.	It includes all types of work, e.g. mechanical, electrical work, etc.
It is considered at constant pressure.	It is considered at constant volume.
Change in free energy of the system (ΔG_{sys}) at constant T and P accounts automatically for the entropy change of the system and surroundings.	Change in work function of the system (ΔA_{sys}) at constant T and V accounts automatically for the entropy change of the system and surroundings.
It is less than work function by amount $P\Delta V$.	It is greater than free energy function.

1.18.1 Free Energy Function is Superior Over Entropy Function

We know that for spontaneous processes the entropy change, $\Delta S_{process}$ (or ΔS_{univ}) is greater than zero, i.e. $\Delta S_{process} = (\Delta S_{sys} + \Delta S_{surr}) > 0$.

In the spontaneous process, the total entropy of the universe increases. It is simple to determine the ΔS_{sys} but in most of the cases it is inconvenient to evaluate the change in surroundings (ΔS_{surr}). This difficulty may be overcome by introducing the free energy of the system alone. The change in free energy function can be evaluated with the help of standard values. These free energy functions are:

(i) Change in Helmholtz free energy of system, i.e. ΔA_{sys} accounts automatically for ΔS_{sys} and ΔS_{surr}
(ii) Change in Gibbs free energy of system, i.e. ΔG_{sys} accounts automatically for ΔS_{sys} and ΔS_{surr}

Hence, the use of ΔG_{sys} is preferred to ΔS_{univ} to decide the nature of physical or chemical processes.

1.19 VARIATION OF FREE ENERGY WITH TEMPERATURE AND PRESSURE

We know that free energy,
$$G = H - TS \tag{1.104}$$

Since $H = E + PV$ $\therefore$ $G = E + PV - TS$ (1.105)

Upon differentiation of Eq. (1.104)
$$dG = dE + PdV + VdP - TdS - SdT \tag{1.106}$$

From the first law of thermodynamics,
$$dE = dq + dw \tag{1.107}$$

Work done by the system (gas) on expansion,
$$dw = -pdV \tag{1.108}$$

And from the definition of entropy,
$$dq = TdS \tag{1.109}$$

$\therefore$ From Eqs. (1.107), (1.108) and (1.109), we get
$$dE = TdS - PdV \tag{1.110}$$

On substituting the value of dE from Eq. (1.110) in Eq. (1.106), we get
$$dG = TdS - PdV + PdV + VdP - TdS - SdT$$

or
$$\boxed{dG = VdP - SdT} \tag{1.111}$$

The Eq. (1.111) shows the variation of free energy with temperature and pressure.
At constant pressure ($dP = 0$) and therefore, $VdP = 0$
Hence, from Eq. (1.111), we get
$$dG = -SdT \tag{1.112}$$

or
$$\left(\frac{\partial G}{\partial T}\right)_P = -S \tag{1.113}$$

At constant temperature, $dT = 0$, $\therefore$ $SdT = 0$
Therefore, from Eq. (1.111)
$$dG = VdP \tag{1.114}$$

or
$$\left(\frac{\partial G}{\partial P}\right)_T = V \tag{1.115}$$

On integrating Eq. (1.115) with the limit $G_1 \to G_2$ and $P_1 \to P_2$

$$\int_{G_1}^{G_2} dG = V \int_{P_1}^{P_2} dP \qquad \left(\because PV = RT \text{ or } V = \frac{RT}{P}\right)$$

$\therefore$

$$\int_{G_1}^{G_2} dG = RT \int_{P_1}^{P_2} \frac{dP}{P}$$

or

$$G_2 - G_1 = RT \ln \frac{T_2}{T_1}$$

or

$$\Delta G = 2.303 \, RT \log \frac{T_2}{T_1} \qquad (1.116)$$

At constant temperature, $P_1 V_1 = P_2 V_2$

or

$$\frac{P_2}{P_1} = \frac{V_1}{V_2}$$

$\therefore$

$$\boxed{\Delta G = 2.303 \, RT \log \left(\frac{V_1}{V_2}\right)} \qquad (1.117)$$

In case of liquid or solid the volume remains constant

$$\int_{G_1}^{G_2} dG = V \int_{P_1}^{P_2} dP$$

or $\quad G_2 - G_1 = V(P_2 - P_1)$

or $\quad \Delta G = V\Delta P \qquad (1.118)$

1.20 PARTIAL MOLAR PROPERTIES

The thermodynamic properties like E, H, $\underline{S}$, A, G, V heat capacity, etc. are extensive properties. In a closed system, these properties depend usually on three parameters, i.e. P, V and T. These parameters are inter-related by an equation of state. Hence, any two of these parameters can be taken as independent variables.

Thus in a closed system, the thermodynamic function can be expressed as the function of any two independent variables. If X denotes any thermodynamic property of the system then we can write

$$K = f(P, T); \, X = f(V, T); \text{ etc.} \qquad (1.119)$$

But in an open system, the thermodynamic property (an extensive property) is a function not only of two independent variables, but also of the number of moles of the various components present in the system as well.

So in an open system if P and T are chosen as two independent variables, any extensive property (say X), may be represented as

$$X = f(P, T, n_1, n_2, n_3, \ldots, n_i) \tag{1.120}$$

where $n_1, n_2, n_3, \ldots, n_i$ represent the different quantities of various components 1, 2, 3, ..., i in the system, usually expressed in gram moles.

1.20.1 Concept of Partial Molar Quantities (Properties)

Let us now consider small changes in all variables of the system including the amounts of components. The net change in thermodynamic property, X will be given by partial differentiation of Eq. (1.120) as under

$$dX = \left(\frac{\partial X}{\partial P}\right)_{T, n_1, n_2, \ldots} dP + \left(\frac{\partial X}{\partial T}\right)_{P, n_1, n_2, \ldots} dT + \left(\frac{\partial X}{\partial n_1}\right)_{P, T, n_2, n_3, \ldots} dn_1$$
$$+ \left(\frac{\partial X}{\partial n_2}\right)_{P, T, n_1, n_3, \ldots} dn_2 + \cdots + \left(\frac{\partial X}{\partial n_i}\right)_{P, T, n_1, n_2, \ldots} dn_i \tag{1.121}$$

the term $\left(\dfrac{\delta X}{\delta n_i}\right)_{P, T, n_1, n_2, \ldots}$ dn_i is called *partial molar property* or *quantity* for the components.

This is usually denoted by putting a bar over the symbol of thermodynamic property. Hence

$$\left(\frac{\delta X}{\delta n_1}\right)_{P, T, n_2, n_3 \ldots} = \bar{X}_1 \tag{1.122}$$

$$\left(\frac{\delta X}{\delta n_2}\right)_{P, T, n_1, n_3 \ldots} = \bar{X}_2 \tag{1.123}$$

$$\left(\frac{\delta X}{\delta n_i}\right)_{P, T, n_1, n_2, n_3, \ldots} = \bar{X}_i \tag{1.124}$$

Thus, $\bar{X}_1, \bar{X}_2, \bar{X}_3, \ldots \bar{X}_i$ denote partial molar properties for the components 1, 2, 3, ..., i, respectively.

Hence, the partial molar property signifies the rate of change of the given thermodynamic extensive property of the system on account of addition of one mole of the given component, when moles of all other components, temperature and pressure remain constant.

The Eq. (1.121) can now be written as

$$dX = \left(\frac{\delta X}{\delta P}\right)_{T, n_1, n_2} dp + \left(\frac{\delta X}{\delta T}\right)_{P, n_1, n_2} dT + \bar{X}_1 dn_1 + \bar{X}_1 dn_2 + \cdots + \bar{X}_i dn_i \tag{1.125}$$

If the system undergoes a change only in mass of its components, at constant temperature ($dT = 0$) and constant pressure ($dP = 0$), then Eq. (1.125) becomes

$$(dX)_{P,T} = \bar{X}_1 dn_1 + \bar{X}_2 dn_2 + \bar{X}_3 dn_3 + \cdots + \bar{X}_i dn_i \tag{1.126}$$

It may be remembered that X stands for any extensive thermodynamic property. Hence on the basis of Eq. (1.126) at constant P and T, the change in entropy may be expressed as follows:

$$(dS)_{P,T} = \bar{S}_1 dn_1 + \bar{S}_2 dn_2 + \bar{S}_3 dn_3 + \cdots + \bar{S}_i dn_i \tag{1.127}$$

$\bar{S}_1$ = Partial molar entropy of component 1
$\bar{S}_2$ = Partial molar entropy of component 2
$\bar{S}_i$ = Partial molar entropy of component i

Partial molar internal energy
$$\bar{E}_i = \left(\frac{\delta E}{\delta n_i}\right)_{P,T,n_1,n_2,\ldots} \tag{1.128}$$

Partial molar enthalpy
$$\bar{H}_i = \left(\frac{\delta H}{\delta n_i}\right)_{P,T,n_1,n_2,\ldots} \tag{1.129}$$

Partial molar Gibb's potential
$$\bar{G}_i = \left(\frac{\delta G}{\delta n_i}\right)_{P,T,n_1,n_2,\ldots} \tag{1.130}$$

Partial molar work function
$$\bar{A}_i = \left(\frac{\delta A}{\delta n_i}\right)_{P,T,n_1,n_2,\ldots} \tag{1.131}$$

Partial molar volume
$$\bar{V}_i = \left(\frac{\delta V}{\delta n_i}\right)_{P,T,n_1,n_2,\ldots} \tag{1.132}$$

1.20.2 Partial Molar Free Energy (Chemical Potential)

The partial molar Gibbs free energy (or partial molar Gibbs potential) is the most important partial molar quantity in physical chemistry.

It may be defined as "the change in free energy of the system that results on adding one mole of that particular substance at a constant temperature and pressure to such a large quantity of the system that there is no considerable change in overall composition of the system".

It is represented by μ. Hence

$$\left(\frac{\partial G}{\partial n_i}\right)_{P,T,n_1,n_2,\ldots} = \mu_i = \bar{G}_i \tag{1.133}$$

In an open system, the Gibbs potential can be expressed as

$$G = f(P, T, n_1, n_2, \ldots, n_i) \tag{1.134}$$

By partial differentiation

$$dG = \left(\frac{\partial G}{\partial P}\right)_{T,n_1,n_2,\ldots} dP + \left(\frac{\partial G}{\partial T}\right)_{P,n_1,n_2,\ldots} dT + \left(\frac{\partial G}{\partial n_1}\right)_{P,T,n_2,n_3,\ldots} dn_1 + \left(\frac{\partial G}{\partial n_2}\right)_{P,T,n_1,n_3,\ldots} dn_2 + \cdots + \left(\frac{\partial G}{\partial n_i}\right)_{P,T,n_1,n_2,\ldots} dn_i \tag{1.135}$$

If the temperature and pressure remain constant (i.e. $dT = 0$ and $dP = 0$) the above equation becomes

$$(dG)_{P,T} = \mu_1 dn_1 + \mu_2 dn_2 + \cdots + \mu_i dn_i \quad (1.136)$$

If the system has a definite composition having n_1, n_2, n_3, ..., n_i moles of the constituents 1, 2, 3, ..., i, respectively then on integrating Eq. (1.136), we get

$$(G)_{P,T,N} = \mu_1 n_1 + \mu_2 n_2 + \cdots + \mu_i n_i \quad (1.137)$$

From Eq. (1.137) the chemical potential may be defined as the contribution per mole of each particular constituent of the mixture to the total free energy of the system at constant temperature and pressure.

For one mole of the pure substance

$$G = \mu \quad (1.138)$$

i.e. the free energy becomes equal to chemical potential.

1.20.3 Physical Significance of Chemical Potential

1. If a small amount (dn_i moles) of the components i is added to a system (keeping P, T and moles of other components, constant), the increase in free energy per mole of ith component added is given by:

$$(dG)_{P,T,n_1,n_2,\ldots} = \mu_i dn_i \quad (1.139)$$

where

$$\mu_i = \left(\frac{\partial G}{\partial n_i}\right)_{P,T,n_1,n_2,\ldots}$$

From Eq. 1.139, it follows that chemical potential of component i is the increase in the Gibbs free energy of the system, which occurs due to the addition of 1 mole of ith component at constant temperature and pressure when other components remain unaltered.

2. The chemical potential is independent of the masses, but is not independent of the composition of the system. Hence chemical potential is an intensive property of the system and must have the same value everywhere within the system in equilibrium.
3. Matter flows spontaneously from a region of high chemical potential to a region of low chemical potential.
4. Another name of chemical potential is escaping tendency, i.e.. μ_i stands for escaping tendency of component i. Thus, if μ_i is high, then component i has large escaping tendency and if μ_i in a system is low, then component i has low escaping tendency.
5. For a pure substance the chemical potential is equal to molar free energy, i.e.

$$\mu = \frac{G}{n}$$

where n = number of moles.

1.20.4 Variation of Chemical Potential with Temperature

We know that

$$\left(\frac{\partial G}{\partial n_i}\right)_{P,T,n_1,n_2,\ldots} = \mu_i = \overline{G}_i \quad (1.140)$$

Differentiating Eq. (1.140) with respect to temperature at constant pressure and composition of other components

$$\left(\frac{\partial \mu_i}{\partial T}\right)_{P,n_1,n_2,...} = \left(\frac{\partial^2 G}{\partial T \partial n_i}\right)_{P,T,n_1,n_2,...} \quad (1.141)$$

From the equation of variation of free energy with temperature and pressure, we know that

$$dG = VdP - SdT$$

Hence, at constant pressure and composition of system

$$dG = -SdT$$

or

$$\left(\frac{\partial G}{\partial T}\right)_{P,n_1,n_2,...} = -\overline{S}_i \quad (1.142)$$

Differentiating this equation w.r.t. number of moles n_i, maintaining other variables constant

$$\left(\frac{\partial^2 G}{\partial n_i \partial T}\right) = \left(\frac{\partial S}{\partial n_i}\right) = -\overline{S}_i \quad (1.143)$$

Combining Eqs. (1.142) and (1.143) we have

$$\left(\frac{\delta^2 \mu_i}{\delta T}\right)_{P,n_1,n_2,...} = \left(\frac{\delta^2 G}{\delta T \delta n_i}\right) = -\overline{S}_i \quad (1.144)$$

This equation gives the variation of the chemical potential of a constituent in a mixture with respect to temperature at constant pressure and composition of the system. S_i is the partial molar entropy of ith component of the mixture. It is clear from the Figure 1.3 that at the melting point (T_m), the chemical potential of solid and liquid phases are equal. Similarly, at the boiling point (T_b) the chemical potential of liquid and gaseous phase are equal.

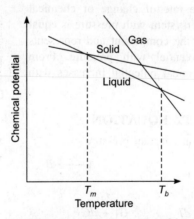

Figure 1.3 Variation of chemical potential with temperature.

From Eq. (1.44) it follows that the chemical potential will decreases with increase in temperature, as the entropy is always a positive quantity. However, the rate of change of chemical potential with temperature will be different for solids, liquids and gases, because

$$S_{(solid)} < S_{(liquid)} < S_{(gas)}$$

1.20.5 Variation of Chemical Potential with Pressure

We know that chemical potential, $\mu_i = \left(\dfrac{\partial G}{\partial n_i}\right)_{P,T,n_1,n_2,...}$

Differentiating this equation with respect to pressure at constant temperature and composition

$$\left(\dfrac{\partial \mu_i}{\partial P n_i}\right)_{T,n_1,n_2,...} = \left(\dfrac{\partial^2 G}{\partial P \partial n_i}\right)_{T,n_1,n_2,...} \quad (1.145)$$

We know that variation of free energy with pressure at constant temperature ($dT = 0$)

$$dG = VdP$$

or

$$\left(\dfrac{\partial G}{\partial P}\right)_T = V \quad (1.146)$$

Differentiating this equation w.r.t. n_i at constant temperature and composition.

$$\left(\dfrac{\partial^2 G}{\partial P \delta n_i}\right)_{T,n_1,n_2,...} = \left(\dfrac{\partial V}{\partial n_i}\right)_{T,n_1,n_2,...} \quad (1.147)$$

Combining Eqs. (1.146) and (1.148), we get

$$\left(\dfrac{\partial \mu_i}{\partial P}\right)_{T,n_1,n_2,...} = \left(\dfrac{\partial V}{\partial n_i}\right)_{T,n_1,n_2,...} \quad (1.148)$$

or

$$\left(\dfrac{\partial \mu_i}{\partial P}\right)_{T,n_1,n_2,...} = \bar{V}_i \quad (1.149)$$

Equation (1.49) states that the rate of change of chemical potential of a component of the system with pressure is equal to the partial molar volume of the component and represents the variation of chemical potential with pressure. From Figure 1.4, it is clear that chemical potential increases with increase in pressure.

1.21 GIBBS–HELMHOLTZ EQUATION

Consider the following system at constant pressure

$$A \longrightarrow B$$

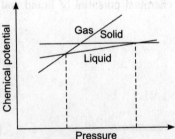

Figure 1.4 Variation of chemical potential with pressure.

The free energy
at temperature	T	G_1	G_2
at temperature	$T + dT$	$G_1 + dG_1$	$G_2 + dG_2$

Now from the equation of variation of free energy with temperature at constant pressure

$$dG = -SdT \quad (1.150)$$

For initial state (A) the free energy change

$$dG_1 = -S_1 dT \quad (1.151)$$

For final state (B) the free energy change

$$dG_2 = -S_2 dT \tag{1.152}$$

where S_1 and S_2 are entropies of the system in initial and final state respectively.

Subtracting Eq. (1.151) from Eq. (1.152)

$$(dG_2 - dG_1) = -(S_2 dT - S_1 dT) \tag{1.153}$$

or

$$d(G_2 - G_1) = -dT(S_2 - S_1)$$

or

$$d(\Delta G) = -dT(\Delta S)$$

or

$$\left(\frac{\partial(\Delta G)}{\partial T}\right)_P = -\Delta S \tag{1.154}$$

We know that

$$\Delta G = \Delta H - T\Delta S$$

or

$$-\Delta S = \frac{\Delta G - \Delta H}{T} \tag{1.155}$$

$$\therefore \left(\frac{\partial(\Delta G)}{\partial T}\right)_P = \frac{\Delta G - \Delta H}{T}$$

or

$$\boxed{\Delta G = \Delta H + T\left(\frac{\partial(\Delta G)}{\partial T}\right)_P} \tag{1.156}$$

Equation (1.156) is known as Gibbs–Helmholtz equation.

At constant volume, Eq. (1.156) may be written as

$$\boxed{\Delta A = \Delta E + T\left(\frac{\partial(\Delta A)}{\partial T}\right)_V} \tag{1.157}$$

where, ΔA = change in free energy at constant volume, ΔE = change in internal energy.

1.21.1 Integrated Form of Gibbs–Helmholtz Equation

Free energy change,

$$\Delta G = \Delta H + T\left(\frac{\partial(\Delta G)}{\partial T}\right)_P$$

Dividing this equation by T^2, both sides

$$\frac{\Delta G}{T^2} = \frac{\Delta H}{T^2} + \frac{1}{T}\left(\frac{\partial(\Delta G)}{\partial T}\right)_P \tag{1.158}$$

Rearranging Eq. (1.158), we get

$$\frac{\Delta G}{T^2} - \frac{1}{T}\left(\frac{\partial(\Delta G)}{\partial T}\right)_P = -\frac{\Delta H}{T^2}$$

or

$$\left(\frac{\partial\left(\Delta G \times \frac{1}{T}\right)}{\partial T}\right)_P = -\frac{\Delta H}{T^2} \tag{1.159}$$

On integrating Eq. (1.159) with limit for temperature $T_1 \to T_2$, we get

$$\frac{\Delta G_2}{T_2} - \frac{\Delta G_1}{T_1} = -\int_{T_1}^{T_2} \frac{\Delta H}{T^2}$$

or
$$\frac{\Delta G_2}{T_2} - \frac{\Delta G_1}{T_1} = -\left[-\frac{\Delta H}{T}\right]_{T_1}^{T_2} \quad (1.160)$$

(The ΔG_1 is the change in free energy at temperature T_1 and ΔG_2 is the change in free energy at temperature T_2.)

or
$$\boxed{\frac{\Delta G_2}{T_2} - \frac{\Delta G_1}{T_1} = \frac{\Delta H_2}{T_2} - \frac{\Delta H_1}{T_1}} \quad (1.161)$$

at constant volume, we can write the Eq. (1.161) as follows:

$$\frac{\Delta A_2}{T_2} - \frac{\Delta A_1}{T_1} = -\int_{T_1}^{T_2} \frac{\Delta E}{T^2}$$

$$\boxed{\frac{\Delta A_2}{T_2} - \frac{\Delta A_1}{T_1} = \frac{\Delta E_2}{T_2} - \frac{\Delta E_1}{T_1}} \quad (1.162)$$

Equations (1.161) and (1.162) are known as integrated forms of Gibbs–Helmholtz equation.

1.21.2 Applications of Gibbs–Helmholtz equation
This equation is used:
1. To calculate ΔH or ΔE, when ΔG or ΔA are given at two temperatures.
2. To calculate the emf of the cell.

   Since $\quad\quad\quad \Delta G = nFE$

   $\therefore \quad\quad\quad E = -\frac{\Delta H}{nF} + T\left(\frac{\partial(\Delta E)}{\partial T}\right)_P \quad (1.163)$

3. In deriving various important equations, e.g. Clausius–Clapeyron equation, van't Hoff isotherm, etc.
4. To calculate entropy change (ΔS) of isothermal process.

1.21.3 Limitations of Gibbs–Helmholtz Equation

1. If $\frac{\partial(\Delta E)}{\partial T} = 0$, Eq. (1.163) becomes $\Delta G = \Delta H$.
2. If $\Delta G \approx 0$, then $\Delta H = -T = \left(\frac{\partial(\Delta G)}{\Delta T}\right)_P$
3. If $T = 0$, then $\Delta G = \Delta H$
4. This equation can be used to calculate ΔH from ΔG but it does not allow the calculation of ΔG from ΔH.

1.22 CLAUSIUS–CLAPEYRON EQUATION

This equation was derived by Clausius and Clapeyron scientists, independently and simultaneously on the basis of second law of thermodynamics for two phase equilibria in one-component systems and is known as *Clausius–Clapeyron equation*.

Following four types of two phase equilibria are possible in one-component systems
1. Solid $\rightleftharpoons$ Liquid (at the melting point of solid)
2. Liquid $\rightleftharpoons$ Vapour (at the boiling point of liquid)
3. Solid $\rightleftharpoons$ Vapour (at the sublimation temperature of solid)
4. One crystalline form $\rightleftharpoons$ another crystalline form at transition temperature e.g. rhombic sulphur (S_R) $\rightleftharpoons$ monoclinic sulphur (S_M)

Consider a system in which two phases are in equilibrium at a given temperature and pressure

$$\text{Phase } A \rightleftharpoons \text{Phase } B$$

Free energy at temperature T, G_A G_B
Free energy at temperature $(T + dT)$, $(G_A + dG_A)$ $(G_B + dG_B)$

Since the system is in equilibrium

$\therefore$ $\Delta G = G_B - G_A = 0$ or $G_B = G_A$ (1.164)

Also $G_B + dG_B = G_A + dG_A$ (1.165)

Subtracting Eq. (1.164) from Eq. (1.165), we get

$$dG_A = dG_B \qquad (1.166)$$

Now, from the equation of variation of free energy with temperature and pressure, we know

$$dG = VdP - SdT \qquad (1.167)$$

for phase A, Eq. (1.167) can be written as

$$dG_A = V_A dP - S_A dT \qquad (1.168)$$

and for phase B

$$dG_B = V_B dP - S_B dT \qquad (1.169)$$

From Eq. (1.166), since $dG_A = dG_B$

$\therefore$ $V_A dP - S_A dT = V_B dP - S_B dT$ (1.170)

On rearranging Eq. (1.170)

$$S_B dT - S_A dT = V_B dP - V_A dP \qquad (1.171)$$

$$dT(S_B - S_A) = dP(V_B - V_A) \qquad (1.172)$$

$\therefore$
$$\frac{dP}{dT} = \frac{\Delta S}{(V_B - V_A)} \qquad (1.173)$$

We know that $\Delta G = \Delta H - T\Delta S$, at equilibrium $\Delta G = 0$

$\therefore$
$$\Delta H = T\Delta S \quad \text{or} \quad \Delta S = \frac{\Delta H}{T} \qquad (1.174)$$

$$\therefore \quad \boxed{\frac{dP}{dT} = \frac{\Delta H}{T(V_B - V_A)}} \quad (1.175)$$

Equation (1.175) is known as *Clausius–Clapeyron equation*. This equation gives variation of equilibrium pressure with temperature for any two phases of a substance.

1.22.1 Integrated Form of Clausius–Clapeyron Equation

This equation can be integrated under the condition of large volume change in the final stage of two-phase equilibria, e.g.

$$\text{Liquid} \rightleftharpoons \text{vapour and solid} \rightleftharpoons \text{vapour equilibria.}$$

Consider the following two phase equilibria:

Volume $\qquad\qquad\qquad$ Liquid $\rightleftharpoons$ Vapour
$\qquad\qquad\qquad\qquad\qquad\quad (V_l) \qquad\quad (V_g)$

For this equilibria the Clausius–Clapeyron equation can be written as

$$\frac{dP}{dT} = \frac{\Delta H_{vap}}{T(V_g - V_l)} \quad (1.176)$$

Since, $V_g \ggg V_l$ (18 ml water at 100°C = 30060 ml vapours), $\therefore V_l$ is neglected.
Hence, the above equation becomes

$$\frac{dP}{dT} = \frac{\Delta H_{vap}}{T \cdot V_g} \quad (1.177)$$

If the vapours behave as ideal gas, then

$$PV_g = RT$$

or $\qquad\qquad\qquad\qquad\qquad V_g = \frac{RT}{P} \quad (1.178)$

$\therefore$ From Eqs. (1.177) and (1.178), we get

$$\frac{dP}{dT} = \frac{\Delta H_{vap}}{RT^2} \cdot P \quad (1.179)$$

On rearranging Eq. (1.179)

$$\frac{dP}{P} = \frac{\Delta H_{vap}}{R} \cdot \frac{dT}{T^2} \quad (1.180)$$

Integrating Eq. (1.180) over a small range of temperature and pressure

$$\int \frac{dP}{P} = \frac{\Delta H_{vap}}{R} \cdot \int \frac{dT}{T^2}$$

or $\qquad\qquad\qquad \boxed{\ln P = \frac{\Delta H_{vap}}{R}\left(\frac{1}{T}\right) + C} \quad (1.181)$

where C is the integration constant.

Equation (1.180) can also be integrated over a certain limit of pressure $P_1 \to P_2$ corresponding to temperature $T_1 \to T_2$

$$\int_{P_1}^{P_2} \frac{dP}{P} = \frac{\Delta H_{vap}}{R} \cdot \int_{T_1}^{T_2} \frac{dT}{T^2}$$

or
$$\ln \frac{P_2}{P_1} = \frac{\Delta H_{vap}}{R} \left[\frac{1}{T_1} - \frac{1}{T_2} \right] \qquad (1.182)$$

or
$$\log \frac{P_2}{P_1} = \frac{\Delta H_{vap}}{2.303\, R} \left[\frac{T_2 - T_1}{T_2 \times T_1} \right] \qquad (1.183)$$

Equation (1.184) is known as *integrated form of Clausius–Claypeyron equation.*

1.22.2 Applications of Clausius–Clapeyron Equation

This equation can be used:
1. To calculate the heat of vaporization, ΔH_{vap}.
2. To study the effect of temperature on vapour pressure of liquid.
3. To study the effect of pressure on melting point of solid or boiling point of liquid.
4. To study the thermodynamics of elevation of boiling point and depression of freezing point.
5. To calculate the latent heat of fusion from vapour pressure data.

1.23 THIRD LAW OF THERMODYNAMICS

"This law states that at the absolute zero of temperature the entropy of perfectly crystalline substance is taken as zero. The entropy of other substances which have zero disorderness of atoms, ions or molecules also possess zero entropy."

1.24 ZEROTH LAW OF THERMODYNAMICS

"This law states that if a body A is in thermodynamic equilibrium with body B and body C then body B and C will also be in thermodynamic equilibrium with each other."

SOLVED NUMERICAL PROBLEMS

1. Five moles of an ideal gas expand reversibly from a volume of 10 dm³ to 100 dm³ at a temperature of 27°C. Calculate the change in entropy.

   Solution: We know that change in entropy
   $$\Delta S = 2.303\, nR \log \left(\frac{V_2}{V_1} \right)$$

   Given that $n = 5$ moles, $V_1 = 10$ dm³, $V_2 = 100$ dm³

   $$\Delta S = 5 \times 8.314 \times 2.303 \log \left(\frac{100}{10} \right)$$

$\Delta S = 41.57 \times 2.303 \times 1 = 95.735$ \quad ($\because$ log 10 = 1)

$\Delta S = \mathbf{95.735\ J\ K^{-1}}$ **Ans.**

2. Calculate the increase in entropy during the evaporation of one mole of water at 100°C in SI unit. Heat of vapourization of water at 100°C = 540 cal/g.

   Solution: Given that, $T_b = 373$ K, $\Delta H = 540$ cal/g = $540 \times 4.184 \times 18 = 40.66848$ J K^{-1} mol^{-1} (where 18 is the molecular mass of H$_2$O)

   $$\Delta S_{vap} = \frac{\Delta H_{vop}}{T_b}$$

   We know that change in entropy during evaporation,

   $$\Delta S = \frac{40.668}{373} = \mathbf{109.03\ J\ K^{-1}\ mol^{-1}}\ \textbf{Ans.}$$

3. One mole of a mono-atomic gas (ideal) expands reversibly from a volume of 15 dm^3 and temperature 298 K to a volume of 25 dm^3 and temperature 250 K. Calculate the entropy change for this process $\left(C_V = \frac{3}{2}R\right)$.

   Solution: From equation $\Delta S = C_V \times 2.303 \log\left(\frac{T_2}{T_1}\right) + R \times 2.303 \log\left(\frac{V_2}{V_1}\right)$

   $\Delta S = 1.5 \times 8.314 + 2.303 \log\left(\frac{250}{298}\right) + 8.314 + 2.303 \log\left(\frac{25}{15}\right)$

   $\Delta S = 28.720 \log 250 - \log 298 + 19.147 \log 25 - \log 15$

   $\Delta S = 28.720 \times (2.3979 - 2.4782) + 19.147 (1.3979 - 1.1760)$

   $\Delta S = 28.720 \times (-0.0803) + 19.147 (0.2219)$

   $\Delta S = -2.3062 + 4.2487$

   $\Delta S = \mathbf{1.9425\ J\ K^{-1}\ mol^{-1}}\ \textbf{Ans.}$

4. One kg of ice melts into water at 0°C. Calculate the change in entropy. Heat of fusion of ice = 80 cal g^{-1}.

   Solution: $\Delta S = \frac{\Delta H_f}{T_f} = \frac{80 \times 4.184 \times 1000}{273} = \mathbf{1226\ J\ K^{-1}\ kg^{-1}}\ \textbf{Ans.}$

5. One mole of water vapourizes at 100°C. Calculate the change in entropy. Heat of vapourization of water at 100°C = 540 cal g^{-1}.

   Solution: $\Delta S = \frac{\Delta H_V}{T_b} = \frac{540 \times 4.184 \times 18}{373} = \mathbf{109.03\ J\ K^{-1}\ mol^{-1}}\ \textbf{Ans.}$

6. Calculate the change in entropy accompanying the heating of one mole of helium gas (ideal gas), from a temperature of 298 K to a temperature of 596 K at constant pressure. Given that $C_V = 3/2\ R$.

   Solution: For ideal gas, $C_P - C_V = R$

   Hence, $C_P = C_V + R = 3/2\ R + R = 2.5\ R$

   At constant pressure, entropy change, $\Delta S = C_P \ln\left(\frac{T_2}{T_1}\right)$

$$\Delta S = 2.303 \times 2.5 \times 8.314 \log\left(\frac{596}{298}\right)$$

$$\Delta S = 47.8055 \log\left(\frac{596}{298}\right) = 47.8055 \times 0.3010 = \textbf{14.3894 J K}^{-1}\textbf{ mol}^{-1} \textbf{ Ans.}$$

7. One mole of an ideal gas is heated from 150 K to 350 K. Calculate the change in entropy if (a) the volume is kept constant and (b) the pressure is kept constant. Given that $C_V = 3/2\ R$.

Solution: Entropy change at constant volume, $\Delta S = C_V \ln\left(\frac{T_2}{T_1}\right)$

$$\Delta S = 1.5 \times 8.314 \times 2.303 \log\left(\frac{350}{150}\right) = 28.7207(2.544 - 2.1760)$$

$$= 28.720 \times 0.386 = \textbf{10.568 J K}^{-1} \textbf{ Ans.}$$

8. One mole of an ideal mono-atomic gas expands reversibly from a volume of 10 dm³ at temperature 298 K to a volume of 20 dm³ at temperature of 250 K. Calculate the change in entropy for this process.

Solution: Given that $C_V = \frac{3}{2} R$.

$$\Delta S = 2.303 \left[C_V \log\left(\frac{T_2}{T_1}\right) + R \log\left(\frac{V_2}{V_1}\right) \right]$$

$$\Delta S = 2.303 \left[1.5 \times 8.314 \log\left(\frac{250}{298}\right) + 8.314 \log\left(\frac{20}{10}\right) \right]$$

$$\Delta S = 2.303\ [1.5 \times 8.314(2.3979 - 2.4742) + 8.314 \times 0.3010]$$

$$= 2.303\ [-2.19 \times 0.0763 + 2.5025]$$

$$= 2.303\ [-0.16709 + 2.5025]$$

$$= 2.303 \times 2.6695 = \textbf{--6.1840 J K}^{-1} \textbf{ mol}^{-1} \textbf{ Ans.}$$

9. One mole of an ideal mono-atomic gas expands reversibly from a temperature of 298 K and pressure 1 atm to a temperature of 273 K and pressure 1.052 atm. Calculate the entropy change for the process. Given that $C_V = 3/2\ R$.

Solution: Entropy change with temperature and pressure,

$$\Delta S = 2.303 \left[C_P \log\left(\frac{T_2}{T_1}\right) + R \log\left(\frac{P_1}{P_2}\right) \right]$$

For ideal gas, $\quad C_P - C_V = R$

Hence, $\quad C_P = C_V + R = \frac{3}{2} R + R = 2.5\ R$

$$\Delta S = 2.303 \left[2.5 \times 8.314 \log\left(\frac{273}{298}\right) + 8.314 \log\left(\frac{760}{800}\right) \right]$$

$$= 2.303[20.785 \times 0.0381 + 8.314\ (2.8808 - 2.9030]$$

$$= 2.303 \times 0.8141 = \textbf{18,748 J K}^{-1} \textbf{ mol}^{-1} \textbf{ Ans.}$$

10. Calculate the free energy change (ΔG) during the isothermal reversible expansion of 1 mole of an ideal gas at 37°C from an initial volume of 55 dm³ to 1000 dm³.

Solution: Given that $n = 1$ mol $T = 373°C$, $V_1 = 50$ dm³

From equation, $\Delta G = 2.303\ nRT \log\left(\dfrac{V_2}{V_1}\right)$

$\Delta G = 2.303 \times 8.314 \times 310 \log\left(\dfrac{55}{1000}\right)$

$\Delta G = 5935.614 \log 55 - \log 1000$

$\Delta G = 2577.34(1.740 - 3)$

$\Delta G = 2577.34 \times -1.26 = -3247.4484$ J mol⁻¹ = **−3.247 kJ mol⁻¹ Ans.**

11. The free energy change (ΔG) accompanying a given process is −85.78 kJ at 298 K and −83.67 kJ at 308 K. Calculate the change in free energy (ΔG) for the process at 303 K.

Solution: ΔG at 298 K = −85.78 kJ and ΔG at 308 K − 83.67 kJ

ΔG at 303 K = $\Delta G = -\left(\dfrac{85.78 + 83.67}{2}\right) = -84.725$ kJ

$\Delta G = \Delta H + T\left(\dfrac{\partial(\Delta G)}{\partial T}\right)$;

Thus $-84.725 = \Delta H + 303\left(\dfrac{\partial(-83.67) - (85.78)}{308 - 298}\right)$

$-84.725 = \Delta H + 303\left(\dfrac{0.211}{10}\right)$

or $-84.725 = \Delta H + 303 \times 0.211$

$\Delta H = -84.725 + (-63.933) = $ **−148.658 kJ Ans.**

12. Calculate the $\dfrac{dT}{dP}$ for the Water $\rightleftharpoons$ Ice system at 273 K. ΔH_f for water is 6007.8 J mol⁻¹ (1 J = 9.87 × 10⁻³ dm³ atm), molar volume of water = 18.00 cm³ and ice = 19.63 cm³.

Solution: Given that $V_1 = 9.00$ cm³ mol⁻¹ = 0.01900 dm³ mol⁻¹,

$V_s = 20.63$ cm³ mol⁻¹ = 0.02063 dm³ mol⁻¹,

$\dfrac{dP}{dT} = \dfrac{\Delta H_f}{T_f(V_1 - V_2)}$

or $\dfrac{dT}{dP} = \dfrac{T_f(V_1 - V_2)}{\Delta H_f}$

$\dfrac{dT}{dP} = \dfrac{273(0.019) - 0.02063)}{6007.8 \times 9.87 \times 10^{-3}} = \dfrac{-0.44499}{59.297}$

= **−7.504 × 10⁻³ K atm⁻¹ Ans.**

Hence, the melting point of ice reduced by 0.0075 K if the pressure is raised by 1 atm.

13. The vapour pressure of water at 373°C is 760 mm. What will be the vapour pressure of water at 363°C? The heat of vapourization of water in this temperature range is 41.27 kJ mol^{-1}.

Solution: From Clausius–Clapeyron equation

$$\log \frac{P_2}{760} = \frac{\Delta H_{vap}}{2.303\,R} \left[\frac{T_2 - T_1}{T_2 \times T_1}\right]$$

$$\log \frac{P_2}{760} = \frac{41.27 \times 10^3}{2.303 \times 8.314} \left[\frac{363 - 373}{368 \times 373}\right]$$

$$\log P_2 - \log 760 = \frac{41.27 \times 10^3}{19.147} \cdot \left[\frac{-10}{137264}\right]$$

$$[\log P_2 - 2.8808 = 2155.429 \times (-7.2855 \times 10^{-5})]$$

$$\log P_2 - 2.8808 = -0.15702$$

$$\log P_2 = -0.15702 + 2.8808$$

$$\log P_2 = 2.723$$

$$P_2 = \text{antilog } 2.723$$

$$P_2 = \mathbf{528.44 \text{ mm Ans.}}$$

14. Ether boils at 306.5 K at 1 atmospheric pressure. At what temperature will it boil at a pressure of 770 mm given that ΔH_{vap} = 369.86 J g^{-1}.

Solution: From Clausius–Clapeyron equation

$$\log \frac{P_2}{P_1} = \frac{\Delta H_{vap}}{2.303\,R} \left[\frac{T_2 - T_1}{T_2 \times T_1}\right], \text{ or (Molecular mass of diethyl ether is 74)}$$

$$\log \frac{750}{760} = \frac{369.86 \times 74}{2.303 \times 8.314}\left[\frac{T_2 - 306.5}{T_2 \times 306.5}\right] \text{ or } \log 750 - \log 760 = \frac{27369.64}{19.1471} \left[\frac{T_2 - 306.5}{T_2 \times 306.5}\right]$$

or $\quad 2.8865 - 2.8808 = \dfrac{27369.64}{19.1471} \left[\dfrac{T_2 - 306.5}{T_2 \times 306.5}\right]$

or $\quad \dfrac{306.5 \times -0.0057}{1429.44} = \left[\dfrac{T_2 - 306.5}{T_2}\right]; \text{ or } \dfrac{-1.74705}{1429.44} = \left[\dfrac{T_2 - 306.5}{T_2}\right]$

or $\quad T_2 = 305.9$ K

Hence, boiling point of ether at **750 mm is 350.9 K Ans.**

15. At what pressure will ice melt at −1.0°C assuming that ΔH_f is independent of pressure and is equal to 6.0095 kJ mol^{-1}? Given that the density of water is 0.9998 g cm^{-3} and that of ice is 0.917 g cm^{-3}.

Solution: From the Clausius–Clapeyron equation

Density of water = 0.9998 g cm^{-3} = 0.9998 × 10^{-3} kg m^{-3}
Density of ice = 0.917 g cm^{-3} = 0.917 × 10^{-3} kg m^{-3}
Molar mass of water = 18 g mol^{-1} = 18 × 10^{-3} kg mol^{-1}

Molar volume of water, $V_l = \dfrac{18 \times 10^{-3}}{0.998 \times 10^3} = 18.0186 \times 10^{-6}$ m^3 mol^{-1}

and molar volume of ice, $V_s = \dfrac{18 \times 10^{-3}}{0.917 \times 10^3} = 19.645 \times 10^{-6}$ m^3 mol^{-1}

$$\Delta H_f = 6.0095 \text{ kJ mol}^{-1} = 6009.5 \text{ J mol}^{-1}$$

$$= \dfrac{6009.5 \text{ J mol}^{-1}}{101325 \text{ Jm}^{-2} \text{ atm}^{-1}} = 5.9309 \times 10^{-2} \text{ m}^3 \text{ atm mol}^{-1}$$

($\because$ 1 atm = 101325 Nm^{-2} = 101325 Jm^{-3})

$$T_1 = 0°C = 273 \text{ K and } T_2 = -1°C = 272 \text{ K}$$

From the Clausius–Clapeyron equation

$$\dfrac{dP}{dT} = \dfrac{\Delta H_f}{T_f \cdot \Delta V_f} \quad \text{or} \quad dP = \dfrac{\Delta H_f}{\Delta V_f} \cdot \dfrac{dT}{T_f}$$

Suppose that the $\dfrac{\Delta H_f}{\Delta V_f}$ is independent of temperature and pressure, then integrating above equation, we get

$$\int_{P_1}^{P_2} dP = \dfrac{\Delta H_f}{\Delta V_f} \int_{T_1}^{T_2} \dfrac{dT}{T} \quad \text{or} \quad (P_2 - P_1) = \dfrac{\Delta H_f}{\Delta V_f} \ln \dfrac{T_2}{T_1}$$

or
$$P_2 - P_1 = 2.303 \dfrac{\Delta H_f}{(V_l - V_s)} \log \dfrac{T_2}{T_1}$$

$$P_2 - P_1 = 2.303 \dfrac{5.9309 \times 10^{-2}}{(18.0186 - 19.645) \times 10^{-6}} \log \dfrac{272}{273}$$

$$P_2 - P_1 = 2.303 \dfrac{5.9309 \times 10^{-2}}{-1.6264 \times 10^{-6}} (2.4345 - 2.4361)$$

$$P_2 - P_1 = \dfrac{13.6588 \times 10^{-2}}{-1.6264 \times 10^{-6}} \times (-0.0016) = 134.37$$

or $\quad P_2 = P_i + 134.37 \quad$ or $\quad P_2 = 1 + 134.37 \quad (\because P_i = 1 \text{ atm})$

or $\quad\quad\quad\quad\quad\quad\quad\quad P_2 = 135.37$ atm **Ans.**

SUMMARY

Thermodynamics: "It is the branch of chemistry which deals with the study of transformation of energy."

OR

"It is the branch of chemistry which involves the study of interrelation of various kinds of energy accompanying physical or chemical changes."

System: A system is the specified part of the universe which is selected for the purpose of study of effect of certain properties such as temperature, pressure, etc.

Surroundings: Rest part of the universe other than the system separated from the system by real or imaginary boundaries is called *surroundings*.

Homogeneous system: The system which is uniform throughout, i.e. all the substances present in it have only one phase is called *homogeneous system*.

Heterogeneous system: The system which is not uniform throughout is called *heterogeneous system*, i.e. it has more than one phases.

State functions: The measurable properties of a system which depends only upon the state of the system (i.e. initial and final state of the system) are known as *state variables*, for example, P, V, T, composition, energy, etc.

State of the system: When the fundamental properties, e.g. pressure, temperature, volume, number of moles and composition have definite values, the system is said to be in a definite state. When there is any change in any one of these properties then system is said to be under a change of state.

Extensive properties: The properties which depend upon amount of the system such as mass, volume, energy, heat, entropy, etc.

Intensive properties: The properties which do not depend upon amount of the system such as temperature, pressure, viscosity, density, refractive index, surface tension, chemical potential, etc.

Work: In thermodynamics, the fundamental physical property is work. Work is done when an object moves against an opposing face. Work is expressed as the product of two factors, i.e. w = intensity factor × capacity factor.

Heat: It may be defined as the quantity of energy which flows between system and surroundings due to the temperature difference. It always flows from high temperature point to low temperature point. Heat $q = m \times s \times t$.

Internal energy: "The internal energy may be the sum of different kind of energies such as translational energy (E_t), rotational energy (E_r), vibrational energy (E_v), kinetic energy (E_k), and potential energy (E_p), electronic energy (E_e), and bonding energy (E_b)."

First law of thermodynamics: "Energy can neither be created nor destroyed, although it can be transformed from one form to another."

OR

"The total energy of the universe is constant".

OR

"The total mass and energy of an isolated system remain unchanged."

Enthalpy or heat content: Enthalpy may be defined as "the sum of internal energy and the product of pressure and volume of the system".

Second law of thermodynamics: Different statements of thermodynamics are:

Kelvin statement: "It is impossible to use a cyclic process to extract heat from a reservoir and to convert it into work without transferring at the same time a certain amount of heat from a hotter to colder part of the body", i.e. it is impossible to construct perpetual motion machine.

Carnot statement: "It is impossible to convert heat into work without compensation."

Generalized concept: "All the natural processes occur spontaneously, i.e. irreversibly it follows that the available energy is decreasing and entropy of universe is increasing continuously."

Spontaneous (irreversible) and Non-spontaneous processes: Spontaneous processes are those processes which proceed by themselves and bring the system closer to the equilibrium. These are also called as positive or natural processes.

Partial molar free energy (chemical potential): "The change in free energy of the system that results on adding one mole of that particular substance at a constant temperature and pressure, to such a large quantity of the system that there is no considerable change in overall composition of the system."

Gibbs–Helmholtz equation:

At constant pressure $\quad\quad$ At constant volume

$$\Delta G = \Delta H + T\left(\frac{\partial(\Delta G)}{\partial T}\right)_P \quad\quad \Delta A = \Delta E + T\left(\frac{\partial(\Delta A)}{\partial T}\right)_V$$

Clausius–Clapeyron equation: This equation was derived by Clausius and Clapeyron independently and simultaneously on the basis of second law of thermodynamics for two-phase equilibria in one-component system and is known as Clausius–Clapeyron equation.

$$\frac{dP}{dT} = \frac{\Delta H}{T(V_B - V_A)}.$$

EXERCISES

1. What is thermodynamics? Give the various objectives and limitations of thermodynamics.
2. What do you understand by the system and surroundings? Explain the various types of systems with examples.
3. Describe the various types of processes in thermodynamics with examples.
4. What do you understand by thermodynamic equilibrium? Explain the various types of equilibrium giving suitable examples.
5. Explain the following terms giving suitable examples: (i) state function. (ii) extensive properties, (iii) intensive properties, (iv) heat, (v) work, (vi) energy, (vii) internal energy, (viii) enthalpy.
6. What is the first law of thermodynamics? Give its mathematical form and explain its limitations.
7. Explain the work done in isothermal reversible and irreversible expansion.
8. What do you understand by spontaneous and non-spontaneous processes? Explain with examples.
9. What is the concept of entropy? Explain the entropy with the help of Carnot cycle.
10. What is entropy? Explain the physical significance of entropy.
11. Derive the equation for the entropy change of ideal gas.
12. What is the entropy change in reversible and irreversible processes? Explain in details.
13. Derive the equation for entropy change for ideal gas with temperature and pressure variables.

14. (a) Give the statements of second law of thermodynamics. Explain the criteria for reversible and irreversible processes in details.

    (b) Give the applications and limitations of second law of thermodynamics.

15. What do you understand by entropy? Give the mathematical statement and units of entropy.

16. What is Helmholtz free energy function?

17. What is the difference between free energy function and work function?

18. The entropy change is positive for a substance involved in transformations. Does it follow that the change is spontaneous?

19. How will you determine the effect of volume on work function?

20. Explain the following:

    (i) What is entropy change in adiabatic process?

    (ii) Prove that entropy is the measure of feasibility/probability.

    (iii) What are the limitations of thermodynamic criteria for determining spontaneity of a process?

    (iv) Give the Boltzmann's definition of thermodynamic probability.

21. What is free energy and work function? Give their physical significance.

22. What do understand by Gibbs free energy? Derive the equation for variation of free energy with temperature and pressure.

23. Prove that $\left(\frac{\partial G}{\partial T}\right)_P = -S$.

24. What is partial molar free energy? Explain the variation of chemical potential with temperature and pressure.

25. Derive the Gibbs–Helmholtz equation and give its various applications.

26. (a) What is the concept of chemical potential? Explain the variation of chemical potential with temperature and pressure.

    (b) Give the physical significance of chemical potential.

27. What is the Clausius–Clapeyron equation? Under what conditions the equation can be integrated?

28. Derive the Clausius–Clapeyron equation for two phase equilibria. Under what conditions this equation can be integrated.

29. Derive the equation for the change of Helmholtz function and Gibbs free energy functions when an ideal gas is expanded isothermally from a higher pressure P_1 to a lower pressure P_2. For this process prove that $\Delta G = \Delta A$.

30. Calculate ΔG and ΔA when one mole of an ideal gas expands isothermally at 300 Kelvin from a pressure of 100 atmospheres to 1 atmosphere.

31. Derive the Gibbs–Helmholtz equation in following different forms:

    (i) $\Delta H = \Delta G - T\left\{\frac{\partial(\Delta G)}{\partial T}\right\}_P$

    (ii) $\left\{\frac{\partial(\Delta G/T)}{\partial T}\right\}_P = -\frac{(\Delta H)}{T^2}$

    (iii) $\left\{\frac{\partial(A/T)}{\partial(1/T)}\right\}_V = E$.

32. Prove that the change in free energy of the system under suitable experimental conditions accounts automatically for the entropy change of the system and surroundings. Discuss the nature of equilibrium systems.
    (i) $(dG)_{T, P} \leq 0$ (ii) $(dA)_{T, V} \leq 0$ (iii) $(dS)_{T, V} \geq 0$.
33. Derive the Clapeyron–Clausius equation for liquid $\rightleftharpoons$ vapour equilibria.
34. Calculate the heat of fusion per mole of benzene at its freezing point, 278 K if the rate of change of melting point with pressure is 0.0128 K atm^{-1} and ΔV is 0.058 cm^{3}g^{-1}.
35. Explain the following:
    (i) Show that the ΔS_{mix} for a binary mixture of ideal gases in an isothermal process is maximum when $x_1 = x_2 = 1/2$, where x are mole fractions?
    (ii) What is the importance of Boltzmann entropy equation?
    (iii) Why is ΔS_{mix} in multi–component system always positive?

    (Hint: $\Delta S_{mix} = -R \sum x_i \ln x_i$, since $x_i < 1$, $\ln x_i$ is negative, ΔS_{mix} hence is positive.)
    (iv) What will be the value of chemical potential for one mole at constant T and P.
    (v) Is it possible to integrate the Gibbs–Helmholtz equation? If yes then find the integrated form of the same.
    (vi) The internal energy of an isolated system is always constant.
    (vii) The decrease in Helmholtz free energy function is equal to the maximum reversible work. Also mention the conditions under which it is true.
    (viii) The decrease in Gibbs free energy at constant temperature is equal to the net over and above the mechanical work.
    (ix) The exothermic reactions are always spontaneous. Give some suitable examples of exothermic reactions which are not spontaneous.
    (x) At equilibrium the free energy must be minimum and entropy should be maximum.
    (xi) Give reason of the following:
        (a) ΔS mixing in a multi-component system is always positive.
        (b) The change in partial molar free energy with temperature and pressure is maximum for gases.
        (c) The Clapeyron–Clausius equation cannot be integrated for solid $\rightleftharpoons$ liquid equlibria?
        (d) All the natural processes proceed spontaneously.
        (e) Entropy is the measure of the disorderness of the system.
        (f) Entropy is the measure of probability?
        (g) The internal energy of an isolated system is always constant. Justify that erntropy change for an adiabatic process is zero.
        (h) Justify that entropy change for an adiabatic process is zero.
        (i) The decrease in Helmholtz free energy function is equal to maximum reversible work.
        (j) The reaction $2H_2(g) + 1/2 O_2(g) \rightarrow 2H_2O(l)$ is thermodynamically physical at room temperature but mixture of H_2 and O_2 at room temperature remains as such almost indefinitely without reactive .
36. What are the criteria for determining the spontaneity of process.

37. What will be value of chemical potential for one mole of substance at constant T and P.
38. Prove that: (i) $\Delta S = C_P \ln T_2/T_1 - R \ln P_2/P$, (ii) $\Delta S = \sum nx_i$ (iii) $\Delta G = RT \ln P_2/P_1$.
39. Find ΔG and ΔA when 1 mole of ideal gas expands isothermally at 300 K from a pressure of 100 atm to 1 atm.
40. Give the Boltzmann definition of thermodynamic probability?
41. Prove that decrease in free energy at constant temperature is equal to the net over and above the mechanical work.
42. Define the entropy change for various phase transitions.
43. What do you understand by unavailable energy? Give its physical significance.
44. Derive the equation $\Delta A = \Delta E + T(\sigma(\Delta A/\Delta T))_V$.
45. What do you understand by chemical potential? Explain the factors affecting the chemical potential.
46. Derive the equation $S = \sum n(C_P \log T - R \log P - R \log X + S_0^-)$.
47. Calculate ΔS for one mole of an mono-atomic ideal gas when it is heated from 27°C to 227°C at constant volume and at constant pressure.
48. Calculate the entropy change if 2 moles of N_2, 3 moles of H_2 and 2 moles of NH_3 are mixed at constant temperature, assume that no chemical reaction occurs.
49. If C_P of water is 4.184 J K^{-1} g^{-1} and that of the alloy is 22 J K^{-1}g^{-1}. Calculate ΔS.
50. Calculate the zero point entropy for CO crystal at 0 K.
51. Calculate ΔA for the process:
    2He (g, 27°C, 1 atm) → 2He (g, 27°C, 5 atm), assuming that the process is reversible.
52. Five moles of an ideal gas expand reversibly from a volume of 8 dm^3 to 80 dm^3 at a temperature of 27°C. Calculate the change in entropy.
53. Calculate ΔG for the process:
    1 mole H_2O (l) 100°C → 1 mole H_2O (V, 100°C) (Hint: for phase change $\Delta G = 0$)
54. Calculate ΔS for one mole of an mono-atomic ideal gas when it is heated from 27°C to 227°C (i) at constant volume and (ii) at constant pressure.
55. One kg of ice melts into water at 0°C. Calculate the ΔS. Heat of fusion of ice = 80 cal g^{-1}. Calculate the change in entropy accompanying the heating of one mole of helium gas (ideal gas) from a temperature of 298 K to a temperature of 1000 K at constant pressure.
56. One mole of an ideal gas is heated from 100 K to 300 K. Calculate the change in entropy if (a) the volume is kept constant, (b) the pressure is kept constant. Given that $C_V = 3/2\ R$.
57. One mole of an ideal mono-atomic gas expand reversibly from a temperature of 298 K and pressure 1 atm to a temperature of 273 K. Calculate the entropy change for the process. Given that $C_V = 3/2\ R$.
58. Calculate the entropy of mixing of one mole of oxygen gas and two moles of hydrogen gas, assuming that no chemical reaction occurs and the gas mixture behaves ideally.
59. Calculate the free energy change, ΔG which occurs when one mole of an ideal gas expands reversibly and isothermally at 37°C from an initial volume of 55 dm^3 to 1000 dm^3.
60. The free energy change (ΔG) accompanying a given process is −85.77 kJ at 298 K and −83.68 kJ at 308 K. Calculate the change in Gibbs free energy change ΔG for the process at 303 K.

61. Calculate the $\frac{dT}{dP}$ for the water $\rightleftharpoons$ ice system at 273 K. ΔH_f for water is 6.007.8 J mol^{-1}. (1 J = 9.87 × 10^{-3} dm^3 atm). Molar volume of water = 18.00 cm^3 and ice = 19.63 cm^3.

62. Ether boils at 306.5 K at 1 atmospheric pressure. At what temperature will it boil at a pressure of 750 mm given that ΔH_{vap} = 369.86 J g^{-1}.

63. At what pressure will ice melt at -1.0 °C assuming that ΔH_f is independent of pressure and is equal to 6.0095 kJ mol^{-1}? Given that the density of water is 0.9998 g cm^{-3} and that of ice is 0.917 g cm^{-3}.

64. The vapour pressure of water at 100 °C is 760 mm. What will be the vapour pressure of water at 95 °C? The ΔH_{vap} of water in this temperature range is 41.27 kJ mol^{-1}.

65. The vapor pressure of Ice $\rightleftharpoons$ Water system at 0.0075°C is 4.58 mm and at 273 K is 759.80 mm of mercury. Calculate the molar heat of fusion of ice. Given that the specific volume of ice and water at 273.0075°C are 1.0907 cc and 1.0001 cc, respectively. Density of mercury at 273.0075 K =13.6 g/cc^{-1}.

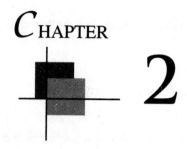

CHAPTER 2

PHASE RULE

2.1 INTRODUCTION

In the physical or chemical systems the different phases can co-exist in equilibrium under certain conditions of variables such as temperature, pressure and compositions. The equilibrium of heterogeneous systems is governed by an important generalization which is called *phase rule*. This rule is a landmark in the study of effect of different variables on the properties of phases of the systems co-existing in *equilibrium*.

2.2 THE PHASE RULE

This rule was discovered by J. Willard Gibbs in 1875 and is known as *Gibbs phase rule*. This is an important generalization used for the study of the heterogeneous equilibria. It predicts quantitatively the effect of temperature, pressure and composition (concentration) on multiple phase equilibria.

It may be stated that "provided that the equilibrium between any number of phases is not influenced by gravitational forces, electrical forces, magnetic forces or by surface action, but only by temperature, pressure and concentration (composition), then the number of degree of freedom (F) of the system is related with the number of component (C) and number of phases (P) by the following equation".

$$\boxed{F + P = C + 2}$$

or

$$\boxed{F = C - P + 2} \tag{2.1}$$

In other words "the sum of number of degrees of freedom and number of phases is equal to number of components plus two". This equation is known as *phase rule equation*.

The above equation does not involve mass of phases as it has no influence on the state of equilibrium of heterogeneous system.

2.2.1 Definition of Terms Used

(a) *Phase (P):* Phase is defined as "a homogeneous, physically distinct and mechanically separable portion of the system which is separated from other such parts of the system by definite boundary surfaces".

For example:
1. A mixture of gases constitutes single phase, since gases are completely miscible.

$$H_2(g) + I_2(g) \rightleftharpoons 2HI(g) \quad P = 1$$

2. Immiscible liquids constitute different phases, e.g. mixture of oil and water, water and benzene, water and CCl_4, etc. form two phases.
3. Miscible liquids constitute single phase as they give uniform solution, e.g. mixture of ethanol and water.
4. In case of mixture of solids, each solid constitutes different phase, e.g. decomposition of $CaCO_3$:

$$CaCO_3(s) \rightleftharpoons CaO(s) + CO_2(g)$$

It consists of three phases (i.e. two solids and one gaseous).

5. True solutions form single phase as they are homogeneous throughout, e.g. aqueous solution of glucose, sugar, NaCl, etc. constitute single phase.
6. In the saturated solution of a solid, the excess solid is in contact with the solution of solid, hence there are two phases, e.g. saturated solution of NaCl ($P = 2$).
7. Colloidal solutions are not physically homogeneous, hence they consist of more than one phases, e.g. gelatin (solid, liquid), clouds (gas, liquid), smoke (gas, liquid).
8. Homogeneous solid solution of a salt constitutes single phase, e.g. Mohr's salt

$[FeSO_4(NH_4)_2SO_4 \cdot 6H_2O]$ constitutes single phase.

(b) *Component (C):* It is defined as "the smallest number of independently variable constituents (molecular species) taking part in the state of equilibrium and by means of which the composition of each phase can be expressed either directly or in terms of chemical equations."

Number of components in any system can be calculated by considering the following points:

1. In water system, ice (s) $\rightleftharpoons$ water (l) $\rightleftharpoons$ vapour (g), the chemical composition of all the three phases can be expressed in terms of H_2O only. Hence, it is a one-component system, $C = 1$.
2. In case of mixture of gases which do not combine chemically with each other, the number of components are equal to the number of constituents, e.g. the mixture of hydrogen, helium and argon is a one phase and three-component system.
3. Aqueous solution of NaCl or any other salt is homogeneous and constitutes one phase two-component system.
4. For a chemically reactive system, the number of components can be calculated by using following equation, i.e.

Number of components, $\quad C = N - a - n - E \quad$ (2.2)

where N = number of species (constituents), a = number of independent equilibrium conditions, n = number of relations due to initial conditions and E = number of independent equations/reactions relating the concentrations of N constituents.

In general, the above equation may be given as $C = N - E$ (2.3)

For ionic reactions $C = N - (E + 1)$ (2.4)

5. In case of gaseous mixture formed by decomposition of any species, the number of component is one.

$$H_2O(g) \rightleftharpoons H_2(g) + 1/2\ O_2(g)$$

For this system only one component is required to specify all the phases, because the information regarding the other two can be obtained from the law of mass action, i.e.

$$\text{Equilibrium constant, } K = \frac{P_{H_2O}}{P_{H_2} \times (P_{O_2})^{1/2}}$$

6. For the system in which gaseous substances reacting with one another to give gaseous product, the number of components are equal to number of reacting gaseous species, e.g.

$$H_2(g) + I_2(g) \rightleftharpoons 2HI(g) \quad (C = 2) \quad \text{as } [H_2] = [I_2]$$

In this system, we require to specify only two components. The concentration of third species can be determined by the equation:

$$K = \frac{[HI]^2}{[H_2] \times [I_2]}$$

7. In the mixture of $N_2(g)$, $H_2(g)$ and $NH_3(g)$ all the three components are required to specify. Hence it is a three-component system.

8. Salt hydrates and water are two-component systems:

$$CuSO_4 \cdot 5H_2O(s) \rightleftharpoons CuSO_4 \cdot 3H_2O(s) + 2H_2O(g),\ P = 3,\ C = 2$$

9. Decomposition of $CaCO_3$

$$CaCO_3(s) \rightleftharpoons CaO(s) + CO_2(g),\ P = 3 \quad \text{and} \quad C = 2$$

Here, only two components are required to specify. The third one can be calculated from the stoichiometry of the reaction. Thus, $P = 3$ and $C = 2$.

In case of mixture of three species, we need to specify all the three constituents, i.e. $C = 3$.

10. *Dissociation of ammonium chloride*, NH_4Cl: there may be two cases:

   (a) When only $NH_4Cl(s)$ is heated in a closed vessel.

$$NH_4Cl(s) \rightleftharpoons NH_3(g) + HCl(g)$$

In this condition, the NH_3 and HCl are present in equimolar ratio, i.e. $[NH_3] = [HCl]$

or $P_{NH_3} = P_{HCl}$, hence $P = 2$ and $C = 1$.

$$K = \frac{[NH_3][HCl]^2}{[NH_4Cl]}$$

The active mass of solid is taken as constant and ignored, therefore

$$K = [NH_3][HCl]$$

(b) When the mixture of $NH_4Cl(s)$, $NH_3(g)$ and $HCl(g)$ is heated in closed vessel, then

$$P_{NH_3} \neq P_{HCl} \quad \text{hence, } P = 2 \quad \text{and} \quad C = 2.$$

11. $Fe(s) + H_2O(g) \rightleftharpoons FeO(s) + H_2(g)$

    There are three phases (two solids and one gaseous phase) hence two solid components, $Fe(s)$ and $FeO(s)$ and one of the two gaseous components must be specified, since the third one can be calculated from the equilibrium constant:

    $$K = \frac{P_{H_2}}{P_{H_2O}}$$

    Hence, $P = 3$ and $C = 3$.

(c) *Degree of Freedom (F):* The degree of freedom of a system is defined as "the minimum number of independent variables such as temperature, pressure and concentration (or composition) which must be specified in order to define the system completely or to represent perfectly the condition of a system". It is also known as *variance*.

The number of degree of freedom can be calculated by using the equation:

$$F = C - P + 2$$

In case of constant temperature or pressure the reduced equation is used, i.e.

$$F = C - P + 1 \tag{2.5}$$

Examples:

1. In case of water system:

    $$Ice(s) \rightleftharpoons Water(l) \rightleftharpoons Vapour(g)$$

    No condition is needed to specify as the three phases can be in equilibrium only at particular temperature and pressure. The system is therefore zero variant or non-variant, i.e. $(F = 0)$.

2. For a gaseous mixture of N_2 and H_2, if the pressure and temperature are fixed the volume automatically becomes definite. Hence, it has two degrees of freedom.

3. Saturated solution of KCl

    $$KCl(s) \rightleftharpoons KCl \text{ (solution)} \rightleftharpoons H_2O \text{ (vapour)}$$

    At a fixed temperature the solubility of solute and vapour pressure have definite values. Hence, the above system can be completely defined by specifying only temperature. Hence, the system is univarient.

2.2.2 Derivation of Phase Rule

Consider a heterogeneous system constitutes of components $C(C_1, C_2, C_3, \ldots C_c)$ distributed over P phases ($\alpha, \beta, \gamma, \delta, \ldots, P$) as shown in Figure 2.1. Assume that passage of a component from one phase to another does not constitute a chemical reaction. The system is in equilibrium and can be completely defined by specifying the two variables, temperature T and pressure P and also by composition of each phase, i.e.

$$T, P \text{ and } (x_1^\alpha, x_2^\alpha, \ldots, x_c^\alpha), (x_1^\beta, x_2^\beta, \ldots, x_c^\beta), (x_1^\gamma, x_2^\gamma, \ldots, x_c^\gamma), (x_1^P, x_2^P, \ldots, x_c^P)$$

(where x = mole fraction of components).

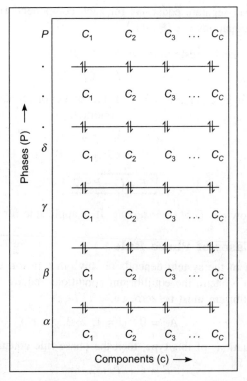

Figure 2.1 Distribution of various components (C) in P phases.

∴ Total variables = pressure + temperature + concentration variables for P phases

$$= 1 + 1 + CP = CP + 2 \tag{2.6}$$

But all $CP + 2$ variables are not essentially be independent variables because of the restrictions imposed by the equilibrium conditions. These restrictions can be calculated as follows:

1. The sum of mole fractions of all the components in any phase is equal to one, i.e.

$$x_1^\alpha + x_2^\alpha + x_3^\alpha + \cdots + x_c^\alpha = x_1^\beta + x_2^\beta + x_3^\beta + \cdots + x_c^\beta = \ldots = x_1^P + x_2^P + x_3^P \cdots x_c^P = 1$$

For one component the number of equation is one, therefore for P phases, total equations will be

$$1 \times P = P \tag{2.7}$$

Thus, there are P relations (or restrictions) due to mole fraction condition.

2. Again for equilibrium state, another thermodynamic requirement is that the chemical potential (μ) of a component must be same in every phase, i.e.

For component 1 : $\mu_1^\alpha = \mu_1^\beta = \mu_1^\gamma = \ldots = \mu_1^P$ (P − 1) equation

For component 2 : $\mu_2^\alpha = \mu_2^\beta = \mu_2^\gamma = \ldots = \mu_2^P$ (P − 1) equation

$\vdots \qquad \vdots \qquad \vdots \qquad \vdots \qquad \vdots \qquad \vdots \qquad \vdots \qquad \vdots$

For component C : $\mu_C^\alpha = \mu_C^\beta = \mu_C^\gamma = \ldots = \mu_C^P$ (P − 1) equation

Thus, there are $(P-1)$ separate equations for each component. Hence, for C components, the number of such equations = $C(P-1)$

Therefore, total number of restrictions = $P + C(P-1)$

$$= P + CP - C \tag{2.8}$$

But, number of degrees of freedom = Total number of variables – Number of restricting conditions

or $$F = CP + 2 - (P + CP - C)$$

or $$F = CP + 2 - P - CP + C$$

or $$\boxed{F = C - P + 2} \tag{2.9}$$

Equation (2.9) is known as *Gibbs phase rule*. It is applicable for nonreactive systems.

2.2.3 Some Special Cases of Phase Rule

Case 1. If the equilibrium of the system depends on any other factor, e.g. chemical affinity (A_f), then for chemically reactive system, the equilibrium conditions require that the chemical affinity, A_f for each reaction at equilibrium must be zero, i.e.

$$A_{f_i} = 0 \quad (i = 1, 2, 3, \ldots, r') \tag{2.10}$$

This means there are r' equations of this type, then the phase rule equation will be

$$F = C - P + 2 - r' \tag{2.11}$$

Case 2. If the equilibrium of system depends on electrical potential (γ), then the phase rule will be

$$F = C - P + 2 + \gamma \tag{2.12}$$

Case 3. If the temperature or pressure remains constant at equilibrium, then the phase rule will be

$$F = C - P + 1$$

Case 4. When one component is absent in one of the phases. Suppose component C_1 is insoluble in α phase, then the total number of equations in equilibrium conditions will be reduced by one, but another condition, the insolubility of component C_1 in α phase will be introduced. Therefore, final expression remains unchanged.

2.2.4 Applications of Phase Rule

1. Phase rule gives a simple method to classify the equilibrium states of the system.
2. By knowing the number of degree of freedom and number of components present in the system we can predict the number of phases co-existing in the system under equilibrium.
3. Thus, it guides to ascertain the existence of a new phase.
4. It can be applied to physical as well as chemical phase reactions.
5. This rule confirms that different systems having the same number of degrees of freedom behave in same manner.

6. This rule explains the behaviour of systems when they are subjected to change in variables such as pressure, temperature, concentration, etc.
7. Since this rule is applicable to macroscopic systems, it is not necessary to have information about molecular structure.
8. This rule is very useful in industries, e.g. metallurgical industry, in the practical fields such as in the isolation of salts from mixture, preparation of alloys, de-silverisation of lead, etc.

2.2.5 Limitations of Phase Rule

1. The phase rule explains the effect of temperature, pressure and composition on phase equilibria. The influence of other variables such as gravitational, magnetic, electrical and surface forces, etc. is ignored.
2. It can be applied only to the heterogeneous systems in equilibrium. It is therefore of no use for such systems which attain the equilibrium state slowly.
3. If liquid or solid phase is finely divided, their vapour pressure will differ from their normal value. It does not give any idea about this changed vapour pressure.
4. Since this rule applies to the system in equilibrium, therefore time is not a variable.
5. This phase rule cannot be applied to such systems which are consisted of two or more solutions separated by semipermeable membranes.
6. This rule is applicable to a single equilibrium system and does not provide any information about the number of other equilibria possible in the system.

2.3 PHASE DIAGRAM

A phase diagram may be defined as the graphical presentation (Figure 2.2), showing interdependence of variables which helps in studying the co-existence of different phases in a multiphase system. The facts of phase diagram are:

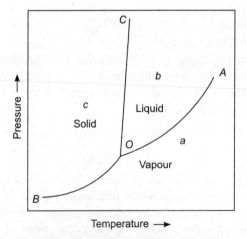

Figure 2.2 A typical phase diagram.

1. The phase diagrams represent the inter-phase relationship of a system.
2. It helps in investigating the conditions of phases which constitute the system.
3. The two-dimensional phase diagram is obtained by considering two variables (i.e. T and P or T and C) on rectangular coordinates.

4. Three-dimensional phase diagram (solid diagram) is obtained by considering three variables (i.e. T, P and C) on three axes to study the multiphase system.
5. The Clausius–Clapeyron equation is used to study the effect of temperature on the vapour pressure of liquid as well as the effect of pressure on transition temperature of a substance.

$$\frac{dP}{dT} = \frac{\Delta H}{T(V_2 - V_1)} = \frac{\Delta S}{\Delta V} \qquad (2.13)$$

where ΔH = change in enthalpy (heat of phase transformations) at temperature T, V_1 = molar volume of the pure substance in phase 1, V_2 = molar volume of the pure substance in phase 2, dP = change in pressure and dT = change in temperature.

6. Transition of the pure substance from one phase to another, i.e. phase transition or transformation can be explained with the help of Clausius–Clapeyron equation.
7. The nature of phase diagram depends on the number of phases and number of components present in the system, i.e. it may be one component, two components, etc.

2.4 ONE-COMPONENT SYSTEM

The system in which all the phases of the system can be defined only by taking one specie, is called one-component system. For example, in water system there are three phases: ice, water and vapour which can be explained only by taking one specie, i.e. H_2O. Similarly, in sulphur system there are four phases, namely, rhombic sulphur, monoclinic sulphur, liquid sulphur and vapour sulphur. All these phases can be explained only by sulphur component.

2.4.1 Water System

In water system the following equilibria can exist:
1. Ice(s) ⇌ water(l)
2. Water(l) ⇌ vapour(g)
3. Ice(s) ⇌ vapour(g)
4. Ice(s) ⇌ water(l) ⇌ vapour(g)

This system can be explained with the help of following $T–P$ phase diagram (Figure 2.3). The number of phases that can co-exist in equilibrium at any point depends upon the conditions of temperature and pressure. The phase diagram of water system is consists of:

1. Three stable curves: OA, OB and OC
2. Three areas a, b and c
3. Triple point O

1. Curves
  (a) *Curve OA:* This curve is called vapour pressure curve of water. It represents the equilibrium between liquid water and water vapours at different temperatures. At 100°C, the vapour pressure of water becomes equal to atmospheric pressure; it is called the *boiling point of water*. This curve starts from point O and extends up to critical temperature A, (374°C and pressure 218 atm).

  Along this curve: liquid ⇌ vapour,

  Degree of freedom,
  $$F = C - P + 2 = 1 - 2 + 2 = 1$$
  (system along this curve is univariant)

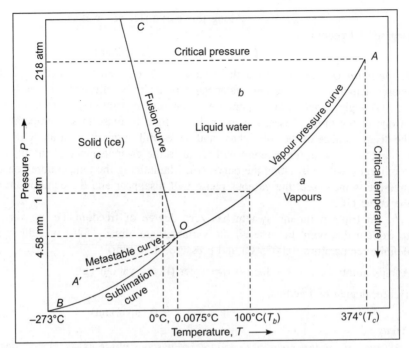

Figure 2.3 *T – P* phase diagram of water system.

(b) *Curve OA' (Metastable curve):* The dotted curve *OA'* is a continuation curve of *OA*. Sometimes it is possible to cool water several degrees below its freezing temperature without converting it into ice. This water is called *super-cooled water*. It represents the vapour pressure of super-cooled water.

The liquid ⇌ vapour equilibrium along this curve is called *metastable equilibrium* and this curve is called *metastable curve*. On a slight disturbance (i.e. adding ice piece or any other solid phase) this curve merges to sublimation curve (*OB*). Along this curve the degree of freedom,

$$F = 1 - 2 + 2 = 1 \text{ (univariant)}$$

Since this curve lies above the sublimation curve, *OB*, thus, the metastable curve has higher vapour pressure than the stable one phase curve at the same temperature.

(c) *Curve OB:* This curve is called *vapour pressure* or *sublimation curve* of ordinary ice. Along this curve, solid ice is in equilibrium with its vapours. It starts at point *O* and ends at point *B*, i.e. absolute zero (−273°C). At this temperature only ice can exist. Along this curve (Ice ⇌ vapour):

Degree of freedom,

$$F = C - P + 2 = 1 - 2 + 2 = 1 \text{ (univariant)}$$

(d) *Curve OC:* This curve is known as *freezing curve* of water or *fusion curve* of ice. This curve shows the temperature at which ice and water are in equilibrium at various pressures. This curve shows the effect of pressure on the melting point of ice, i.e. lowering of melting point of ice with increase in pressure. The curve slops only slightly towards the pressure axis as the effect of pressure on melting point is small.

Along this curve (Ice ⇌ water).

Degree of freedom,

$$F = 1 - 2 + 2 = 1 \text{ (univariant)}.$$

2. **Triple point O:** At this point three curves OA, OB and OC meet, i.e. at this point three phases namely, ice, water and water vapours are in equilibrium with each other.

A slight variation in temperature or pressure at point O may result in the disappearance of one of the three phases. For example, a slight increase in the temperature will result in the disappearance of the solid phase and the equilibrium will shift along the curve OA while a slight decrease in temperature will result in the disappearance of the liquid phase and the equilibrium will shift along the curve OB. Maintaining the temperature constant, if only the pressure is increased, the gaseous phase will disappear and the equilibrium will shift along the curve OC.

At triple point the system has zero degree of freedom, i.e. no variable is required to explain this point because all the variables are already fixed, i.e. this point occurs at a definite temperature (0.0075°C) and pressure (4.58 mm).

At this point: Ice (s) ⇌ water (l) ⇌ vapour (g)

Hence, degree of freedom,

$$F = 1 - 3 + 2 = 0 \text{ (invariant)}$$

3. **Areas**

   (i) Area *a*: In this area only vapours, i.e. gaseous phase exists. Hence, degree of freedom:

$$F = 1 - 1 + 2 = 2 \text{ (bivariant)}$$

   (ii) Area *b*: In this area only water, i.e. liquid phase exists. Hence, degree of freedom:

$$F = 1 - 1 + 2 = 2 \text{ (bivariant)}$$

   (iii) Area *c*: In this area only ice, i.e. solid phase exists. Hence, degree of freedom:

$$F = 1 - 1 + 2 = 2 \text{ (bivariant)}$$

The water system is summarized in Table 2.1.

Table 2.1 Brief description of water system

System	Specification	Phases in equilibrium	Degree of Freedom (variance)
I Curves			
1. Curve OA	Vapour pressure curve of water	Water ⇌ vapour	1
2. Curve OB	Sublimation curve of ice	Ice ⇌ Vapour	1
3. Curve OC	Fusion curve of ice	Ice ⇌ Water	1
4. Curve OA'	Meta stable curve	Supercooled Water ⇌ Vapour	1
II Point O	Triple point ($T = 0.0075°C$, $P = 4.58$ mm)	Ice ⇌ water ⇌ vapour	0
III Areas			
1. Area A	AOB	Vapour (gas phase)	2
2. Area b	AOC	Water (liquid phase)	2
3. Area c	BOC	Ice (solid phase)	2

Applications of water system

1. *Ice skating:* As it is clear from the phase diagram of water system that on increasing the pressure the freezing point of ice is lowered. So due to the pressure of skater the ice melts into water. This water layer acts as a lubricant between the ice and the pad of scatter and makes the skating easier.
2. *Sliding/flow of glacier:* The glaciers may be joined together in the form of huge pieces. So due to the high pressure of upper glacier piece the ice of lower piece melts which acts as a lubricant. So due to the separation of glaciers by water layer it slides and floats over water surface due to less density of ice than water.
3. This system helps to study of the polymorphism of water.

2.4.2 Carbon Dioxide System

In carbon dioxide system the following equilibria can exist:

1. Solid CO_2 ⇌ liquid CO_2
2. Liquid CO_2 ⇌ gas CO_2
3. Solid CO_2 ⇌ gas CO_2
4. Solid CO_2 ⇌ liquid CO_2 ⇌ gas CO_2

The number of phases that can co-exist in equilibrium at any point depends upon the conditions of temperature and pressure. The phase diagram of carbon dioxide system consists of:

1. Three stable curves—*OA*, *OB* and *OC*
2. Three areas—*a*, *b* and *c*
3. Triple point—*O*.

This system can be explained with the help of following $T - P$ phase diagram (Figure 2.4).

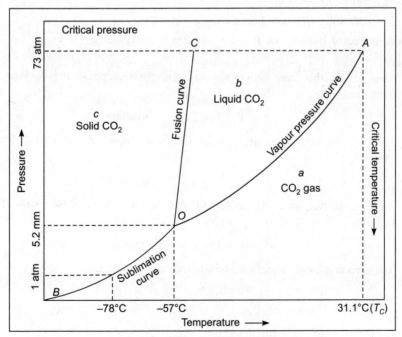

Figure 2.4 $T - P$ phase diagram of CO_2 system.

1. **Curves**
   (a) *Curve OA:* This curve is called vapour pressure curve of liquid CO_2. It represents the equilibrium between liquid CO_2 and CO_2 gas at different temperatures. This curve starts from point O and extends up to critical temperature A, (31.1°C and pressure 73.0 atm). Along this curve: liquid $CO_2 \rightleftharpoons CO_2$ gas

   Thus, the degree of freedom is $F = C - P + 2 = 1 - 2 + 2 = 1$ (univariant).

   (b) *Curve OB:* This curve is called vapour pressure or sublimation curve of solid CO_2 (dry ice). Along this curve solid CO_2 is in equilibrium with its vapours. It starts at point O and ends at point B. Along this curve: solid $CO_2 \rightleftharpoons$ vapour CO_2

   Thus, the degree of freedom is: $F = C - P + 2 = 1 - 2 + 2 = 1$ (univariant).

   (c) *Curve OC:* This is known as fusion curve of solid CO_2 or freezing curve of liquid CO_2. This curve shows the temperature at which solid CO_2 and liquid CO_2 are in equilibrium at various pressures. It shows the effect of pressure on melting point of dry ice, i.e. increase of melting point of dry ice (solid CO_2) with rise in pressure. The curve slopes only slightly away from the pressure axis as the effect of pressure on melting point of solid CO_2 is small.

   Along this curve: solid $CO_2 \rightleftharpoons$ liquid CO_2,

   Thus, the degree of freedom is: $F = 1 - 2 + 2 = 1$ (univariant).

2. **Triple point O:** At this point three curves OA, OB and OC meets and three phases namely, solid CO_2, liquid CO_2 and CO_2 gas are in equilibrium with each other.

   At triple point O, the system has zero degree of freedom, i.e. no variable is required to explain this point because all the variables are already fixed, i.e. this point occurs at a definite temperature (–57°C) and pressure (5.2 atm).

   At this point: Solid $CO_2 \rightleftharpoons$ liquid $CO_2 \rightleftharpoons$ gas CO_2

   Hence, degree of freedom is: $F = 1 - 3 + 2 = 0$ (invariant).

3. **Areas**
   (a) *Area a:* In this area only CO_2 gas, i.e. gaseous phase exists. Hence, degree of freedom,

   $$F = 1 - 1 + 2 = 2 \quad \text{(bivariant)}.$$

   (b) *Area b:* In this area only liquid CO_2, i.e. liquid phase exists. The degree of freedom,

   $$F = 1 - 1 + 2 = 2 \quad \text{(bivariant)}.$$

   (c) *Area c:* In this area only solid CO_2 (dry ice) i.e. solid phase exists. The degree of freedom,

   $$F = 1 - 1 + 2 = 2 \quad \text{(bivariant)}.$$

   The carbon dioxide system is summarized in Table 2.2.

Table 2.2 Brief description of carbon dioxide (CO_2) system

System	Specification	Phases in equilibrium	Degree of Freedom (F)
I Curves:			
1. Curve OA	Vapour pressure curve of liquid CO_2	Liquid $CO_2 \rightleftharpoons CO_2$ gas	1
2. Curve OB	Sublimation curve of solid CO_2	Solid $CO_2 \rightleftharpoons CO_2$ gas	1
3. Curve OC	Fusion curve of solid CO_2	Solid $CO_2 \rightleftharpoons$ liquid CO_2	1
II Point O	Triple point ($T = -57°C$, $P = 5.2$ atm)	$CO_2(s) \rightleftharpoons CO_2(l) \rightleftharpoons CO_2(g)$	0
III Areas:			
1. Area a	AOB	CO_2 gas (gas phase)	2
2. Area b	AOC	liquid CO_2 (liquid phase)	2
3. Area c	BOC	Solid CO_2 (solid phase)	2

Applications of carbon dioxide system

1. In the separation of lipids and phosphides by super critical fluid chromatography (SFC).
2. In the extraction of coffee from green beans.

Water system and carbon dioxide system differences (Table 2.3):

Table 2.3 Differences between water system and carbon dioxide system

	Water system		Carbon dioxide system
1.	The fusion curve slopes towards the pressure axis. This indicates that increase of pressure raises the melting point of solid carbon dioxide.	1.	The fusion curve slopes away from the pressure axis. This indicates that increase of pressure raises the melting point of solid CO_2.
2.	The factor dT/dP of the Clausius–Clapeyron equation is positive indicating thereby that $V_l < V_s$, i.e. specific volume of liquid water is less than that of solid ice.	2.	The factor dT/dP of the Clausius–Clapeyron equation is positive indicating that $V_l > V_s$, i.e. specific volume of liquid carbon dioxide is greater than that of solid carbon dioxide.
3.	In water system, ice and water can exist in equilibrium even at a low pressure equal to 4.58 mm.	3.	Solid carbon dioxide can exist in equilibrium with its liquid (i.e. it can melt) only at a very high pressure equal to 5.2 atm.
4.	The vapour pressure of solid carbon dioxide at extremely low temperature is very low (4.58 mm).	4.	The vapour pressure of solid carbon dioxide even at extremely low temperature is very high (higher than that of ice).
5.	Water can be solidified without the appearance of liquid phase, merely by cooling below 0°C.	5.	Even at atmospheric pressure (1 atm), CO_2 gas can be directly solidified without the appearance of liquid phase by cooling to –78°C.

Solid carbon dioxide is known as dry ice

Even at atmospheric pressure (1 atm), carbon dioxide gas can be directly solidified without the appearance of liquid phase merely on cooling to –78°C. It is, therefore, commonly known as *dry ice*.

2.5 POLYMORPHISM

The existence of a given substance in more than one crystalline form possessing different physical properties is known as *polymorphism*. This phenomenon occurs in a number of elements and compounds. When it occurs in elements, it is frequently referred to as *allotropy*.

Each polymorphic form comprises a separate phase. The temperature at which one form changes into another, at a given pressure, is known as the *transition temperature*. For example, in case of rhombic sulphur when it is heated at a pressure of one atmosphere, changes into monoclinic sulphur at 95.6°C and if the sulphur is cooled under a pressure of one atmosphere it changes into rhombic sulphur at 95.6°C (transition temperature of sulphur). Polymorphic forms which can transform reversibly from one form to another at the transition temperature are known as *enantiotropic forms* and the phenomenon is called *enantiotropy*.

2.5.1 Polymorphism of Water at High Pressure

According to Tamman and Bridgman the water under high pressure up to 50,000 atm shows several forms of ice other than ordinary ice. These forms are stable under different conditions of temperature and pressure. These modifications of ice are shown in Figure 2.5. The ordinary ice is shown as ice–I, other modifications of ice which are stable at very high pressures are designated as ice–II, ice–III, ice–IV, ice–V, ice–VI and ice–VII. These different form of ice differ in crystalline structure, density and other physical properties. When ice–I is compressed, its melting point is lowered and reaches –22°C at a pressure of about 2240 atm.

On further increasing the pressure ice–I transforms into ice–III. Ice–IV has not been isolated so it is not shown in the phase diagram. Ice–VII is the extreme high pressure modification which melts at about 100°C and 25,000 atm. This melting ice at 100°C is so hot and really it is a surprising fact. The regions marked by I, II, III, V, VI and VII indicate that these forms are stable in these regions. Except ice–II all other forms of ice can co-exist with water. There are six triple points C, D, E, F, G and H other than normal triple point O. All these points are invariant as maximum three phases can co-exist in equilibrium.

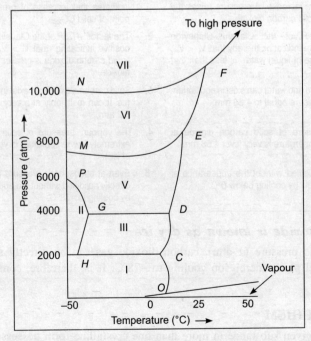

Figure 2.5 High pressure diagram of water system (polymorphic forms of water).

2.6 TWO-COMPONENT SYSTEMS

"The system, in which the composition of all phases is expressed in terms of two independent constituents, is called *two-component system*."

In a two-component system, when $C = 2$, degree of freedom has the highest value, $F = C - P + 2 = 2 - 1 + 2 = 3$, i.e. maximum number of degree of freedom is three and for the lowest number of degree of freedom (i.e. 0), the maximum number of phases that can exist in equilibrium is four, i.e. $P = 2 - 0 + 2 = 4$.

A solid–liquid equilibrium of an alloy has practically no gas phase and the effect of pressure is small in this type of equilibrium. Therefore, experiments are usually conducted under atmospheric pressure. Thus, keeping the pressure constant of a system in which vapour phase is not considered is known as *condensed system*.

The phase rule becomes, $$F = C - P + 1 \tag{2.14}$$

This equation is known as *condensed phase rule equation*.

Reduced phase rule equation

In the study of two-component system, usually, one of the three variables is kept constant and, therefore, the degree of freedom is reduced by one.

Hence, the phase rule equation for two components reduces to $F = C - P + 1$, and called *reduced phase rule equation*.

This condensed phase rule has two variables, namely temperature and concentration (or composition) of the constituents. Therefore, the solid–liquid equilibria are represented on temperature–composition diagram.

2.7 EUTECTIC SYSTEM

A binary system which consists of two substances, miscible in all proportions in the liquid phase but do not react chemically, is known as the eutectic system, e.g. a mixture of lead and silver comprises of such a system.

1. *Eutectic mixture:* It is a solid solution of two or more substances showing the lowest freezing point of all the possible mixtures of components. The mixture of metal alloys is of low melting point than the pure metal which generally form eutectic mixtures.
2. *Eutectic point:* The minimum freezing point attainable corresponding to the eutectic mixture is called eutectic point (lowest melting point).
   For example—Bi (m.p. 273°C) and Cd (m.p. 323°C) form eutectic mixture at composition of (60% Bi + 40% Cd) and eutectic temperature 140°C.
3. *Applications of Eutectic System*
   (a) To study the properties of alloys of different mixtures.
   (b) To study the lowest melting or freezing point mixture.
   (c) To study the strength of alloys.
   (d) To prepare the alloys of lowest melting points used in safety devices.
   (e) To study the extraction and purification of metals.

2.7.1 Lead–Silver (Pb–Ag) System

This system constitutes of two components (Pb and Ag) and three phases Pb(s), Ag(s), solution of molten Pb and Ag. Since the boiling point of Pb and Ag is very high so the vapour phase is practically absent and hence there is no effect of pressure on the system, so the pressure variable is neglected and condensed phase rule equation $(F = C - P + 1)$ is used.

In such a case, the system can be explained with the help of T–C phase diagram considering only two variables, i.e. temperature and composition (Figure 2.6).

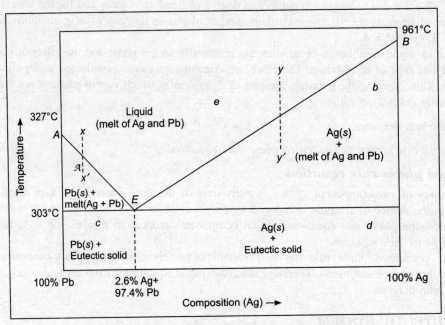

Figure 2.6 T– C Phase diagram of Pb–Ag system.

Salient features of Pb–Ag system

1. **Curves:** There are two curves:
    (a) *Curve AE:* This curve is called *fusion curve* or *freezing point curve* of lead and shows the effect on melting point of lead (327°C) on adding silver in it. Along this curve two phases i.e. Pb(s) and solution of Pb and Ag melt are in equilibrium. Then, using the condensed phase rule equation, we have

    $$F = C - P + 1 = 2 - 2 + 1 = 1 \quad \text{(univariant)}.$$

    (b) *Curve BE:* This curve is called *fusion curve* or *freezing point curve* of silver and shows the effect on melting point of silver (961°C) on adding lead in it. Along this curve two phases, i.e. Ag(s) and solution of Ag, and Pb melt are in equilibrium. Then, using the condensed phase rule equation, we have

    $$F = C - P + 1 = 2 - 2 + 1 = 1 \quad \text{(univariant)}.$$

2. **Point E:** This point is called *eutectic point*. At this point two curves, AE and BE meet. This point is obtained at the fixed temperature (303°C) and composition (2.6% Ag and 97.4% Pb). If the temperature is raised above eutectic point, the solid phase disappears and if the temperature is lowered below this point, the liquid phase disappears. Using condensed phase rule equation, we have

$$F = C - P + 1 = 2 - 2 + 1 = 1 \quad \text{(invariant)}$$

3. *Areas*

   (a) *Area a:* This area contains Pb(s) and solution of molten Pb and Ag, and represents the two-phase system.
   Thus,
   $$F = C - P + 1 = 2 - 2 + 1 = 1 \quad \text{(univariant)}.$$

   (b) *Area b:* This area contains Ag(s) and solution of molten Ag and Pb, and represents the two-phase system.
   Thus,
   $$F = C - P + 1 = 2 - 2 + 1 = 1 \quad \text{(univariant)}.$$

   (c) *Area c:* This area lies below all the melting points, hence contains only solid phases, i.e. Pb(s) and eutectic solid, and represents two-phase system.
   Thus,
   $$F = C - P + 1 = 2 - 2 + 1 = 1 \quad \text{(univariant)}.$$

   (d) *Area d:* This area also lies below all the melting points, hence contains only solid phases, i.e. Ag(s) and eutectic solid, and represents two-phase system.
   Thus,
   $$F = C - P + 1 = 2 - 2 + 1 = 1 \quad \text{(univariant)}.$$

   (e) *Area e:* This area lies above all the melting points and contains only solution of molten Ag and Pb, and represents the single phase system.
   Thus,
   $$F = C - P + 1 = 2 - 1 + 1 = 2 \quad \text{(bivariant)}.$$

The lead–silver system is summarized in Table 2.4.

Table 2.4 Brief description of Pb–Ag system

System	Specification	Phases in equilibrium	Degree of freedom (variance, F)
Curves:			
1. Curve AE	Fusion curve of lead	Pb(s) $\rightleftharpoons$ solution (Pb and Ag melt)	1
2. Curve BE	Fusion curve of silver	Ag(s) $\rightleftharpoons$ solution (Pb and Ag melt)	1
Point E	Eutectic poin (T = 303°C, C = 2.6 % Ag + 97.4% Pb)	Pb(s) $\rightleftharpoons$ solution Ag(s)	0
Areas:			
Area a	Below curve AE	Pb(s) $\rightleftharpoons$ solution (Pb & Ag melt)	1
Area b	Below curve BE	Ag(s) $\rightleftharpoons$ solution (Pb & Ag melt)	1
Area c	Below all melting points	Pb(s) and eutectic solid	1
Area d	Below all melting points	Ag(s) and eutectic solid	1
Area e	Above all melting points	Single phase [solution (Pb and Ag melt)]	2

Application of Pb–Ag system in the desilverisation of lead (Pattinson's process)

If a sample of argentiferrous lead containing less than 2.6% Ag is allowed to cool gradually along line xx' (Figure 2.6), lead will separate out leaving behind the solution richer in Ag till the percentage of Ag reaches 2.6%. On further cooling the whole mass will solidify as such. On the other hand, if the Pb–Ag alloy solution containing Ag greater than 2.6% is cooled along line yy' the pure silver is separated out till the eutectic point E (2.6% Ag) is reached. In this way, the pure silver is separated out from the mixture of lead and silver.

2.8 CONGRUENT SYSTEM

The system in which two components at a certain composition combine together to form a stable compound which melts congruently, i.e. melts sharply at a certain temperature without change in composition in the liquid state. Example, zinc–magnesium system.

Characteristics of such systems

The binary systems, forming chemical compounds with congruent and incongruent melting points possess the following characteristics:

1. The components of these systems, under suitable conditions of temperature and composition, enter into chemical combination, producing compounds which are frequently stable and melt without decomposition. Such melting is called *congruent melting*.
2. At the melting point of the compound the liquid so formed has the same composition of solid and liquid phases and is called a *congruent melting compound*.
3. The compound having incongruent melting point, separates out as a new solid phase from the solution.
4. The phase diagram of such cases shows a peak in the liquidus at its middle with an eutectic on either side. If the peak is sharp (a well-defined maxima), the compound formed is stable and its melting point is at the maxima.
5. In case of flat maxima the compound formed appears to be less stable.
6. When more than one compounds are formed, the number of peaks in the liquidus will also be greater.
7. Formation of congruent melting compound is confirmed by constructing the phase diagram of the system.
8. The examples of such type of systems are intermetallic compounds (when the two components are metals), double compounds (when the two components are organic compounds), salt hydrates (when the two components are salt and water), etc.

The specific cases are: (i) zinc–magnesium system, (ii) aluminium–magnesium system, mercury–thorium system, (iv) phenol–aniline system (v) ferric chloride–water system.

2.8.1 Zinc–Magnesium (Zn–Mg) System

The T–C phase diagram of Zn–Mg system is shown in Figure 2.7.

Salient features of Zn–Mg system

The phase diagram (Figure 2.7) shows that it consists of two simple eutectic diagrams joined together at point C. The diagram to the left represents the eutectic system AE_1C, while that on the right the other eutectic system BE_2C.

1. **Curves:**
    (a) *Curve AE_1:* It is the fusion curve of Zn. It shows that when Mg is added to Zn, the freezing point of Zn is lowered. Along this curve, two phases, i.e. solid Zn and solution of Zn and Mg are in equilibrium with each other. Applying condensed phase rule equation,

    $$F = C - P + 1 = 2 - 2 + 1 = 1$$

    Thus, the system is univariant.

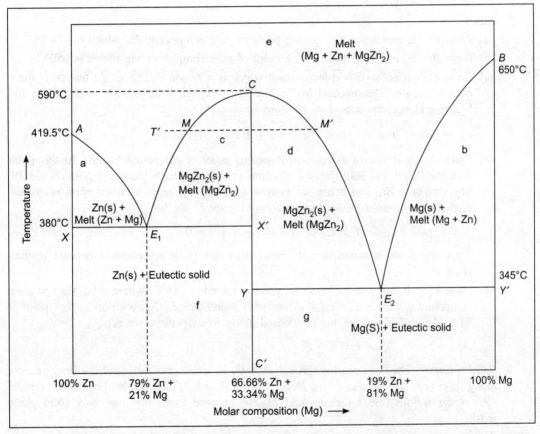

Figure 2.7   Phase diagram of Zn–Mg system showing the formation of compound with congruent melting point (P = constant).

(b) *Curve BE_2:* It is the fusion curve of Mg. It expresses that when Zn is added to Mg, the freezing point of Mg is lowered. Along this curve, the solid Mg is in equilibrium with solution of Zn and Mg. Hence according to condensed phase rule equation,

$$F = C - P + 1 = 2 - 2 + 1 = 1 \quad \text{(univariant)}.$$

(c) *Curve CE_1 and CE_2:* Curve CE_1 and CE_2 are the freezing or fusion curves of congruent compound $MgZn_2$. Along these curves, the solid compound $MgZn_2$ is in equilibrium with the liquid phase of Zn and Mg along CE_1 and CE_2, respectively at different temperatures. Hence, along these curves the degree of freedom,

$$F = C - P + 1 = 2 - 2 + 1 = 1 \quad \text{(univariant)}.$$

It can be observed that at a certain temperature T the liquid phase can have two compositions M and M' with the same solid $MgZn_2$ i.e. the compound $MgZn_2$ can have two solubilities at same temperature. This is because the compound $MgZn_2$ is in equilibrium with different components i.e. Zn and Mg at two different points M and M' respectively.

2. **Points**
   (a) *Point A:* It represents the melting point of pure component Zn, which is 419.5°C.
   (b) *Point B:* It represents the melting point of pure component Mg which is 650°C.
   (c) *Point E_1:* It is the first eutectic point which is obtained at 380 °C. At this point three phases, i.e. Zn(s), compound $MgZn_2$ are in quilibrium with solution (melt of Zn, and $MgZn_2$). Hence, the degree of freedom is:

   $$F = C - P + 1 = 2 - 3 + 1 = 0 \quad \text{(invarient)}.$$

   (d) *Point C:* It is known as *congruent melting point* of compound $MgZn_2$. At this point both the liquid and solid phases of compound $MgZn_2$ have same composition (33.3% Mg + 66.66% Zn) and hence the *system at this point becomes one-component system.* Applying condensed phase rule equation to point C, we get

   $$F = C - P + 1 = 1 - 2 + 1 = 0$$

   Thus, the point is nonvariant. It means that this point represents a definite melting point.

   (e) *Point E_2:* It is the second eutectic point obtained at 345°C, where solid Mg and solid compound $MgZn_2$ are in equilibrium with liquid phase. Composition at this point is 81% Mg–19% Zn. Applying condensed phase rule equation, we get

   $$F = C - P + 1 = 1 - 2 + 1 = 0.$$

3. **Solidus lines:** The lines passing through eutectic E_1 and E_2 are called solidus lines because along them only solid phase exists. Hence lines XX' and YY' are solidus lines. The vertical CC' is the solidus line for compound $MgZn_2$ because along this line only solid phase exist.

4. **Areas**
   (a) *Area a:* It represents the area below curve AE_1. This area consists of two phases, namely solid Zn and liquid melt. Hence,

   $$F = C - P + 1 = 2 - 2 + 1 = 1 \quad \text{(univariant)}.$$

   (b) *Area b:* It represents the area below curve AE_2 and consists of two phases—solid Mg and liquid melt. Hence,

   $$F = C - P + 1 = 2 - 2 + 1 = 1 \quad \text{(univariant)}.$$

   (c) *Area c:* (below curve CE_1): It consists of two phases, i.e. solid compound $MgZn_2$ and liquid melt and hence,

   $$F = C - P + 1 = 2 - 2 + 1 = 1.$$

(d) *Area d* (below curve CE_2): It represents the co-existence of two phases, namely solid compound $MgZn_2$ and liquid melt. Hence,

$$F = C - P + 1 = 2 - 2 + 1 = 1 \quad \text{(univariant)}.$$

(e) *Area e* (below line XX'): It represents the coexistence of solid Zn and solid compound $MgZn_2$. Hence,

$$F = C - P + 1 = 2 - 2 + 1 = 1 \quad \text{(univariant)}.$$

(f) *Area f* (below line YY'): It represent the coexistence of solid Mg and solid compound $MgZn_2$. Hence system is univariant.

(g) *Area g* (above all melting points, AE_1CE_2B): It represent the existence of only one phase, i.e. liquid solutions. Hence, $F = C - P + 1 = 2 - 1 + 1 = 2$, system is bivariant.

The complete Zn–Mg system is summarized in Table 2.5.

Table 2.5 Brief description of Zn–Mg system

System	Specification	Phases in equilibrium	Degree of freedom (F)
Curves:			
Curve AE_1	Fusion curve of zinc	$Zn(s) \rightleftharpoons$ solution (melt of Zn and Mg)	1
Curve CE_1	Fusion curve of congruent compound $MgZn_2$	$MgZn_2(s) \rightleftharpoons$ melt ($MgZn_2$ and Zn)	1
Curve CE_2	Fusion curve of congruent compound $MgZn_2$	$MgZn_2(s) \rightleftharpoons$ melt ($MgZn_2$ and Mg)	1
Curve BE_2	Fusion curve of magnesium	$Mg(s) \rightleftharpoons$ solution (melt of Mg and Zn)	1
Points:			
Point A	Melting point of Zn (419.5°C)	Pure Zn	0
Point E_1	First eutectic point (380°C)	$Zn(s) \rightleftharpoons$ solution $\rightleftharpoons MgZn_2(s)$	0
Point C	Melting point of congruent compound $MgZn_2$ (590°C)	$MgZn_2(s) \rightleftharpoons$ Melt of $MgZn_2$ (this point is one-component system)	0
Point E_2	Second eutectic point (347°C)	$Mg(s) \rightleftharpoons$ Melt of $MgZn_2 \rightleftharpoons MgZn_2(s)$	0
Point B	Melting point of Mg (650°C)	Pure Mg	0
Areas:			
Area a	Below curve AE_1	$Zn(s) \rightleftharpoons$ solution (Zn+Mg)	1
Area b	Below curve BE_2	$Mg(s) \rightleftharpoons$ solution (Mg + Zn)	1
Area c	Below curve CE_1	$MgZn_2(s) \rightleftharpoons$ solution ($MgZn_2$ + Zn)	1
Area d	Below curve CE_2	$MgZn_2(s) \rightleftharpoons$ solution ($MgZn_2$+Mg)	1
Area e	Below line XX' (below all melting points)	Zn(s) + solid eutectic	1
Area f	Below line YY' (below all melting points)	Mg(s) + solid eutectic	1
Area g	Above line AE_1CE_2B (above all melting points)	Single phase (solution of Zn, Mg and $MgZn_2$)	2

Applications of congruent system

1. To study the chemical affinity (reaction) between two metals.
2. To study the formation of inter-metallic compounds or double compounds.
3. To study the composition of inter-metallic or double compounds at different temperatures.
4. To study the thermal stability of such compounds.
5. To study the composition of such compounds at melting temperature.

2.9 INCONGRUENT SYSTEMS

In these type of systems, two components combine chemically to form a compound (incongruent compound) which melts incongruently, i.e. melts giving a new solid phase. The composition of solution (melt) is different from that of solid phase. These type of compounds having incongruent melting points are called *incongruent melting compounds*.

$$\underset{\text{Original compound (solid)}}{AB} \rightleftharpoons \underset{\text{New solid}}{C} + \text{Solution (or melt)}$$

This reaction is called *transition reaction* or peritectic or meritectic reaction and the temperature is called *transition temperature* or *peritectic* or *meritectic* temperature (or point).

2.9.1 Sodium–Potassium (Na–K) System

In this system, the two components, sodium and potassium combine together chemically to form a compound KNa_2 which melts incongruently giving a new solid phase:

$$KNa_2 \rightleftharpoons \text{New solid phase} + \text{Melt}$$

General characteristics of sodium–potassium system

The phase diagram (Figure 2.8) shows that it consists of:

1. *Curves:* There are mainly three curves in the system. Fourth one is hypothetical curve.

    (a) *Curve AE:* This curve is called *fusion curve* or *freezing curve of potassium* and shows the effect on melting point (lowering/depression of m.p.) of potassium (63.8°C) on adding sodium in it. Along this curve, two phases, i.e. K(s) and solution of K and Na melt are in equilibrium. Hence, using the condensed phase rule equation, we have

    $$F = C - P + 1 = 2 - 2 + 1 = 1 \quad \text{(univariant)}.$$

    (b) *Curve BI:* This curve is called *fusion curve* or *freezing point curve* of sodium and shows the effect on melting point of sodium (97.8°C) on adding potassium in it. Along this curve two phases, i.e. Na(s) and solution of Na and K melt are in equilibrium. Hence, we have,

    $$F = C - P + 1 = 2 - 2 + 1 = 1 \quad \text{(univariant)}.$$

    (c) *Curve IE:* This curve is called *fusion curve* or *freezing point curve* of incongruent compound KNa_2 and shows the effect on melting point (lowering/depression of m.p.) of KNa_2 on adding potassium in it. Along this curve two phases, i.e. KNa_2(s) and solution of KNa_2 and K melt are in equilibrium.

Hence, we have,

$$F = C - P + 1 = 2 - 2 + 1 = 1 \quad \text{(univariant)}.$$

(d) *Curve IC:* It is the hypothetical congruent point. If the compound KNa_2 does not melt incongruently at point *I* then it will melt at point *C* congruently, i.e. without change in composition. Along this curve the solid KNa_2 and $Na(s)$ will be in equilibrium with solution. Hence, we have,

$$F = C - P + 1 = 2 - 2 + 1 = 1 \quad \text{(univariant)}.$$

2. **Points**
   (a) *Point A:* Melting point of pure K (63.8°C).
   (b) *Point B:* Melting point of pure Na (97.8°C).
   (c) *Point E:* It is the eutectic point. At this point two phases: $K(s)$ and $KNa_2(s)$ are in equilibrium with solution (melt of K and KNa_2). Hence, the degree of freedom at this point,

$$F = C - P + 1 = 2 - 3 + 1 = 0 \quad \text{(invariant)}.$$

   (d) *Point I:* It is the incongruent point at which the compound KNa_2 melts giving a new solid phase. There are three phases $KNa_2(s)$, new solid phase and solution are in equilibrium. Hence we have,

$$F = C - P + 1 = 2 - 3 + 1 = 0 \quad \text{(invariant)}.$$

   (e) *Point C:* It is the hypothetical congruent point. At this point the compound KNa_2 may be in equilibrium with its melt.

$$F = C - P + 1 = 1 - 2 + 1 = 0 \quad \text{(invariant)}.$$

3. **Areas**
   (a) *Area a:* It represents the area below curve AE_1. This area consists of two phases, namely solid K and melt. Hence,

$$F = C - P + 1 = 2 - 2 + 1 = 1 \quad \text{(univariant)}.$$

   (b) *Area b:* It represents the area below curve *AI* and consists of two phases–solid Na and melt. Hence,

$$F = C - P + 1 = 2 - 2 + 1 = 1 \quad \text{(univariant)}.$$

   (c) *Area c* (below curve *IE*): It consists of two phases solid compound KNa_2 and melt and hence,

$$F = C - P + 1 = 2 - 2 + 1 = 1 \quad \text{(univariant)}.$$

   (d) *Area d* (below line *DD'*): It represents co-existence of two phases, namely solid K and compound $KNa_2(s)$. Hence,

$$F = C - P + 1 = 2 - 2 + 1 = 1 \quad \text{(univariant)}.$$

(e) *Area e* (below line *LL'*): It represents the coexistence of solid K and solid compound KNa_2. Hence,

$$F = C - P + 1 = 2 - 2 + 1 = 1 \quad \text{(univariant)}.$$

(f) *Area f* (above all melting points, *AEIB*): It represents the existence of only one phase, i.e. liquid solutions. Hence,

$$F = C - P + 1 = 2 - 1 + 1 = 2 \quad \text{(bivariant)}.$$

The complete summary of K–Na system is given in Table 2.6.

Table 2.6 Brief description of Na–K system

System	Specification	Phases in equilibrium	Degree of freedom (F)
Curves:			
Curve *AE*	Fusion curve of potassium	K(s) ⇌ solution (melt of K and Na)	1
Curve *IE*	Fusion curve of incongruent compound KNa_2	KNa_2 (s) ⇌ melt (KNa_2 and K)	1
Curve *BI*	Fusion curve of sodium	Na(s) ⇌ solution (melt of Na and K)	1
Curve *IC*	Hypothetical congruent curve of compound KNa_2	KNa_2 (s) ⇌ solution	1
Points:			
Point *A*	Melting point of K (63.8°C)	Pure K	0
Point *E*	Eutectic point (T_E°C)	K(s) ⇌ solution ⇌ KNa_2(s)	0
Point *I*	Melting point of incongruent compound KNa_2 (T_I°C)	KNa_2(s) ⇌ solution ⇌ new solid phase	0
Point *B*	Melting point of Na (97.8°C)	Pure Na	0
Point *C*	Hypothetical congruent point	KNa_2(s) ⇌ KNa_2(melt)	0
Areas:			
Area *a*	Below curve *AE*	K(s) ⇌ solution (melt of K and Na)	1
Area *b*	Below curve *BE*	Na(s) ⇌ solution (melt of Na and K)	1
Area *c*	Below curve *IE*	KNa_2(s) ⇌ melt (KNa_2 and K)	1
Area *d*	Below line *DD'* (below all melting points)	Na(s) + KNa_2(s)	1
Area *e*	Below line *LL'* (below all melting points)	K(s) + KNa_2(s)	1
Area *f*	Above line *AEIB* (above all melting points)	Single phase (solution of K, Na and KNa_2)	2

Study of cooling effect on curves

Consider a point *X* in liquid area above the curve *AE*. If the liquid is cooled along line *XX'* (Figure 2.8), then there is no change in composition of liquid up to point *A'* but on further cooling it follows the path *A'C* with the separation of solid K, i.e. its composition changes up to the formation of eutectic compound at point *E*.

We observe the similar trend on cooling the liquid along line *YY'*, above the area *BI*. On cooling of liquid there is no change in composition up to the point *B'*. Cooling of liquid after point *B'* leads to change in composition with separation of solid Na and it follows the path *B'I*.

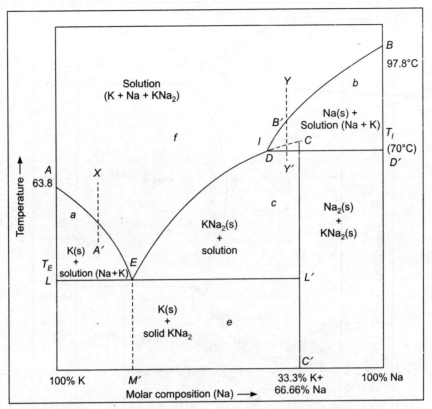

Figure 2.8 Phase diagram of sodium–potassium (Na–K) system.

Applications of incongruent system

1. To study the chemical affinity and formation of intermetallic compounds between the two metals.
2. To study the composition of inter metallic or double compounds at different temperatures.
3. To study the thermal stability of such compounds.
4. To study the composition of such compounds at melting temperature.
5. To study the formation of new solid phase at melting temperature of such compounds.

2.9.2 Sodium Sulphate–Water System

This system is an example of incongruent system. Sodium sulphate forms two hydrates $Na_2SO_4 \cdot 10H_2O$ and $Na_2SO_4 \cdot 7H_2O$. Also, the anhydrous salt can exist in two anisotropic crystalline forms, namely, rhombic and monoclinic. Thus in this system, there are five solid phases in all, including ice in the system. The phase diagram of the system is shown in Figure 2.9.

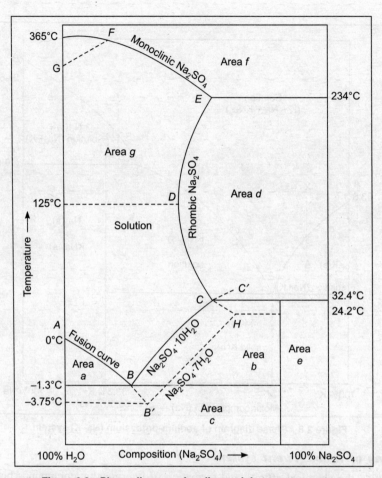

Figure 2.9 Phase diagram of sodium sulphate–water system.

The sodium sulphate–water system consists of eight phases as given in Table 2.7.

Table 2.7 Phases (with composition) present in sodium sulphate–water system

S. No.	Phase	Composition	S.No.	Phase	Composition
(i)	$Na_2SO_4 \cdot 10H_2O$ (s)	$Na_2SO_4 + 10H_2O$	(v)	Ice (s)	H_2O
(ii)	$Na_2SO_4 \cdot 7H_2O$ (s)	$Na_2SO_4 + 7H_2O$	(vi)	Water (l)	H_2O
(iii)	$Na_2SO_4 \cdot$ (Rhombic)	Na_2SO_4	(vii)	$Na_2SO_4 \cdot$ (Solution)	$Na_2SO_4 \cdot xH_2O$
(iv)	$Na_2SO_4 \cdot$ (Monoclinic)	Na_2SO_4	(viii)	Vapour	H_2O

Since all the measurements are made at constant pressure, hence vapour phase is not considered in the phase diagram:

The phase–diagram of $Na_2SO_4 \cdot H_2O$ system can be explained with the help of following T–C phase diagram:

1. ***Point A:*** It is the freezing point of pure water (0°C temperature).
2. ***Curve AB:*** It is the fusion curve of ice, which represents the depression in freezing point

of water on adding Na_2SO_4 in it. Two phases viz, ice and solution of Na_2SO_4 in water are in equilibrium.

Hence, $\qquad F = C - P + 1 = 2 - 2 + 1 = 1 \quad$ (univariant).

3. **Point B:** It is the eutectic (cryohydric point, $-1.3°C$ temperature). At this point new solid phase $(Na_2SO_4 \cdot 10H_2O)$ begins to appear. At this point three phases, viz., $Na_2SO_4 \cdot 10H_2O$, ice and solution co-exist.

Hence, $\qquad F = C - P + 1 = 2 - 3 + 1 = 0 \quad$ (invariant).

4. **Curve BC:** It is the solubility curve of $Na_2SO_4 \cdot 10H_2O$, which represents the effect of temperature on solubility of Na_2SO_4. In presence of excess of $Na_2SO_4 \cdot 10H_2O$ salt, ice starts melting and salt gets dissolved to give solution. Along this curve, two phases, solid $Na_2SO_4 \cdot 10H_2O$ and its solution are in equilibrium.

Hence, $\qquad F = C - P + 1 = 2 - 2 + 1 = 1 \quad$ (the system is univariant).

5. **Point C:** It is the incongruent melting point of decahydrate $Na_2SO_4 \cdot 10H_2O$. At this point $Na_2SO_4 \cdot 10H_2O$ melts incongruently (i.e. melts with change in composition) giving a new solid phase, i.e. Na_2SO_4.

$$Na_2SO_4 \cdot 10H_2O(s) \rightleftharpoons \text{Rhombic } Na_2SO_4 \text{ (s) + Solution}$$

At this point three phases viz., solid $Na_2SO_4 \cdot 10H_2O$, rhombic Na_2SO_4 and solution co-exist in equilibrium.

Hence, $\qquad F = C - P + 1 = 2 - 3 + 1 = 0 \quad$ (invariant).

6. **Curve CDE:** It is the solubility curve of rhombic Na_2SO_4, which shows the effect of temperature on solubility of rhombic Na_2SO_4. Along this curve anhydrous rhombic Na_2SO_4 is in equilibrium with solution and hence $F = C - P + 1 = 2 - 2 + 1 = 1$ (system is univariant). The inclination of this curve towards temperature axis up to D shows that the solubility of rhombic Na_2SO_4 decreases up to temperature $125°C$ and then increases slightly with further rise in temperature up to $234°C$ (point E).

7. **Point E:** This point is called as *transition point* (temperature $234°C$). At this point enantiotropic transformation of rhombic Na_2SO_4 to monoclinic Na_2SO_4 takes place and three phases are in equilibrium.

$$\text{Rhombic } Na_2SO_4 \rightleftharpoons \text{Monoclinic } Na_2SO_4 \rightleftharpoons \text{Solution}$$

Hence, $\qquad F = C - P + 1 = 2 - 3 + 1 = 0 \quad$ (invariant or non-variant).

8. **Curve EF:** It is the solubility curve of anhydrous monoclinic Na_2SO_4, which shows the effect of temperature on solubility of monoclinic Na_2SO_4. Inclination of curve towards the temperature axis shows that the solubility of monoclinic Na_2SO_4 decreases with rise in temperature up to $365°C$. Along this curve anhydrous monoclinic Na_2SO_4 is in equilibrium with solution and hence $F = C - P + 1 = 2 - 2 + 1 = 1$ (system is univariant).

9. **Point G:** This point gives the maximum temperature limit up to which the system can be studied.

10. *Metastable equilibria*
    (a) *Curve BB′ and point B′:* Curve BB′ is the metastable fusion curve of ice. If $Na_2SO_4 \cdot 10H_2O$ fails to appear at point B then the curve AB extends to B′, where $Na_2SO_4 \cdot 7H_2O$ appears. Along curve BB′ ice is in equilibrium with solution and degree of freedom is one. At *metastable point B′* (−3.75°C) three phases, $Na_2SO_4 \cdot 7H_2O$, ice and solution are in equilibrium.

    Hence, $F = C - P + 1 = 2 - 3 + 1 = 0$ (system is invariant or nonvariant).

    (b) *Curve CC′:* It is the metastable solubility curve of $Na_2SO_4 \cdot 10H_2O$. Along this curve two phases $Na_2SO_4 \cdot 10H_2O$ and solution are in metastable equilibrium.

    Hence, $F = C - P + 1 = 2 - 2 + 1 = 1$ (system is univariant).

    (c) Curve *CH′:* It is the metastable solubility curve of heptahydrate, $Na_2SO_4 \cdot 7H_2O$ and phases co-existing along this curve are:

    $$Na_2SO_4.7H_2O \ (s) \rightleftharpoons Na_2SO_4 \cdot 7H_2O \ \text{(solution)}$$

    Hence, $F = 2 - 2 + 1 = 1$ (univariant)

    (d) *Point H:* It is the metastable transition point of heptahydrate to anhydrous rhombic Na_2SO_4 at 24.2°C temperature. Along this curve:

    $$Na_2SO_4 \cdot 10H_2O \rightleftharpoons \text{Rhombic } Na_2SO_4 \rightleftharpoons \text{Solution}$$

    Hence, $F = 2 - 3 + 1 = 0$ (nonvariant).

    (e) *Curve FG:* It is the metastable solubility curve of monoclinic Na_2SO_4. It shows the solubility of monoclinic Na_2SO_4 in vapour phase. Solubility of Na_2SO_4 decreases with decrease in temperature along FG. At point G solubility becomes zero. Along this curve:

    $$\text{Monoclinic } Na_2SO_4(s) \rightleftharpoons \text{Monoclinic } Na_2SO_4 \ \text{(solution in vapours)}$$

    Hence, $F = 2 - 2 + 1 = 1$ (univariant).

11. *Area*
    (a) *Area a:* In this area, ice is in equilibrium with solution, i.e. Ice $\rightleftharpoons$ solution of $Na_2SO_4 \cdot 10H_2O$. Hence, $F = 2 - 2 + 1 = 1$ (univariant).
    (b) *Area b:* In this area solid phase, $Na_2SO_4 \cdot 10H_2O$ and its solution are in equilibrium. Hence, $F = 2 - 2 + 1 = 1$ (univariant).
    (c) *Area c:* Ice $\rightleftharpoons Na_2SO_4 \cdot 10H_2O$. Hence, $F = 2 - 2 + 1 = 1$ (univariant).
    (d) *Area d:* In this area rhombic $Na_2SO_4 \cdot 10H_2O$ and its solution are in equilibrium. Hence, $F = 2 - 2 + 1 = 1$ (univariant).
    (e) *Area e:* In this area $Na_2SO_4 \cdot 10 \ H_2O(s)$ and $Na_2SO_4 \cdot 7 \ H_2O(s)$ are in equilibrium. Hence, $F = 2 - 2 + 1 = 1$ (univariant).
    (f) *Area f:* In this area monoclinic $Na_2SO_4(M)$ and its solution are in equilibrium. Hence, $F = 2 - 2 + 1 = 1$ (univariant).
    (g) *Area g:* In this area there is only unsaturated solution of various species present. Hence, $F = 2 - 1 + 1 = 2$ (bivariant).

The brief summary of sodium sulphate–water system is given in Table 2.8.

Table 2.8 Brief summary of sodium sulphate-water system

System	Specification	Phases in equilibrium	Degree of freedom
Points:			
Point A	Freezing point of pure water (or m.p. of ice (0°C)	Pure ice	0
Point B	Eutectic (cryohydric) point (–1.3°C)	$Na_2SO_4 \cdot 10H_2O \rightleftharpoons$ Ice $\rightleftharpoons$ Solution	0
Point C	Incongruent point of $Na_2SO_4 \cdot 10H_2O$ (32.4°C)	$Na_2SO_4 \cdot 10H_2O \rightleftharpoons$ Rhombic Na_2SO_4 $\rightleftharpoons$ Solution	0
Point E	Transition point (240°C)	$Na_2SO_4 \rightleftharpoons$ Solution	0
Point B′	Metastable point (–3.75°C)	$Na_2SO_4 \cdot 7H_2O \rightleftharpoons$ Ice $\rightleftharpoons$ Solution	0
Point H	Metastable transition point (24.2°C)	$Na_2SO_4 \cdot 7H_2O \rightleftharpoons Na_2SO_4$ (s) $\rightleftharpoons$ Solution	
Point F	Critical point (365°C)	Water $\rightleftharpoons$ Vapour	0
Curves:			
Curve AB	Fusion curve of ice	Ice $\rightleftharpoons$ Solution of Na_2SO_4	1
Curve BC	Solubility curve of $Na_2SO_4 \cdot 10H_2O$	$Na_2SO_4 \cdot 10H_2O \rightleftharpoons$ Solution	1
Curve CE	Solubility curve of rhombic $Na_2SO_4 \cdot 10H_2O$	Rhombic $Na_2SO_4 \rightleftharpoons$ Solution Na_2SO_4	1
Curve EF	Solubility curve of monoclinic Na_2SO_4	Monoclinic $Na_2SO_4 \rightleftharpoons$ Solution	1
Metastable Curves:			
Curve BB′	Metastable fusion curve of ice	Ice $\rightleftharpoons$ Solution of $Na_2SO_4 \cdot 10H_2O$	1
Curve B′H	Metastable solubility curve of $Na_2SO_4 \cdot 7H_2O$	$Na_2SO_4 \cdot 7H_2O \rightleftharpoons$ Solution	1
Curve CC′	Metastable solubility curve of $Na_2SO_4 \cdot 10H_2O$	$Na_2SO_4 \cdot 10H_2O \rightleftharpoons$ Solution	1
Curve CH	Metastable solubility curve of rhombic Na_2SO_4	Rhombic $Na_2SO_4 \rightleftharpoons$ Solution	1
Curve FG	Metastable solubility curve of monoclinic Na_2SO_4 in vapours	Monoclinic $Na_2SO_4 \rightleftharpoons$ Solution in vapour	1
Areas:			
Area a	–	Ice $\rightleftharpoons Na_2SO_4 \cdot 10H_2O$ (solution)	1
Area b	–	$Na_2SO_4 \cdot 10H_2O$(s) $\rightleftharpoons Na_2SO_4 \cdot 10H_2O$ (solution)	1
Area c	–	Ice $\rightleftharpoons Na_2SO_4 \cdot 10H_2O$	1
Area d	–	Na_2SO_4 rhombic $\rightleftharpoons Na_2SO_4$ (solution)	1
Area e	–	Mixture of $Na_2SO_4 \cdot 10H_2O$(s) and $Na_2SO_4 \cdot 7H_2O$(s)	1
Area f	–	Na_2SO_4(M) $\rightleftharpoons Na_2SO_4$ (M solution)	1
Area g	–	Unsaturated solution	2

Applications of incongruent (salt hydrate) system

1. To study the formation of salt hydrate of inorganic compounds.
2. To study the compotation and stability of salt hydrates at different temperatures.
3. To study the dehydration temperature of salt hydrates.
4. To study the enantiotropic transformation of salts.
5. To study the solubility of different enantiotrops of salt in water.

2.10 THERMAL ANALYSIS–COOLING CURVES

When a pure substance or a mixture of two constituents of known composition is heated in an inert atmosphere to obtain the homogeneous melt or liquid phase and then melt or liquid is allowed to cool reversibly at slow rate without change of state, a temperature–time $(T-t)$ curve is obtained which is continuous and exponential in the form. During cooling the temperature of the standardized thermocouple is recorded as a function of time till the melt is completely solidified. This process is repeated for the other compositions of the pure substance within the range of 0–100%.

"The temperature–time plots for each amount of pure substance or each composition of mixture are called *cooling curves*. These plots are used to detect the various transformations and phase transitions which take place during cooling.

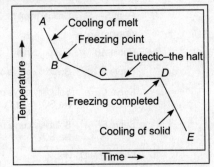

This can be explained by plotting the fall of temperature with time as shown in Figure 2.10. It shows that along the curve *AB* the rate of cooling is very fast and no solid phase separates out. At point *B* one of the solids start separating out. Therefore, the rate of cooling along curve *BC* is slowed down. The cooling curve at *B* shows a distinct break. At *C* eutectic point is reached and hence the temperature remains constant along *CD* until the complete solidification takes place.

Figure 2.10 Cooling curve of a two-component system forming an eutectic.

Now the system becomes invariant as the liquid phase disappears completely and there is only solid phase. Further cooling results in the fall of temperature. In total, there are three breaks in the cooling curve (i) at the freezing point (*B*) of the mixture, where the solid first commence to appears (ii) at eutectic point (*C*) and (iii) at complete solidification of the mixture.

The nature of cooling curve for a mixture of two-component system is shown in Figure 2.11.

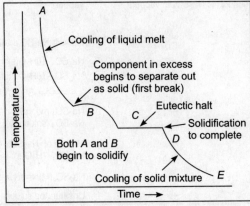

Figure 2.11 Cooling curve for a mixture of two components.

Phase Rule ♦ 73

SOLVED NUMERICAL PROBLEMS

1. Calculate the number of component (C) and number of degree of freedom (F) for the following systems:
   (i) Aqueous solution of glucose
   (ii) Dissociation of $NH_4Cl(s)$ in a sealed tube
   (iii) Solid camphor in equilibrium with its vapour
   (iv) $I_2(s) \rightleftharpoons I_2(H_2O) \rightleftharpoons I_2(benzene)$
   (v) $Na_2SO_4 \cdot 10H_2O(s) \rightleftharpoons Na_2SO_4(s) + 10H_2O(l)$
   (vi) Water at its freezing point.
   (vii) $Fe(s) + H_2O(g) \rightleftharpoons Fe(s) + H_2(g)$
   (viii) $H_2(g) + 1/2 O_2(g) \rightleftharpoons H_2O(g)$ when $P_{H_2} = P_{O_2}$
   (ix) $PCl_5(g) \rightleftharpoons PCl_3(g) + Cl_2(g)$, when $P_{Cl_3} = P_{Cl_2}$ and when $P_{Cl_3} \neq P_{Cl_2}$
   (x) $N_2O_4(g) \rightleftharpoons 2NO_2(g)$
   (xi) $H_2O(g) \rightleftharpoons H_2(g) + O_2(g)$
   (xii) $CaCl_2 \cdot 6H_2O(s)$, $Ca^{2+}(aq)$, $Cl^-(aq)$, $H_2O(l)$, $H_2O(g)$
   (xiii) $NH_4Cl(s)$, $NH_4^+(aq)$, $Cl^-(aq)$, $H_2O(l)$, $H_3O^+(aq)$, $H_2O(g)$, $NH_3(g)$, $OH^-(aq)$, $NH_4OH(aq)$
   (xiv) $CH_3COONH_4(s)$, $CH_3COO^-(aq)$, $NH_4^+(aq)$, $H_3O^+(aq)$, $NH_3(g)$, $OH^-(aq)$,
   (xv) $CH_3COOH(aq)$, $H_2O(l)$, $H_2O(g)$ assuming hydrolysis to take place.
   (xvi) Br_2 dissolved in CCl_4.

Solution:
   (i) Aqueous solution of sugar
       In this case, $C = 2$, i.e. water (H_2O) and sugar, $P = 1$ (liquid), hence,
       $$F = C - P + 2 = 2 - 1 + 2 = 3$$
       Thus, three variables (T, P and concentration) are required to define the system.
   (ii) Dissociation of $NH_4Cl(s)$ in a sealed tube. There may be two cases:
        (a) When only NH_4Cl is heated in a closed vessel.
            $$NH_4Cl(s) \rightleftharpoons NH_3(g) + HCl(g)$$
            In this condition the NH_3 and HCl are present in equimolar ratio, i.e.
            $[NH_3] = [HCl]$ or $P_{NH_3} = P_{HCl}$, hence $P = 2$, $C = 1$.
            $$K = \frac{[NH_3][HCl]}{[NH_4Cl]}$$
            The active mass of solid is taken as constant and ignored, therefore
            $$K = [NH_3][HCl]$$
        (b) When the mixture of $NH_4Cl(s)$, $NH_3(g)$ and $HCl(g)$ are heated in closed vessel, then
            $P_{NH_3} \neq P_{HCl}$, hence $P = 2$, $C = 2$.

(iii) Solid camphor in equilibrium with its vapour

$$C = 1 \text{ (camphor only)}, \quad P = 2 \text{ (solid camphor and its vapour)}$$

(iv) $I_2(s) \rightleftharpoons I_2(H_2O) \rightleftharpoons I_2$ (benzene)

$$C = 3 (I_2, H_2O \text{ and benzene}),$$
$$P = 3 [I_2 \text{ (solid)}, H_2O \text{ (liquid) and benzene (liquid) immiscible liquids}]$$

Hence, $F = C - P + 2 = 3 - 3 + 2 = 2$

(v) $Na_2SO_4 \cdot 10H_2O(s) \rightleftharpoons Na_2SO_4(s) + 10H_2O(l)$

$$C = 2 (Na_2SO_4 \cdot 10H_2O \text{ and } Na_2SO_4 \text{ or } H_2O)$$
$$P = 3 (Na_2SO_4 \cdot 10H_2O(s), Na_2SO_4(s) \text{ and } H_2O(l))$$

(vi) Water at its freezing point

$$C = 1 (H_2O \text{ only}), P = 3 \text{ (ice, water and vapours)}$$

Hence, $\quad F = C - P + 2 = 1 - 3 + 2 = 0$ (invariant system).

(vii) $H_2(g) + 1/2 O_2(g) \rightleftharpoons H_2O(g)$ when $P_{H_2} = P_{O_2}$

$$C = 2 (H_2 \text{ or } O_2 \text{ and } H_2O), P = 1 \text{ (gaseous mixture)}$$

Hence, $\quad F = C - P + 2 = 2 - 1 + 2 = 3.$

when $\quad P_{H_2} \neq P_{O_2}, C = 3 (H_2, O_2 \text{ and } H_2O), P = 1,$

$$F = C - P + 2 = 3 - 1 + 2 = 4.$$

(viii) $PCl_5(g) \rightleftharpoons PCl_3(g) + Cl_2(g),$

When, $\quad P_{PCl_3} = P_{Cl_2}, C = 2 (PCl_5, PCl_3 \text{ or } Cl_2), P = 3, F = 2 - 3 + 2 = 1$

and when $\quad P_{PCl_3} \neq P_{Cl_2}, C = 3 (PCl_5, PCl_3 \text{ and } Cl_2), P = 3, F = 3 - 3 + 2 = 2$

(ix) $N_2O_4(g) \rightleftharpoons 2NO_2(g)$

$$C = 1 N_2O_4(g) \text{ or } 2NO_2(g), P = 1. \text{ Hence, } F = C - P + 2 = 1 - 1 + 2 = 2.$$

(x) $H_2O(g) \rightleftharpoons H_2(g) + O_2(g)$ when $P_{H_2} = P_{O_2}$

$$C = 1 (H_2O \text{ only}), P = 1 \text{ (Gaseous mixture)}$$

Hence, $\quad F = C - P + 2 = 1 - 1 + 2 = 2.$

when $\quad P_{H_2} \neq P_{O_2}, C = 2 (H_2O \text{ and } H_2 \text{ or } O_2), P = 1,$

$$F = C - P + 2 = 2 - 1 + 2 = 3.$$

(xi) Br_2 dissolved in CCl_4

$$C = 2 (Br_2 \text{ and } CCl_4), P = 2.$$

Hence, $\quad F = C - P + 2 = 2 - 2 + 2 = 2.$

(xii) $CaCl_2 \cdot 6H_2O(s)$, $Ca^{2+}(aq)$, $Cl^-(aq)$, $H_2O(l)$, $H_2O(g)$

No. of species, $N = 4$ ($CaCl_2 \cdot 6H_2O$, Ca^{2+}, Cl^- and H_2O)

No. of equations, $E = 2$, as shown below:
1. $CaCl_2 \cdot 6H_2O(s) \rightleftharpoons Ca^{2+}(aq) + Cl^-(aq) + 6H_2O(l/g)$
2. Electro-neutrality of solution

∴ $C = N - E = 4 - 2 = 2$, $P = 3$ ($CaCl_2 \cdot 6H_2O(s)$, liquid ($Ca^{2+}(aq) + Cl^-(aq) + 6H_2O$) and gas ($H_2O(g)$))

Hence, $F = C - P + 2 = 2 - 3 + 2 = 1$.

(xiii) $NH_4Cl(s)$, $NH_4^+(aq)$, $Cl^-(aq)$, $H_2O(l)$, $H_3O^+(aq)$, $H_2O(g)$, $NH_3(g)$, $OH^-(aq)$, $NH_4OH\ (aq)$

Number of species, $N = 8$ (NH_4Cl, NH_4^+, Cl^-, H_2O, H_3O^+, NH_3, OH^-, NH_4OH)

Number of equations, $E = 5$, as shown below:
1. $NH_4Cl(s) \rightleftharpoons NH_4^+(aq) + Cl^-(aq)$
2. $NH_4^+(aq) + H_2O(l) \rightleftharpoons NH_3(g) + H_3O^+(aq)$
3. $NH_3(g) + H_2O(l) \rightleftharpoons NH_4OH(aq)$
4. $2H_2O(l) \rightleftharpoons H_3O^+(aq) + OH^-(aq)$
5. Electro-neutrality of solution

∴ $C = 8 - 5 = 3$, $P = 3$ [$NH_4Cl(s)$, liquid ($NH_4^+(aq)$, $Cl^-(aq)$, $H_2O(l)$, $H_3O^+(aq)$ and gas ($NH_3(g)$, $H_2O(g)$)] $NH_3(g)$, $OH^-(aq)$, $NH_4OH\ (aq)$.

Hence, $F = C - P + 2 = 3 - 3 + 2 = 1$.

(xiv) $CH_3COONH_4(s)$, $CH_3COO^-(aq)$, $NH_4^+(aq)$, $H_3O^+(aq)$, $NH_3(g)$, $OH^-(aq)$,

Number of species, $N = 8$ (CH_3COONH_4, CH_3COO^-, CH_3COOH, NH_4^+, NH_3, OH^-, H_3O^+, H_2O)

Number of equations, $E = 5$, as shown below:
1. $CH_3COONH_4\ (s) \rightleftharpoons CH_3COO^-(aq) + NH_4^+(aq)$
2. $CH_3COOH(l) + H_2O(l) \rightleftharpoons CH_3COO^-(aq) + H_3O^+(aq)$
3. $CH_3COO^-(aq) + H_2O(l) \rightleftharpoons CH_3COOH(l) + OH^-(aq)$
4. $NH_4^+(aq) + H_2O(l) \rightleftharpoons NH_3(g) + H_3O^+(aq)$
5. Electro-neutrality of solution

∴ $C = 8 - 5 = 3$, $P = 3$ [$CH_3COONH_4(s)$, liquid ($CH_3COOH(l)$, $CH_3COO^-(aq)$, $NH_4^+(aq)$, $H_3O^+(aq)$ $H_2O(l)$, $OH^-(aq)$) and gas (($NH_3(g)$, $H_2O(g)$)]

Hence, $F = C - P + 2 = 3 - 3 + 2 = 2$.

(xv) $CH_3COOH(aq)$, $H_2O(l)$, $H_2O(g)$ assuming hydrolysis to take place.

Number of species, $N = 3$ (CH_3COOH, CH_3COO^-, H_3O^+)

Number of equations, $E = 5$, as shown below:
1. $CH_3COOH\ (l) + H_2O(l) \rightleftharpoons CH_3COO^-(aq) + H_3O^+(aq)$
2. Electro-neutrality of solution

∴ $C = 3 - 2 = 1$, $P = 2$ [miscible liquids $CH_3COOH(l)$, $H_2O(l)$, $H_2O(gl)$]

Hence, $F = C - P + 2 = 3 - 2 + 2 = 3$.

(xvi) Br_2 dissolved in CCl_4

$C = 2$ (Br_2 and CCl_4), $P = 1$ (liquid phase)

Hence, $F = C - P + 2 = 2 - 1 + 2 = 3$.

2. Calculate the dT/dP for the water $\rightleftharpoons$ ice system at 273 K. ΔH_f for water is 6007.8 J mol^{-1} (1 J = 9.87 × 10^{-3} dm^3 atm), molar volume of water = 18.00 cm^3 and ice = 19.63 cm^3.

Solution: Given that $V_l = 19.00$ cm^3 mol^{-1} = 0.01900 dm^3 mol^{-1},

$V_s = 20.63$ cm^3 mol^{-1} = 0.02063 dm^3 mol^{-1},

$$\frac{dP}{dT} = \frac{\Delta H_f}{T_f(V_1 - V_2)}$$

or

$$\frac{dT}{dP} = \frac{T_f(V_1 - V_2)}{\Delta H_f}$$

$$\frac{dT}{dP} = \frac{273(0.019 - 0.02063)}{6007.8 \times 9.87 \times 10^{-3}} = \frac{-0.44499}{59.297} = -7.504 \times 10^{-3} \text{ K atm}^{-1}$$

Hence, the melting point of ice reduced by 0.0075 K if the pressure is raised by 1 atm.

3. The vapour pressure of water at 373°C is 760 mm. What will be the vapour pressure of water at 363°C? The heat of vapourization of water in this temperature range is 41.27 kJ mol^{-1}.

Solution: From Clausius–Clapeyron equation

$$\log \frac{P_2}{P_1} = \frac{\Delta H_{Vap}}{2.303 R} \left[\frac{T_2 - T_1}{T_2 \times T_1}\right];$$

$$\log \frac{P_2}{760} = \frac{41.27 \times 10^3}{2.303 \times 8.314} \left[\frac{363 - 373}{368 \times 373}\right]$$

$$\log P_2 - \log 760 = \frac{41.27 \times 10^3}{19.147} \left[\frac{-10}{135399}\right]$$

$$\log P_2 - 2.8808 = 2155.429 \times 7.3855 \times 10^{-5}$$

$$\log P_2 - 2.8808 = -0.15918$$

$$\log P_2 = -0.15918 + 2.8808$$

$$\log P_2 = 2.72162$$

$$P_2 = \text{anti-log } 2.72162$$

$$\mathbf{P_2 = 526.768 \text{ mm Ans.}}$$

4. At what pressure will ice melt at $-1.0\,°C$ assuming that ΔH_f is independent of pressure and is equal to 6.0095 kJ mol^{-1}? Given that the density of water is 0.9998 g cm^{-3} and that of ice is 0.917 g cm^{-3}.

Solution: From the Clausius–Clapeyron equation

Density of water = 0.9998 g cm^{-3} = 0.9998×10^{-3} kg m^{-3}

Density of ice = 0.917 g cm^{-3} = 0.917×10^{-3} kg m^{-3}

Molar mass of water = 18 g mol^{-1} = 18×10^{-3} kg mol^{-1}

$$\text{Molar volume of water, } V_l = \frac{18 \times 10^{-3}}{0.998 \times 10^3} = 18.0186 \times 10^{-6} \text{ m}^3 \text{ mol}^{-1}$$

$$\text{and molar volume of ice, } V_s = \frac{18 \times 10^{-3}}{0.917 \times 10^3} = 19.645 \times 10^{-6} \text{ m}^3 \text{ mol}^{-1}$$

$$\Delta H_f = 6.0095 \text{ kJ mol}^{-1} = 6009.5 \text{ J mol}^{-1}$$

$$= \frac{6009.5 \text{ J mol}^{-1}}{101325 \text{ Jm}^{-2} \text{ atm}^{-1}} = 5.9309 \times 10^{-2} \text{ m}^3 \text{ atm mol}^{-1}$$

$(\because\ 1 \text{ atm} = 101325 \text{ Nm}^{-2} = 101325 \text{ Jm}^{-3})$

$T_1 = 0°C = 273$ K and $T_2 = -1°C = 272$ K

From the Clausius–Clapeyron equation

$$\frac{dP}{dT} = \frac{\Delta H_f}{T_f \cdot \Delta V_f} \quad \text{or} \quad dP = \frac{\Delta H_f}{\Delta V_f} \cdot \frac{dT}{T_f}$$

Suppose that the $\dfrac{\Delta H_f}{\Delta V_f}$ is independent of temperature and pressure, then integrating above equation, we get

$$\int_{P_1}^{P_2} dP = \frac{\Delta H_f}{\Delta V_f} \int_{T_1}^{T_2} \frac{dT}{T} \quad \text{or} \quad (P_2 - P_1) = \frac{\Delta H_f}{\Delta V_f} \ln \frac{T_2}{T_1}$$

or

$$P_2 - P_1 = 2.303 \frac{\Delta H_f}{(V_l - V_s)} \log \frac{T_2}{T_1}$$

$$P_2 - P_1 = 2.303 \frac{5.9309 \times 10^{-2}}{(18.0186 - 19.645) \times 10^{-6}} \log \frac{272}{273}$$

$$P_2 - P_1 = 2.303 \frac{5.9309 \times 10^{-2}}{-1.6264 \times 10^{-6}} (2.4345 - 2.4361)$$

$$P_2 - P_1 = \frac{13.6588 \times 10^{-2}}{-1.6264 \times 10^{-6}} \times (-0.0016) = 134.37$$

or $\quad P_2 = P_i + 134.37 \quad$ or $\quad P_2 = 1 + 134.37 \quad (\because\ P_i = 1 \text{ atm})$

or $\quad\quad\quad$ **$P_2 = 135.37$ atm Ans.**

SUMMARY

Phase rule: It may be stated that "provided that the equilibrium between any number of phases is not influenced by gravitational forces, electrical forces, magnetic forces or by surface action, but only by temperature, pressure and concentration (composition), then the number of degree of freedom (F) of the system is related with the number of component (C) and number of phases (P) by the following equation."

$$F = C - P + 2$$

Phase (P): A phase is defined as "a homogeneous, physically distinct and mechanically separable portion of the system which is separated from other such parts of the system by definite boundary surfaces."

Component (C): It is defined as "the smallest number of independently variable constituents (molecular species) taking part in the state of equilibrium and by means of which the composition of each phase can be expressed either directly or in terms of chemical equations."

Degree of freedom (F): The degree of freedom of a system is defined as "the minimum number of independent variables such as temperature, pressure and concentration (or composition) which must be specified in order to define the system completely or to represent perfectly the condition of a system". It is also known as *variance*.

Applications of phase rule: Phase rule gives a simple method to classify the equilibrium states of the systems, to predict the number of phases co-existing in the system under equilibrium, to study the behaviour of systems when they are subjected to change in variables, very useful in industries, e.g. metallurgical industry, etc.

Limitations of phase rule: The phase rule does not explains the effect of other variables such as gravitational, magnetic, electric and surface forces, etc. for finely divided liquid or solid phase, does not give any idea about this changed vapour pressure, cannot be applied to such systems which consists of two or more solutions separated by semipermeable membranes.

Phase diagram: A phase diagram may be defined as the graphical presentation showing interdependence of variables which helps in studying the co-existence of different phases in a multiphase system.

Metastable curve: Sometimes it is possible to cool water several degree below its freezing temperature without converting it into ice. This water is called *super-cooled water*.

The liquid $\rightleftharpoons$ vapour equilibrium along this curve is called *metastable equilibrium*.

Triple point: The point at which three phases are in equilibrium with each other.

Polymorphism: "The polymorphism may be defined as the phenomenon in which a substance can exist in two or more crystalline forms, having different physical or chemical properties."

Two-component systems: "The system, in which the composition of all phases is expressed in terms of two independent constituents, is called two-component system."

The phase rule becomes, $F = C - P + 1$ and it is known as condensed phase rule equation.

Reduced phase rule equation: In the study of two-component system, usually, one of the three variables is kept constant and therefore, the degree of freedom is reduced by one, i.e.

$$F = C - P + 1.$$

Eutectic system: A binary system which consists of two substances, miscible in all proportions in the liquid phase but do not react chemically, is known as the eutectic system, e.g. a mixture of lead and silver comprises of such a system.

Eutectic mixture: It is a solid solution of two or more substance showing the lowest freezing point of all the possible mixtures of components. The mixture of metal alloys is of low melting point than pure metal which are generally forms eutectic mixtures.

Eutectic point: The minimum freezing point attainable corresponding to the eutectic.

Congruent system: The system in which two components at a certain composition combine together to form a stable compound which melts congruently, i.e. melts sharply at a constant temperature without change in composition in the liquid state. Example zinc–magnesium system.

Incongruent system: In this type of system, two components combine chemically to form a compound which melts incongruently, i.e. melts giving a new solid phase. The composition of solution (melt) is different from that of solid phase. These type of compounds having incongruent melting point are called incongruent melting compounds.

Salt hydrate system: Sodium sulphate forms two hydrated $Na_2SO_4 \cdot 10H_2O$ and $Na_2SO_4 \cdot 7H_2O$. Also, the anhydrous salt can exist into anisotropic crystalline forms, namely, rhombic and monoclinic thus there are five solid phases in all, including ice in the system.

Cooling curves: The temperature–time plots for each amount of pure substance or each composition of mixture are called *cooling curves*.

EXERCISES

1. State the Gibbs phase rule and explain the various terms used.
2. What are the conditions for thermal, mechanical and chemical equilibria for a two-phase system.
3. Derive the phase rule equation, $F = C - P + 2$ and give different conditions under which this equation can be used.
4. What is phase diagram? How is it helpful in studying the conditions of the system?
5. Describe the applications and limitations of Gibbs phase rule.
6. Explain the following terms:
   (i) phase, (ii) number of components, (iii) Degree of freedom, (iv) triple point, (v) meta-stable equilibrium, (vi) eutectic point, (vii) congruent melting point, (viii) incongruent or peritectic point or meritectic point, (ix) solid solution, (x) transition point, (xi) critical point, (xii) cryohydric temperature.
7. What do you mean by one-component system? Explain the water system with the help of $T-P$ phase diagram.
8. What is meta-stable state? How does it differ from normal state?
9. How is the phase diagram of water helpful to explain: (i) ice skating and (ii) flow of glaciers?
10. What is super-cooled water? Under what conditions it can be obtained?
11. Describe the carbon dioxide system with the help of $T-P$ phase diagram and give its applications.
12. Distinguish between water system and carbon dioxide system.
13. What is condensed system? Why in such a case the phase rule equation used is $F = C - P + 1$.
14. What is eutectic system? Explain the Pb–Ag system with the help of $T-C$ phase diagram.
15. Discuss the applications of phase rule to Pb–Ag phase equilibrium diagram and show how it is used to obtain desilverized lead.

16. Give the maximum number of phases that can co-exist in equilibrium in one-component system.
17. What is transition temperature? Give the value of transition temperature for water and carbon dioxide.
18. Discuss and apply phase rule to two-component system taking the example of Na–K system.
19. Draw and explain the phase diagram of a binary system of components forming an congruent compound.
20. What is eutectic system? Explain the Pb–Ag system with the help of $T-C$ phase diagram. Give various applications of eutectic system.
21. Describe the applications of lead–silver system in desilverisation of lead (Patinson's process).
22. Water containing KCl and NaCl is a three-component system while water containing KCl and NaBr is a four-component system. Explain it.
23. 100 kg of argentiferrous lead contains 0.2% silver, is melted and allowed to cool. If eutectic mixture contains 2.6% Ag. What mass of lead will separate out?
24. What do you understand by congruent system? Explain its general characteristics.
25. Describe the Zn–Mg system with the help of $T-C$ phase diagram. Give its applications.
26. What do you understand by salt hydrate system? Explain the general characteristics of salt hydrate system.
27. What do you understand by incongruent system? Explain its general characteristics.
28. Explain the sodium–potassium system by $T-C$ phase diagram. Give its applications.
29. Describe the sodium sulphate–water system with the help of $T-C$ phase diagram. Give its various applications.
30. Write short notes on the following:

    (i) Polymorphism, (ii) eutectic system, (iii) meta stable equilibrium, (iv) thermal analysis, (v) cooling curves, (vii) phase diagram, (viii) salt hydrate system, (ix) incongruent or peritactic or meritectic or syndiotectic point.
31. Draw the phase diagram of water system and explain its applications.
32. Distinguish between:

    (i) A stable equilibrium from an unstable equilibrium.

    (ii) A eutectic point from a peritectic point.

    (iii) Eutectic mixture from a solid solution.

    (iv) A congruent melting compound from an incongruent melting compound.

    (v) Congruent point and incongruent system.
33. Give reason of the following:

    (i) An eutectic mixture has a definite composition and compound.

    (ii) The lowest temperature attained in a system of salt and ice is the cryohydric temperature.

    (iii) In water, a large number of molecular species like H_2O, $(H_2O)_2$, $(H_2O)_3$ etc. exist, yet the number of component is only one.

    (iv) Why in the case of one-component system the solid–vapour and liquid–vapour curves have always a positive slope, while the solid–liquid curve may have a negative or positive slope.

(v) In the water system three phases are in equilibrium at the freezing point and also at the triple point. The triple point is invariant, but the freezing point varies with pressure. How would you explain this?

34. Suppose four distinct phases were observed in a laboratory specimen of a binary alloy. Is such an observation possible? Explain.
35. Draw the phase diagrams for: (a) water, (b) Pb–Ag and (c) Zn–Mg systems.
36. Explain the applications of phase rule in: (i) desilverization of lead and (ii) purification of metals.
37. Two metals A (m.p. 631°C) and B (m.p. 327°C) forms a eutectic with 80 mole per cent of B and melting point 246°C. There is no solid–solid solubility and the liquid A is completely miscible with liquid B. Give the phase diagram of the system. If a molar mixture of equimolar A and B is allowed to cool down, draw the approximate nature of the cooling curve.

Calculate the number of component (C) and number of degree of freedom for the following systems:

(i) Aqueous solution of glucose.
(ii) Dissociation of $NH_4Cl(s)$ in a sealed tube.
(iii) Solid camphor in equilibrium with its vapour.
(iv) $I_2(s) \rightleftharpoons I_2(H_2O) \rightleftharpoons I_2(benzene)$
(v) $Na_2SO_4 \cdot 10H_2O \rightleftharpoons Na_2SO_4 + 10H_2O$
(vi) Water at its freezing point.
(vii) Water at its boiling point.
(viii) $Fe(s) + H_2O(g) \rightleftharpoons Fe(s) + H_2(g)$
(ix) $H_2(g) + 1/2O_2(g) \rightleftharpoons H_2O(g)$ when $P_{H_2} = P_{O_2}$
(x) $PCl_5(g) \rightleftharpoons PCl_3(g) + Cl_2(g)$, when $P_{C_3} = P_{C_2}$ and when $P_{C_3} \neq P_{C_2}$
(xi) $N_2O_4(g) \rightleftharpoons 2NO_2(g)$
(xii) $H_2O(g) \rightleftharpoons H_2(g) + O_2(g)$
(xiii) $CaCl_2 \cdot 6H_2O(s)$, $Ca^{2+}(aq)$, $Cl^-(aq)$, $H_2O(l)$, $H_2O(g)$
(xiv) $NH_4Cl(s)$, $NH_4^+(aq)$, $Cl^-(aq)$, $H_2O(l)$, $H_3O^+(aq)$, $H_2O(g)$, $NH_3(g)$, $OH^-(aq)$, $NH_4OH(aq)$
(xv) $CH_3COONH_4(s)$, $CH_3COO^-(aq)$, $NH_4^+(aq)$, $H_3O^+(aq)$, $NH_3(g)$, $OH^-(aq)$
(xvi) $CH_3COOH(aq)$, $H_2O(l)$, $H_2O(g)$ assuming hydrolysis to take place
(xvii) Br_2 dissolved in CCl_4
(xviii) $NH_4Cl(s) \rightleftharpoons NH_3(g) + HCl(g)$.
When $P_{NH_3} = P_{HCl}$, when $P_{NH_3} \neq P_{HCl}$
(xix) Sugar solution contained in an open beaker.
(xx) A beaker containing beaker water and sand.
(xxi) Binary azeotropic liquid mixture.
(xxii) A substance at critical temperature (point).
(xxiii) A substance below critical temperature.
(xxiv) A substance above critical temperature.
(xxv) A mixture of $H_2(g)$, $O_2(g)$ and $H_2O(g)$.
(xxvi) A mixture of $CaCO_3(s)$, $CaO(s)$ and $CO_2(g)$.
(xxvii) $2KClO_3(s) \rightleftharpoons 2KCl(s) + 3O_2(g)$.

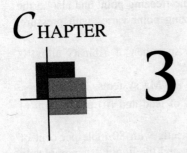

CHAPTER 3

CATALYSIS

3.1 INTRODUCTION

Some reactions which completed by their own within days, months and even years but some materials play the amazing role in completing these reactions very fast. These materials are known as *catalysts*.

The systematic study of the effect of various foreign substances on the rates of chemical reactions was made first by J.J. Berzelius, in 1835. He suggested the name catalyst for such substances.

3.2 CATALYST AND CATALYSIS

Catalyst is defined as "a substance which alters the rate of reaction without undergoing any change and can be recovered as such at the completion of the reaction. The phenomenon of altering the rate of reaction with the help of a catalyst is known as *catalysis*."

3.2.1 Role of the Catalyst

In general, the catalyst increases the rate of reaction. Regarding the role of the catalyst in increasing the rate, it is believed that it provides a new faster path through which a reaction can proceed. If a catalyst participates in a reaction mechanism, it must be regenerated in the end. Consider the following general catalyzed reaction:

$$R_1 + C \longrightarrow I$$
$$I + R_2 \longrightarrow P_2 + C$$

Overall reaction $R_1 + R_2 \longrightarrow P_2$

In above reactions, R_1 and R_2 are the two reactants, C is the catalyst and P_2 is the product. In the first step, the catalyst is attached to R_1 to form the intermediate I, which then reacts with R_2 to give the product P_2 and regenerates the catalyst. The catalyst is always regenerated in the final step of a series of reactions.

3.2.2 Activity of Catalyst

The catalyst increases the rate of reaction. A catalyst provides new faster path for the reaction which has lower activation energy E_a. This can be explained by the activated complex theory. This theory gives the basis of the decrease in Gibbs free energy of activation. According to this theory:

Reaction rate
$$k_f = \left(\frac{k_B T}{h}\right)^{-(\Delta G_f^0/RT)} \tag{3.1}$$

where k_f is the rate constant for the forward reaction, k_B is the Boltzmann constant and (ΔG_f^0) is the standard Gibbs free energy of activation.

In the presence of the catalyst we can write

$$k_f' = \left(\frac{k_B T}{h}\right)^{-(\Delta G_f^0/RT)'} \tag{3.2}$$

where prime (') denotes a catalyzed reaction. Since $(\Delta G_f^0)'$ is greater than (ΔG_f^0), also the $k_{f'}$ is greater than k_f. This can be seen in the Figure 3.1. The catalyst lowers the free energy of activation for both the forward and the backward reactions without altering the overall free energy change of the reaction. It means a catalyst changes the rates of both the forward and the backward reactions.

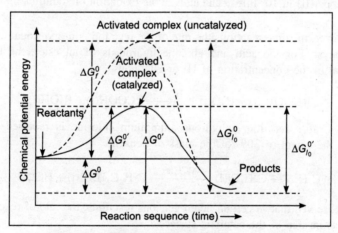

Figure 3.1 Change in rate of reaction by catalyst.

3.2.3 Catalyst and Equilibrium Constant

Mathematically, for the forward and the backward reactions in the presence of catalyst, we can write:

Rate of forward reaction, $\qquad k_f' = (k_B T/h)^{-(\Delta G_f^0)'/RT} \tag{3.3}$

Rate of backward reaction, $\qquad k_b' = (k_B T/h)^{-(\Delta G_b^0)'/RT}$

Therefore, $\qquad K_{eq} = \dfrac{k_f'}{k_b'} = \dfrac{(k_B T/h)^{-(\Delta G_f^0)'/RT}}{(k_B T/h)^{-(\Delta G_b^0)'/RT}} \tag{3.4}$

$$= \exp\left[-(\Delta G_f^0)'/RT - (-(\Delta G_b^0)'/RT)\right] = \exp(-\Delta G^0/RT) \tag{3.5}$$

where,
$$\Delta G^0 = (\Delta G_b^0)' - (\Delta G_f^0)'$$

Since ΔG^0 does not change, the equilibrium constant K_{eq} remains unchanged in the presence of the catalyst ($-\Delta G^0 = RT \ln K$). It shows that a catalyst facilitates in attaining the equilibrium position rapidly but does not alter the relative proportion of reactants and products at equilibrium.

3.3 GENERAL CHARACTERISTICS OF CATALYST/CATALYSIS

Following are the general characteristics of catalyst and catalysis:

1. *A catalyst remains unchanged in mass and chemical composition at the end of the reaction:* The amount of the catalyst found at the end of the reaction is the same as taken in the beginning. Its composition also remains the same as before. However, in some cases it may undergo some physical changes. For example, manganese dioxide used in the granular form as a catalyst in the decomposition of $KClO_3$ is left as a fine powder at the completion of the reaction.

2. *Only a small quantity of the catalyst is generally required to catalyze unlimited reaction:* For example, in the decomposition of hydrogen peroxide, one gram of colloidal platinum can catalyze 10^8 litres of hydrogen peroxide. Similarly, very low concentration, i.e. only one mole of copper(II) in 10^6 litres can catalyze the oxidation of sodium sulphite by atmospheric oxygen.

   However, in some reactions the rate of the reaction is proportional to the concentration of the catalyst. For the acid and alkaline hydrolysis of an ester, the rate of reaction is proportional to the concentration of H^+ or OH^- ions

$$RCOOR'(l) + H_2O(l) \xrightarrow{H^+ \text{ or } OH^-} RCOOH(l) + R'OH(l)$$

   In Friedel–Craft's reaction, anhydrous aluminium chloride is required in relatively large amount to the extent of 30% of the mass of benzene.

$$C_6H_6(l) + C_2H_5Cl(l) \xrightarrow{AlCl_3(s)} C_6H_6C_2H_5Cl(l) + HCl(l)$$

   It is also observed that in certain heterogeneous reactions, the rate of reaction increases with the increase in area of the catalytic surface.

3. *The catalyst does not initiate the reaction:* The role of the catalyst is to alter the speed of the reaction rather than to start it. The reaction in the presence of a positive catalyst adopts some alternative path which requires less amount of activation energy.

   However, there are certain instances where it is observed that the reaction cannot be started in the absence of a catalyst. For example, there is no reaction between H_2 and O_2 at room temperature, but the reaction occurs very rapidly in presence of platinum black.

$$2H_2(g) + O_2(g) \xrightarrow{\text{Room temp.}} \text{No reaction}$$
$$2H_2(g) + O_2(g) \xrightarrow{\text{Pt black}} 2H_2O(g)$$

4. *The catalyst is generally specific in nature:* A substance, which acts as a catalyst for a particular reaction, fails to catalyze the other reaction. For example, manganese dioxide which acts as a catalyst for the decomposition of potassium chlorate fails to catalyze the decomposition of potassium per chlorate.

$$2KClO_3(s) \xrightarrow[270°C]{MnO_2(s)} 2KCl(s) + 3O_2(g)$$

$$2KClO_4(s) \xrightarrow{MnO_2(s)} \text{No reaction}$$

Also the different catalysts for the same reactants may form different products. For example, ethanol yields ethene when passed over alumina, but in presence of hot copper, an acetaldehyde is formed.

$$C_2H_5OH(l) \begin{cases} \xrightarrow{Al_2O_3(s)} C_2H_4(g) + H_2O(l) \text{ (dehydration)} \\ \xrightarrow{Cu(s)} CH_3CHO(l) + H_2(g) \text{ (dehydrogenation)} \end{cases}$$

5. *The catalyst cannot change the position of equilibrium in a reversible reaction:* In case of reversible reactions, if the equilibrium has been established, the concentrations of the products and reactants cannot be affected by the catalyst. However, the use of a catalyst can help to achieve the equilibrium state in lesser time as forward and backward reactions are influenced to the same extent by the catalyst.

For example, the use of platinized asbestos as a catalyst in the reaction of sulphur dioxide and oxygen, causes an appreciable increase in the rate of the reaction but it does not, in anyway, increase the yield of sulphur trioxide under the given conditions of temperature and pressure.

$$2SO_2(g) \xrightarrow{\text{Platinized asbestos}} 2SO_3(g)$$

6. *The catalyst cannot alter the nature of the products in the reaction:* Under suitable conditions, the combination of nitrogen and hydrogen results always in the formation of ammonia whether manganese dioxide is added or not. There are, however, a few exceptions. For example, carbon monoxide and hydrogen combine to form methane, methyl alcohol or formaldehyde depending upon the nature of the catalyst used:

$$CO(g) + 3H_2(g) \xrightarrow{\text{Nickel}} CH_4(g) + H_2O(g)$$

$$CO(g) + 2H_2(g) \xrightarrow{ZnO + Cr_2O_3} CH_3OH(g)$$

$$CO(g) + H_2(g) \xrightarrow{Cu} HCHO(g)$$

7. *A catalyst is poisoned by certain substances:* It has been found that some impurities, even if present in small amounts, inhibit or retard the rate of catalyzed reactions to a large extent. These impurities are, therefore, called *catalytic poisons*. For example, in the contact process, the rate of combination of sulphur dioxide and oxygen is slowed down considerably if some arsenic compounds are present even in traces. Another example: the activity of iron catalyst is destroyed by the presence of H_2S or CO in the synthesis of ammonia by Haber's process.

3.4 CLASSIFICATION OF CATALYTIC REACTIONS (CATALYSIS)

The catalysis and catalytic reactions can be broadly divided into four categories:

3.4.1 Homogeneous Catalysis

When the reactants and the catalyst are in the same phase (i.e. solid, liquid or gas), the catalysis is said to be homogeneous. The following are the examples of homogeneous catalysis:

1. Oxidation of sulphur dioxide into sulphur trioxide with oxygen in the presence of oxides of nitrogen as the catalyst in the lead chamber process.

$$2SO_2(g) + O_2(g) \xrightarrow{NO(g)} 2SO_3(g)$$

The reactants sulphur dioxide and oxygen are in gaseous phase. The catalyst nitric oxide is also in gaseous phase, i.e. all are in the same phase.

2. Hydrolysis of methyl acetate is catalyzed by H⁺ ions furnished by hydrochloric acid.

$$CH_3COOCH_3(l) + H_2OH(l) \xrightarrow{HCl(l)} CH_3COOH(l) + CH_3OH(l)$$

Both the reactants and the catalyst are in the same phase.

3. Hydrolysis of sugar is catalyzed by H⁺ ions furnished by sulphuric acid.

$$C_{12}H_{12}O_{11}(l) + H_2O(l) \xrightarrow{H_2SO_4(l)} C_6H_{12}O_6(l) + C_6H_{12}O_6(l)$$

Both the reactants and catalyst are in the same phase.

3.4.2 Heterogeneous Catalysis

The catalytic process in which the reactants and the catalyst are in the different phases is known as *heterogeneous catalysis*. Some of the examples of heterogeneous catalysis are given below:

1. Oxidation of sulphur dioxide into sulphur trioxide in the presence of platinum metal or vanadium pentaoxide as a catalyst in the contact process for the manufacture of sulphuric acid.

$$2SO_2(g) + O_2(g) \xrightarrow{Pt(s)} 2SO_3(g)$$

The reactants are in gaseous state while the catalyst is in the solid state.

2. Combination between nitrogen and hydrogen to form ammonia in the presence of finely divided iron in Haber's process.

$$N_2(g) + 3H_2(g) \xrightarrow{Fe(s)} 2NH_3(g)$$

The reactants are in gaseous state while the catalyst is in solid state.

3. Oxidation of ammonia into nitric oxide in the presence of platinum gauze as a catalyst in Ostwald's process.

$$4NH_3(g) + 5O_2(g) \xrightarrow{Pt(s)} 4NO(g) + 6H_2O(g)$$

The reactants are in gaseous phase while catalyst is in solid phase.

4. Hydrogenation of vegetable oils in the presence of finely divided nickel as catalyst.

$$\text{Vegetable oil}(l) + H_2(g) \xrightarrow{Ni(s)} \text{Vegetable ghee}(g)$$

One of the reactants is in liquid state and the other in gaseous phase while the catalyst is in solid phase.

3.5 TYPES OF CATALYST

There are following types of catalyst:

3.5.1 Positive Catalyst

The foreign materials which accelerate the rate of the reaction is called *positive catalyst* and the phenomenon is called *positive catalysis*.

Examples of positive catalysis are:

1. Oxidation of ammonia in the presence of platinum gauze:

$$4\,NH_3(g) + 5\,O_2(g) \xrightarrow[300°C]{Pt(s)} 2\,NO(g) + 6\,H_2O(g)$$

2. Decomposition of H_2O_2 in the presence of colloidal platinum:

$$2\,H_2O_2(l) \xrightarrow{Pt(colloidal)} 2\,H_2O(g) + O_2(g)$$

3. Decomposition of $KClO_3$ in the presence of manganese dioxide.

$$2\,KClO_3(s) \xrightarrow[270°C]{MnO_2(s)} 2\,KCl(s) + 3\,O_2(g)$$

4. Synthesis of ammonia by Haber process in the presence of a mixture of iron

$$N_2(g) + 3\,H_2(g) \xrightarrow[450-500°C]{Fe(s)} 2\,NH_3(g)$$

3.5.2 Negative Catalyst

There are certain substances which, when added to the reaction mixture, slow down the reaction rate instead of increasing it. These are called *negative catalysts* or *inhibitors* and the phenomenon is known as *negative catalysis*.

Examples of negative catalysis are:

1. The oxidation of sodium sulphite by air is retarded by alcohol

$$2\,Na_2SO_3(s) + O_2(g) \xrightarrow{Alcohol(l)} 2\,Na_2SO_4(s)$$

Hence, alcohol acts as a negative catalyst.

2. Trichloromethane (chloroform) is oxidized in air to form toxic carbonyl chloride

$$4\,CHCl_3 + 3\,O_2 \xrightarrow{Alcohol(l)} 4\,COCl_2 + 2\,H_2O + 2\,Cl_2$$

But, when 2% ethyl alcohol is added to chloroform its oxidation is stopped, hence ethyl alcohol acts as catalytic poison.

3. Tetraethyl lead (TEL) acts as an anti-knocking agent in the case of gasoline combustion. Thus, it decreases knocking of petrol and acts as a negative catalyst.

3.5.3 Catalytic Promoter or Activator

The substances which themselves are not catalysts but when mixed in a small quantity with the catalyst enhance their efficiency are called *catalytic promoter* or *activator*.

For example:
1. In the synthesis of ammonia by Haber process, traces of Mo increases the activity of iron catalyst:

$$N_2(g) + 3H_2(g) \xrightarrow[450-500°C]{Fe(s) + Mo(s)} 2NH_3(g)$$

2. In the hydrogenation of oil, the activity of nickel catalyst is increased on adding small amount of copper.

3.5.4 Catalytic Poisons

Substances which retard or destroy the activity of the catalyst by their presence are known as *catalytic poisons*. Some of the examples are:

1. The presence of traces of arsenious oxide (As_2O_3) in the reacting gases reduces the activity of platinized asbestos which is used as a catalyst in contact process for the manufacture of sulphuric acid.
2. The activity of iron catalyst is destroyed by the presence of H_2S or CO in the synthesis of ammonia by Haber's process.

$$N_2(g) + 3H_2(g) \xrightarrow[H_2S \text{ or } CO, \text{ cat. poison}]{Fe(s)} 2NH_3(g)$$

3. The platinum catalyst used in the oxidation of hydrogen peroxide is poisoned by CO:

$$2H_2O_2(l) \xrightarrow[CO, \text{ Cat. poison}]{Pt \text{ Catalyst}} 2H_2O(g) + O_2(g)$$

The poisoning of the catalyst is probably due to the preferential adsorption of poison on the surface of the catalyst, thus reducing the space available for the adsorption of reacting molecules.

3.5.5 Auto Catalyst

In certain reactions, one of the products acts as a catalyst. In the initial stage, the reaction is slow but as soon as the products come into existence, the reaction rate increases. This type of phenomenon in which one of the products itself acts as a catalyst, is known as *auto-catalysis*. Examples of auto-catalysis are:

1. The rate of oxidation of oxalic acid by acidified potassium permanganate increases as the reaction progresses. This acceleration is due to the presence of Mn^{2+} ions which are formed during reaction. Thus, Mn^{2+} ions act as auto-catalyst.

$$5H_2C_2O_4(s) + 2KMnO_4(s) + 3H_2SO_4(l) \rightarrow 2MnSO_4(s) + K_2SO_4(s) + 10CO_2(g) + 8H_2O(l)$$

2. When nitric acid is poured on copper, the reaction is very slow in the beginning but gradually the reaction becomes faster due to the formation of nitrous acid during the reaction which acts as an auto-catalyst.

3.5.6 Induced Catalyst

When two reactions are carried out in same container, one reaction for other acts as catalyst which is known as *induced catalyst* and the phenomenon in which one reaction influences the rate of other

reaction, which does not occur under ordinary conditions, is known as *induced catalysis*. Examples of induced catalysis are as follows:
1. The reduction of mercuric chloride ($HgCl_2$) with oxalic acid is very slow, but potassium permanganate is reduced readily with oxalic acid. If, however, oxalic acid is added to a mixture of potassium permanganate and mercuric chloride, both are reduced simultaneously. The reduction of potassium permanganate, thus, induces the reduction of mercuric chloride.
2. Sodium arsenite solution is not oxidized by air. If, however, air is passed through a mixture of the solution of sodium arsenite and sodium sulphite, both of them undergo simultaneous oxidation. The oxidation of sodium sulphite, thus, induces the oxidation of sodium arsenite.

3.6 THEORIES OF CATALYSIS

It is not possible to give a uniform explanation for the mechanism of the phenomenon of catalysis as catalytic reactions are of varied nature. Two broad theories of catalytic action have been proposed. These theories are—intermediate compound formation theory and adsorption theory.

3.6.1 Intermediate Compound Formation Theory

This theory was proposed by Clement and Desormes in 1806 to explain the mechanism of homogeneous catalytic reactions.

This theory can be explained by considering the following steps:
1. The catalyst first forms an intermediate with one of the reactants with less energy consumption than needed for the actual reaction.
2. The intermediate so formed, being unstable, combines with other reactant to form the desired product.
3. The catalyst is regenerated at the end of the reaction.

For example, a reaction of the type $A + B \xrightarrow{K} AB$ may take place as

$$A + \underset{\text{Catalyst}}{C} \longrightarrow \underset{\text{Intermediate}}{AC}$$

$$AC + B \longrightarrow \underset{\text{Product}}{AB} + \underset{\text{Catalyst}}{C}$$

Many catalytic reactions can be explained on the basis of this theory.
1. The catalytic oxidation of sulphur dioxide to sulphur trioxide in the lead chamber process probably takes place as:

$$2NO + O_2 \longrightarrow \underset{\text{Intermediate}}{2NO_2}$$
$$\underset{\text{Catalyst}}{}$$

$$NO_2 + SO_2 \longrightarrow \underset{\text{Product}}{SO_3} + \underset{\text{Catalyst}}{NO}$$

2. The formation of diethyl ether from ethyl alcohol using sulphuric acid as a catalyst can be explained as:

$$C_2H_5OH + \underset{\text{Catalyst}}{H_2SO_4} \longrightarrow \underset{\text{Intermediate}}{[C_2H_5HSO_4]}$$

$$C_2H_5HSO_4 + C_2H_5OH \longrightarrow \underset{\text{Product}}{C_2H_5OC_2H_5} + \underset{\text{Catalyst}}{H_2SO_4}$$

Merits and limitations of intermediate compound theory

This theory explains why the catalyst remains unchanged in mass and chemical composition at the end of the reaction and is effective even in small quantities. However, this theory is limited to only homogeneous catalysis as the formation of intermediate is possible in the case of homogeneous catalysis. It also fails to explain the action of catalytic promoters, catalytic poisons and action of finely divided catalysts.

3.6.2 Adsorption Theory

This theory explains the mechanism of heterogeneous catalysis. According to old view, when the catalyst is in solid state and the reactants are in gaseous state or in solutions, the molecules of the reactants are adsorbed on the surface of the catalyst. The increase in concentration of the reactants on the surface influences the rate of reaction (law of mass action). The adsorption being an exothermic process, the heat of adsorption is taken up by the surface of catalyst and utilized in enhancing the chemical reaction of the molecules.

The adsorption is broadly of two types, i.e. physical adsorption and chemical adsorption. The chemical adsorption is specific and involves chemical combination on the surface of the catalyst.

The modern adsorption theory is the combination of intermediate compound formation theory and the old adsorption theory. The catalytic activity is localized on the surface of the catalyst (Figure 3.2). The mechanism involves following steps:

1. *Diffusion and adsorption of reactants to the surface of the catalyst:* The reactant molecules A and B strike the catalyst surface and held up by weak van der Waal's forces (in case of physical adsorption) or by chemical bonds (in case of chemical adsorption).

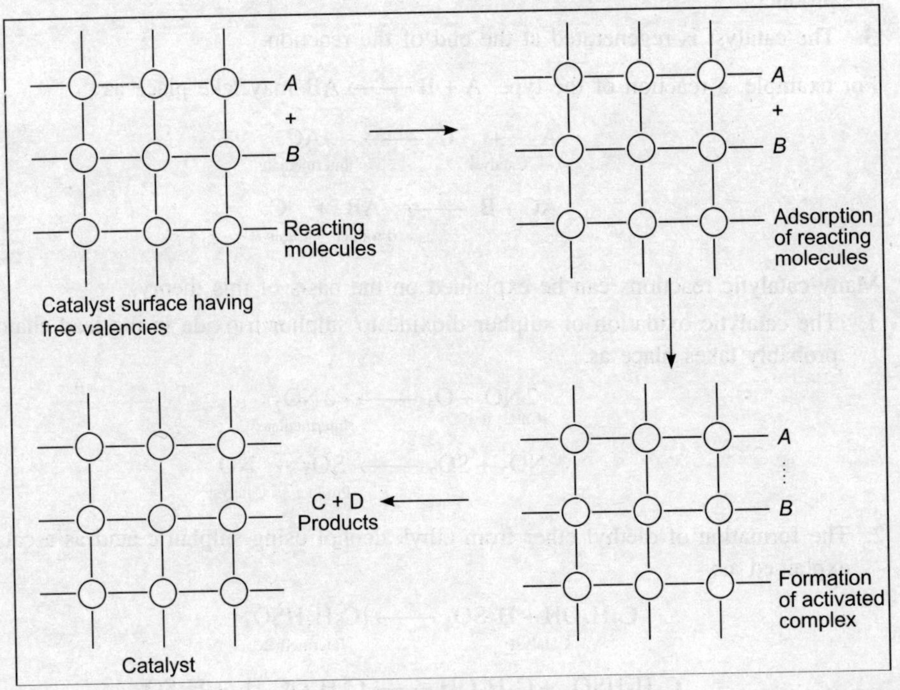

Figure 3.2 Formation of activated complex and product after adsorption on catalyst surface.

2. *Reaction between the reactant molecules to form activated complex:* The reactant molecules combine to form intermediate activated complex by using the heat of adsorption from catalyst surface.
3. *Decomposition of activated complex:* The activated complex then decomposes to give the products C and D and held up on the catalyst surface by weak forces.
4. *Desorption and diffusion of products away from the catalyst surface:* The product molecules then desorbed and defuse away from the catalyst surface and can exist as independent entity.

This theory can be explained by taking the example of hydrogenation of ethene in presence of nickel catalyst. The catalyst, e.g. Ni surface is a seat of chemical forces of attraction. There are valencies on its surface. When H_2 gas comes in contact with such a surface, its molecules are held up by weak chemical combination. The ethene is also adsorbed side by side and reacts with hydrogen to form activated complex on the surface of nickel catalyst. This activated complex then decomposes to give the ethane which defuses away (escapes) from the surface of the catalyst leaving the way for the fresh reactant (ethane and hydrogen) molecules (Figure 3.2).

In case of free valencies, if these are responsible for the catalytic activity, it follows that with the increase in the number of these valencies on the surface of a catalyst, the catalytic activity will be greatly enhanced.

The free valencies can be increased in the following two ways:
1. *By sub-division of the catalyst:* This is carried out by grinding the catalyst into finely powdered or colloidal form, which results in the increase of surface area [Figure 3.3(a)].
2. *By rough surface of the catalyst:* There are a number of active spots in the form of edges, corners, cracks and peaks on the rough surface. They give rise to an increase in the number of free valencies. These active spots enhance the adsorption and thereby increase the catalytic efficiency of the catalyst [Figure 3.3(b)].

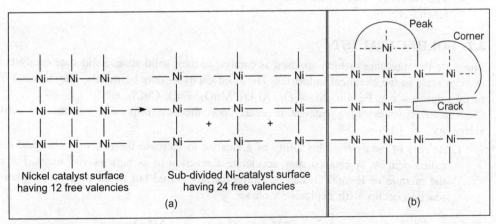

Figure 3.3 Increase in number of free valencies due to sub-division of catalyst. Active centres (peak, corner and crack) on catalyst surface.

3.6.3 Facts of Heterogeneous Catalysis on the Basis of Adsorption Theory

Following are the facts of heterogeneous catalysis:
1. The surface of the catalyst is used again and again due to alternate adsorption and desorption. Thus, a small quantity of the catalyst can catalyze large amounts of reactants.

2. Chemical adsorption depends on the nature of the adsorbent and adsorbate. Hence, catalysts are specific in action.
3. Desorption leaves the catalytic surface unchanged. Thus, the catalyst at the end of reaction remains same in mass and composition.
4. The energy of adsorption compensates the activation energy of the reacting molecules to some extent. Thus, reactions occur at faster rate.
5. Efficiency of the catalyst is greater in finely divided state and rough surface.
6. It adequately explains the poisoning of catalysts. The poisons are preferentially adsorbed at the active centres of the catalyst. This effect reduces the free valencies for the reacting molecules and, thus, the catalytic activity decreases.
7. Promoters are responsible for increasing the roughness of the surface of the catalyst. This effect increases the free valencies for the reacting molecules and thus, the activity of the catalyst increases.

3.6.4 Role of Heterogeneous Catalysts on the Basis of Adsorption Theory

The role of heterogeneous catalysts can be explained in terms of adsorption of reactants on the surface of the catalyst. The adsorption helps the reaction in the following ways:

1. Adsorption increases the concentration of reactants on the surface of the catalyst. Due to increased concentration of the reactants, the reactions proceed rapidly.
2. Adsorbed molecules get dissociated to form active species like free radicals which react faster than molecules.
3. The adsorbed molecules are not free to move about and, therefore, they collide with other molecules on the surface.
4. The heat of adsorption evolved acts as an energy of activation for the reaction (chemisorptions).

3.7 SOLID CATALYST

These are the substances which are used as catalyst in their solid state. Solid state catalysts are used at large scale in the chemical industries. The solid catalysts may be metals, alloys, metal oxides or metal sulphides, e.g. Fe, Cu, Ni, V_2O_5, Al_2O_3, MnO_2, ZnO, $CuCl_2$, etc.

The effectiveness of a catalyst depends upon the two important aspects, i.e. activity and selectivity.

1. *Activity of catalyst:* The ability of a catalyst to increase the rate of a chemical reaction is called *activity*. A catalyst may accelerate a reaction to as high as 10^{10} times. For example, the mixture of H_2 and O_2 can be stored for any period but in the presence of platinum, the reaction occurs with explosive violence.

$$2H_2(g) + O_2(g) \xrightarrow{Pt} 2H_2O(g)$$

2. *Selectivity of catalyst:* The ability of the catalyst to direct a reaction to give a particular product is called *selectivity*. For example; n-heptane in the presence of Pt catalyst selectively gives toluene.

$$nC_7H_{16} \xrightarrow{Pt} C_6H_5CH_3$$

Propylene ($CH_3CH{=}CH_2$) reacts with O_2 in the presence of bismuth molybdate catalyst to selectively give acrolein ($CH_2{=}CHCHO$).

$$CH_3CH{=}CH_2 + O_2 \xrightarrow{\text{Bismuth molybdate}} CH_2{=}CHCHO$$

Similarly, acetylene on reaction with H_2 in the presence of Pt catalyst gives ethane while in the presence of Lindlar's catalyst (palladium and $BaSO_4$ poisoned with quinoline or sulphur) gives ethylene.

The action of catalyst is highly specific. A given catalyst can act as catalyst only in a particular reaction and not in all reactions. Thus, catalyst is highly selective in nature.

3.7.1 Shape-selective Catalysts (Zeolites)

Zeolites are microporous aluminosilicates of the general formula $Mx/n[(AlO_2)_x (SiO_2)_y] \cdot mH_2O$. These are three-dimensional networks of silicates in which some silicon atoms are replaced by aluminium atoms. These are porous and have cavities of molecular dimensions. These are used in petrochemical industries for cracking of hydrocarbons and isomerization. The reactions in zeolites depends upon the size of the cavities (cages) and pores (apertures) present in them. The most remarkable feature of zeolite catalysis is the shape selectivity. Therefore, the selectivity of the catalyst depends on the pores structure. It has been observed that the pore size in zeolites generally varies between 260 picometre and 740 picometre. Depending upon the size of the molecules of reactants and products, and the sizes of the pores of zeolites, the reactions proceed in specific manner.

For example, zeolite catalyst known as *ZSM-5* converts alcohols to gasoline. The alcohol is dehydrated in the cavities and the hydrocarbons are formed. The shape selectivity in the reactions can be judged from the conversion of methanol and 1-heptanol to hydrocarbon mixtures (Table 3.1).

Table 3.1 The shape selectivity in the reactions for the conversion of methanol and 1-heptanol to hydrocarbon mixtures

Product	Starting with CH_3OH (%)	Starting with $n\text{-}C_7H_{15}OH$ (%)
Methane	1.0	0.0
Ethane	0.6	0.3
n-butane	5.6	11.0
Iso-butane	18.7	19.3
Iso-pentane	7.8	8.7
Benzene	1.7	3.4
Toluene	10.5	14.3
Xylene	17.2	11.6

The composition of the product mixture depends on the ability of pores to accommodate linear and *iso*-alkenes as well as benzene derivatives.

3.8 ACID–BASE CATALYSIS

Generally, homogeneous catalysis in solution is carried out by acids or bases or both. On the basis of studies done by Arrhenius and Ostwald in the hydrolysis of esters and nitrates, it was established that in acid–base catalysis, it is the hydrogen ion or hydroxyl ion which acts as a catalyst. A reaction which is catalyzed by proton (H^+ ions) but not by other Bronsted acids (i.e. proton donors) is called *specifically proton-catalyzed reaction*. Examples: keto–enol transformation, solvolysis of esters, etc. While the reaction which is catalyzed by OH⁻ ions is called *specifically base-catalyzed reaction*. Example: decomposition of nitramide. On the other hand, the reaction which is catalyzed by Bronsted base is called *general base catalysis*.

1. Hydrolysis of an ester:

$$C_2H_5COOC_2H_5(l) + H_2O(l) \xrightarrow{H^+ \text{ or } OH^-} C_2H_5COOH(l) + C_2H_5OH(l)$$

2. Conversion of acetone into diacetone alcohol (keto-enol transformation):

$$CH_3COCH_3(l) + CH_3COCH_3(l) \xrightarrow{OH^-} CH_3COCH_2C(CH_3)_2OH(l)$$

3. Decomposition of nitramide

$$NH_2 \cdot NO_2(l) \xrightarrow{OH^-} N_2O(g) + H_2O(l)$$

Recently, it has been proved that all the substances which have the tendency to loose or to gain protons can show catalytic activity, i.e. all Bronsted acids and bases act as acid–base catalysts.

3.8.1 Mechanism of Acid–Base Catalysis

1. *Acid catalyzed reaction:* In acid catalysis, the proton given by Bronsted acid forms an intermediate complex with the reactant which then reacts to give back the proton (H^+).

   For example, the hydrolysis of ester occurs in the following manner:

$$CH_3-CO-C_2H_5 + H^+ \rightarrow CH_3-\underset{+}{\overset{OH}{C}}-OC_2H_5 \rightarrow CH_3-\underset{\underset{H}{\overset{+}{O}}\overset{}{H}}{\overset{OH}{C}}-OC_2H_5 \rightarrow CH_3-\overset{OH}{\underset{O}{C}} + C_2H_5OH + H^+$$

2. *Base-catalyzed reaction (hydrolysis):* In this types of reactions the OH⁻ ion of any Bronsted base accepts a proton from the reactant to form an intermediate complex which then decomposes to regenerate the OH⁻ or Bronsted base. In presence of OH⁻ ions, the decomposition of nitroamine can be shown as

$$NH_2 \cdot NO_2 + OH^- \rightarrow NHNO_2^- + H_2O$$
$$\downarrow$$
$$N_2O + OH^-$$

or in the presence of CH_3COO^- ions,

$$OH^- + CH_3COOH \rightarrow CH_3COO^- + H_2O$$

3.8.2 Kinetics of Acid-Base Catalyzed Reactions

This can be illustrated by considering the following two types of catalytic mechanisms:

1. *First mechanism:* Consider the following acid-catalyzed reaction:

$$S + AH^+ \underset{k_1'}{\overset{k_1}{\rightleftharpoons}} SH^+ + A \qquad (3.6)$$

$$SH^+ + H_2O \xrightarrow{k_2} P + H_3O^+ \qquad (3.7)$$

Here, we assume that a proton is transferred from an acid AH^+ to the substrate S. The acid form of the substrate then reacts with a water molecule to form the product P.

Reaction (3.7) is slow and hence, it is the rate-determining step. The rate constant for the $[SH^+]$ in terms of formation of product can be given as:

$$\frac{d[P]}{dt} = k_2[SH^+] \qquad (3.8)$$

The rate of reaction of SH^+ with time, $d[SH^+]/dt$ (Rate of formation of SH^+ from S and HA − Rate of reaction of SH^+ and H_2O)

$$= k_1[S][AH^+] - k_1'[A][SH^+] - k_2[SH^+] \qquad (3.9)$$

Applying steady state approximation (SSA) for $[SH^+]$, we have

$$d[SH^+]/dt = 0$$

And also
$$k_1[S][AH^+] - k_1'[A][SH^+] - k_2[SH^+] = 0 \qquad (3.10)$$

Since, we work with very dilute solutions, the concentration of H_2O remains almost constant so that in Eq. (3.10) the last term on the right-hand side is written as $k_2[SH^+]$ rather than $k_2[SH^+] H_2O$.

Solving Eq. (3.10) for $[SH^+]$, we find that

$$k_1[S][AH^+] - [SH^+]\{k_1'[A] + k_2\} = 0 \qquad (3.11)$$

or
$$k_1[S][AH^+] = [SH^+]\{k_1'[A] + k_2\} \qquad (3.12)$$

or
$$[SH^+] = \frac{k_1[S][AH^+]}{k_1'[A] + k_2} \qquad (3.13)$$

Putting the value of $[SH^+]$ in Eq. (3.8), we get

$$\frac{d[P]}{dt} = \frac{k_1 k_2[S][AH^+]}{k_2 + k_1'[A]} \qquad (3.14)$$

If $k_2 \gg k'[A]$, then
$$\frac{d[P]}{dt} = k_1[S][AH^+] \qquad (3.15)$$

Thus, we can say that the reaction is general acid catalyzed.
If, however, $k_2 \gg k_1'[A]$, then

$$\frac{d[P]}{dt} = \frac{k_1 k_2[S][AH^+]}{k_2 + k_1'[A]} \qquad (3.16)$$

Now, consider the dissociation of weak acid (HA),

$$HA + H_2O \xrightarrow{K} H_3O^+ + A^-$$

$$K = \frac{[A][H_3O^+]}{[AH^+]}$$

or

$$\frac{[AH^+]}{[A]} = \frac{[H_3O^+]}{K} \tag{3.17}$$

Thus, from Eqs. (3.16) and (3.17), we get

$$\frac{d[P]}{dt} = \frac{k_1 k_2 [H_3O^+]}{k_1' K} \tag{3.18}$$

In this case we say that the reaction is specifically hydrogen-ion catalyzed.

2. *Second mechanism:* Suppose the acid form of the substrate reacts with base A instead of water molecule,

$$S + AH^+ \underset{k_1'}{\overset{k_1}{\rightleftharpoons}} SH^+ + A$$

$$SH^+ + A \xrightarrow{k_2} P + AH^+$$

Then, Eq. (3.13) can be given as

$$[SH^+] = \frac{k_1 [S][AH^+]}{k_1' + k_2 [A]} \tag{3.19}$$

Hence, from Eqs. (3.13) and (3.19), we get

$$\frac{d[P]}{dt} = k_2 [SH^+][A] = \frac{k_1 k_2 [S][AH^+]}{k_1' + k_2} \tag{3.20}$$

This is an example of general acid catalysis.

3.9 ENZYME CATALYSIS

"Enzymes are complex nitrogeneous organic compounds (i.e. protein molecules of high molecular mass), produced by living plants and animals which increase the rate of reaction without undergoing any change in their composition and nature." They form colloidal solutions in water and are very effective catalysts.

The enzymes are known as *biochemical catalysts* as they catalyze numerous reactions, especially those associated with natural processes and in the body of living organism. Each enzyme can catalyze a specific reaction.

Many enzymes have been obtained in pure crystalline state from living cells. However, the first enzyme was synthesized in the laboratory in 1969. The following are some of the examples of enzyme-catalyzed reactions:

1. *Inversion of cane sugar:* The invertase enzyme converts cane sugar into glucose and fructose:

$$C_{12}H_{22}O_{11}(s) + H_2O(l) \xrightarrow{\text{Invertase}} C_6H_{12}O_6(s) + C_6H_{12}O_6(s)$$

2. *Conversion of glucose into ethyl alcohol:* The zymase enzyme converts glucose into ethyl alcohol and carbon dioxide:

$$C_6H_{12}O_6(s) \xrightarrow{Zymase} 2C_2H_5OH(l) + 2CO_2(g)$$

3. *Conversion of starch into maltose:* The diastase enzyme converts starch into maltose.

$$2(C_6H_{10}O_5)_n(s) + nH_2O(l) \xrightarrow{Diastase} nC_{12}H_{22}O_{11}(s)$$
$$\text{Starch} \qquad\qquad\qquad\qquad\qquad \text{Maltose}$$

4. *Decomposition of urea into ammonia and carbon dioxide:* The enzyme urease catalyze this decomposition:

$$NH_2CONH_2(s) + H_2O(l) \xrightarrow{Urease} 2NH_3(g) + CO_2(g)$$

5. *Conversion of milk into curd:* It is an enzymatic reaction brought about by lactic bacilli enzyme present in curd.

Some important enzymes, their source and the reaction which they catalyze are shown in Table 3.2.

Table 3.2 Enzymes as catalyst used in different reactions

Enzyme	Source	Enzymatic reactions
Invertase	Yeast	Sucrose → glucose and fructose
Zymase	Yeast	Glucose → ethyl alcohol and carbon dioxide
Diastase	Malt	Starch → maltose
Maltase	Yeast	Maltose → glucose
Urease	Soyabean	Urea → ammonia and carbon dioxide
Pepsin	Stomach	Protein → amino acids
Trypsin	Intestine	Protein → amino acids
Amylase	Saliva	Starch → glucose
Lactic bacilli	Curd	Fermentation of milk
Mycoderma aceti	Vinegar	Ethyl alcohol → acetic acids
Lipase	Castor seeds	Fat → glycerol
Ptylin	Saliva	Starch → sugar

3.10 KINETICS OF ENZYME-CATALYZED REACTION

It has been observed that there are a number of cavities present on the surface of the enzymes. These cavities are of characteristic shape and possess active groups such as $-NH_2$, $-COOH$, $-SH$, $-OH$, etc. These are in fact the active centres on the surface of enzyme particles. When an enzyme comes in contact with the molecules of the reactant (substrate), which have complementary shape, fits into these cavities just like a key fits into a lock and activated complex is formed which then yields the products as shown in Figure 3.4.

Michaelis and M. Menten proposed the following mechanism of enzyme catalysis:

Step 1: Binding of an enzyme to substrate to form an activated complex:

$$\underset{\text{Enzyme}}{E} + \underset{\text{Subrate}}{S} \underset{k_2}{\overset{k_1}{\rightleftharpoons}} \underset{\text{Activated complex}}{ES} \qquad \text{(Fast)}$$

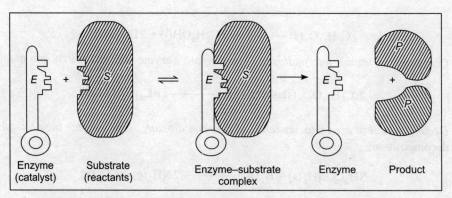

Figure 3.4 Key-lock model of enzyme-catalyzed reaction mechanism.

Step 2: Formation of product in the activated complex:

$$ES \rightarrow EP$$

Step 3: Decomposition of activated complex (EP) into product and enzyme again:

$$EP \rightarrow E + P \text{ (slow)}$$

In the overall reaction, the enzyme is consumed in Step 1 and regenerated in Step 3.

This can be explained by using *equilibrium approximation* or *steady state approximation*.

On the basis of experimental observations it was found that equilibrium is not achieved in the first step (fast) but in the third step (slow) the enzyme–substrate complex is converted into product releasing the enzyme free. This mechanism accounts for the high specificity of enzymatic reactions.

3.10.1 Michaelis–Menten Equation

In 1913, B.L. Michaelis and M. Menten proposed a mechanism for the kinetics of enzyme catalyzed reactions which predicts the following steps:

Step 1: Formation of the enzyme–substrate complex:

$$E + S \underset{k_1}{\overset{k_1}{\rightleftharpoons}} ES \text{ (complex)} \quad \text{(fast)} \tag{3.21}$$

Step 2: Decomposition of the complex:

$$ES \xrightarrow{k_2} P + E \quad \text{(slow)} \tag{3.22}$$

where E is the free enzyme, S is the substrate (reactants) and ES is the enzyme–substrate complex and P is the product.

In the overall reaction (S → P), the enzyme is consumed in Step 1 and regenerated in Step 2. According to the slow step, i.e. rate-determining step, the rate constant for the [SH$^+$] in terms of formation of the product can be given as

$$r = \frac{d[P]}{dt} = k_2[ES] \tag{3.23}$$

Using steady state approximation, the rate of reaction of SH$^+$ with time can be given as

$$\frac{d[ES]}{dt} = k_1[E][S] - k_1'[ES] - k_2[ES] = 0 \tag{3.24}$$

The equilibrium between the free and bound enzyme can be given by the enzyme conservation equation, viz.,

$$[E_0] = [E] + [ES]$$

or
$$[E] = [E_0] - [ES] \tag{3.25}$$

Substituting the value of E in Eq. (3.24), we get

$$\frac{d[ES]}{dt} = 0$$

and also
$$k_1\{[E_0] - [ES]\}[S] - k_1'[ES] - k_2[ES] = 0$$

or
$$k_1[E_0][S] = \{k_1' + k_2 + k_1[S]\}[ES] \tag{3.26}$$

Simplifying Eq. (3.26) for [ES], we find that

$$[ES] = \frac{k_1[E_0][S]}{k_1' + k_2 + k_1[S]} \tag{3.27}$$

Substituting the value of [ES] in Eq. (3.23), we get

$$r = \frac{d[P]}{dt} = \frac{k_2 k_1[E_0][S]}{k_1' + k_2 + k_1[S]} \tag{3.28}$$

Dividing both the numerator and denominator by k_1 of Eq. (3.28), we get

$$r = \frac{\dfrac{k_1 k_2[E_0][S]}{k_1}}{\dfrac{k_1' + k_2}{k_1} + [S]} \tag{3.29}$$

or
$$\boxed{r = \frac{k_2[E_0][S]}{K_m + [S]}} \tag{3.30}$$

This equation (3.30) is called Michaelis–Mentan equation, where the new constant K_m is called Michaelis–Menten constant, which is given by

$$K_m = \frac{k_1' + k_2}{k_1} \tag{3.31}$$

when the enzymes reacted completely with the substrate at high concentration, the reaction will hit maximum rate and no free enzyme will left. Hence,

$$[E_0] = [ES] \tag{3.32}$$

Therefore, from Eq. (3.23), we get

$$r_{\max} = V_{\max} = k_2[E_0] \tag{3.33}$$

where V_{max} is the maximum rate, using the notation of enzymology, the Michaelis–Menten Eq. (3.30) can now be written as

$$r_{max} = \frac{V_{max}[S]}{K_m + [S]} \quad \text{(where, } V_{max} = k_2/[E_0]) \quad (3.34)$$

From Eq. (3.34) the following two cases may arise:

(i) If, $K_m \gg [S]$, then [S] can be neglected in the denominator and Eq. (3.34) can be written as

$$r_{max} = \frac{V_{max}[S]}{K_m} = k'[S] \quad \text{(first order reaction)} \quad (3.35)$$

(ii) If, $[S] \gg K_m$, the K_m can be neglected in the denominator and Eq. (3.34) can be written as
$r_{max} = V_{max} = $ constant (zero order reaction) \quad (3.36)

If, $\quad\quad\quad\quad K_m = [S], \; r_{max} = \frac{1}{2} V_{max}$ \quad\quad\quad (3.37)

These two cases are shown diagrammatically in Figure 3.5(a) and (b).

It is difficult to determine the value of V_{max} directly from the plot of r against [S]. So, Eq. (3.34) can be rearranged to determine the V_{max} easily as follows:

$$\frac{1}{r} = \frac{K_m + [S]}{V_{max}[S]} = \frac{K_m}{V_{max}} \times \frac{1}{[S]} + \frac{1}{V_{max}} \quad (3.38)$$

This equation can be treated as straight line equation, i.e. $y = m.x + c$

A plot of $1/r$ against $1/[S]$ gives a straight line whose intercept on x-axis and y-axis are $-1/K_m$ and $1/V_{max}$, respectively and slope is K_m/V_{max} as indicated in Figure 3.5(a) and (b).

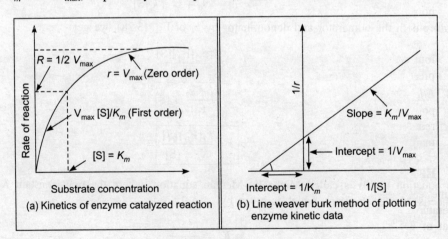

Figure 3.5

3.11 CHARACTERISTICS OF ENZYME CATALYST

Enzyme catalysis is intermediate between homogeneous and heterogeneous catalysis. In general, it is similar to inorganic heterogeneous catalysis and sometimes it is called *microheterogeneous catalysis*. The following characteristics are shown by enzyme catalysts:

1. *Most efficient catalysts:* The enzyme-catalyzed reactions are very fast in comparison to the reactions catalyzed by inorganic substances. This is due to the fact that activation energy of the reaction in presence of an enzyme is low. One molecule of an enzyme may transform one million molecules of the reactants per minute.
2. *Highly specific in nature:* Each enzyme is specific for a given reaction, i.e. one enzyme cannot catalyze more than one reaction. For example, the enzyme urease catalyses the hydrolysis of urea only, it does not catalyze any other amide, not even methylol urea.
3. *Temperature dependence:* The rate of an enzyme-catalyzed reaction depends on the temperature. The enzyme activity rises rapidly with temperature and becomes maximum at a definite temperature which is called *optimum temperature*. Beyond the optimum temperature the enzyme activity decreases and ultimately becomes zero. The enzyme activity is destroyed at about 70°C. The optimum temperature of enzyme reactions occurring in the human body is 37°C. At higher temperatures (fever), the enzyme activity becomes less. The favourable temperature range for enzymatic activity is 25–37°C.
4. pH *dependence:* The rate of an enzyme-catalyzed reaction varies with pH of the system. The enzyme activity is maximum at a particular pH called *optimum* pH. The optimum pH of enzyme reactions occurring in human body is 7.4. The favourable pH range for enzymatic reactions is 5–7.
5. *No effect on equilibrium state:* Like ordinary catalysts, enzymes cannot disturb the final state of equilibrium of reversible reactions.
6. *Colloidal nature:* Enzymes form colloidal solutions in water. Their efficiency is retarded in presence of large quantity of electrolytes. Enzymes are destroyed by ultraviolet rays.
7. *Effect of activators or co-enzymes:* The enzymatic activity is increased in the presence of certain substances, known as co-enzymes. It has been observed that when a small amount of non-protein (i.e. vitamin) is present along with an enzyme, the catalytic activity is enhanced considerably.

    Activators are generally metal ions such as Na^+, Mn^{2+}, Co^{2+}, Cu^{2+}, etc. These metal ions when weakly bonded to enzyme molecules increase their catalytic activity. Amylase in presence of sodium chloride, i.e. Na^+ ions are catalytically very active.
8. *Effect of inhibitors and poisons:* Like ordinary catalysts, enzymes are also inhibited or poisoned. They interact with the active functional groups on the enzyme surface and often reduce or completely destroy the catalytic activity of the enzymes. For example, the use of many drugs is related to their action as enzyme inhibitors in the body.

3.12 EFFECT OF TEMPERATURE ON ENZYME CATALYSIS

Enzyme catalysis is greatly affected by temperature. Their activity is enhanced with rise in temperature. It is because the enzymes decreases the activation energy of a given reaction similar to that of chemical catalysts. The decrease in activation energy by enzymes is far greater than that by the chemical catalysts, but at very high temperatures, the activity of enzyme decreases and even becomes zero (Figure 3.6). It is because enzymes are very sensitive to high temperatures as they are protenious in nature and are denatured at high temperatures. This results into the decrease in effective concentration of enzyme and hence decreases the rate of reaction. Hence, we can say that the catalytic activity of enzymes is effective up to a certain temperature limit (≈ 45°C) and above this temperature the thermal denaturation of enzymes becomes significantly high. Above 70°C the enzymes get destroyed completely. The optimum temperature range for enzyme activity is 25–37°C.

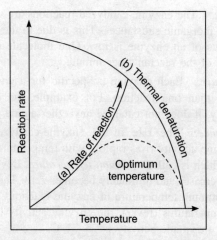

Figure 3.6 Effect of temperature on enzyme-catalyzed reaction.

3.13 pH DEPENDENCE OF RATE CONSTANT OF CATALYZED REACTION

3.13.1 Acid-Catalyzed Reaction

Consider the following acid-catalyzed reaction:

$$S + H^+ \rightarrow P + H^+ \tag{3.39}$$

where S is the substrate and P is the product. The rate of reaction in terms of decomposition of reactants can be given as

$$\frac{-d[S]}{dt} = k_{H^+}[S][H^+] \tag{3.40}$$

The concentration of H^+ ions remains same as they are not consumed. Hence, the reaction can be said to be of pseudo-first order. Thus, at constant pH the rate of reaction

$$\frac{-d[S]}{dt} = k_{app}[S] \tag{3.41}$$

where k_{app} is the apparent rate constant (i.e. pseudo-first order). Therefore, from Eqs. (3.40) and (3.41), we get

$$k_{app} = k_{H^+}[H^+] \tag{3.42}$$

or
$$\log k_{app} = \log k_{H^+} + \log[H^+] \quad (-\log[H^+] = pH)$$

$$= \log k_{app} = \log k_{H^+} - pH \tag{3.43}$$

The effect of pH and k_{app} on the rate of reaction (3.40) is slow and hence, it is the rate determining step. Example: hydrolysis of acetal.

The variation of k_{app} and $\log k_{app}$ with pH for a specific enzyme catalyzed reaction is shown in Figure 3.7(a) and (b).

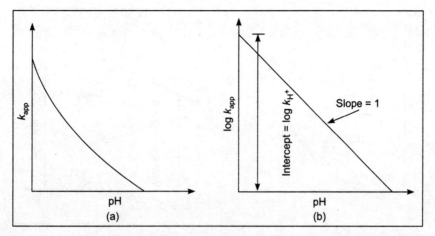

Figure 3.7 (a) and (b) pH dependence of rate constant of specific acid-catalyzed reaction.

3.13.2 Base-Catalyzed Reaction

$$S + OH^- \rightarrow P + OH^-$$

$$\text{Rate of decomposition} = \frac{-d[S]}{dt} = k_{app}[S][OH^-] \quad (3.44)$$

If $[OH^-]$ is constant. Then

$$-\frac{d[S]}{dt} = k_{app}[S] \quad (3.45)$$

where k_{app} is the apparent rate constant (i.e. pseudo first order). Therefore from Eqs. (3.44) and (3.45), we get

$$k_{app} = k_{OH^-}[OH^-] \quad (3.46)$$

On increasing pH, OH^- ions increase. Then the ionic product of water can be given as

$$K_w = [H^+][OH^-] = 10^{-14} \quad \text{(at 25°C)}$$

or

$$[OH^-] = k_w/[H] \quad (3.47)$$

Thus, from Eqs. (3.46) and (3.47), we have

$$k_{app} = k_{OH^-}\left(\frac{k_w}{[H^+]}\right) \quad (3.48)$$

$$\log k_{app} = \log k_{OH^-} + \log k_w + \log \frac{1}{[H^+]} \quad (3.49)$$

$$\log k_{app} = \log k_{OH^-} - 14 + \text{pH} \quad (3.50)$$

($\because \log k_w = -14$)

Example: Hydrolysis of esters and amides.

The variation of log k_{app} with pH and log k_{app} with pH for a specific base-catalyzed reaction is shown in Figure 3.8(a) and (b).

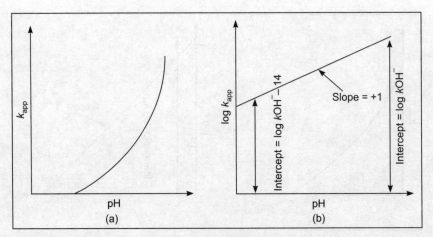

Figure 3.8 (a and b) pH dependence of rate constant of specific base catalyzed reaction.

How the catalytic promoter increases the activity of catalyst?

It is believed that the catalytic promoter enhances the lattice spacing of catalyst which allows the absorption of more number of reactant molecules on the catalyst surface. Moreover, the promoter increases the peaks and cracks, i.e. the surface area to increase the concentration of reactant molecules and hence increases, the rate of reaction.

How the catalytic poison retards or destroys the activity of catalyst?

The catalytic poison covers the surface of catalyst, i.e. blocks the pores (reacting cites). Sometimes the catalytic poison can react chemically with the catalyst and makes the catalyst surface unavailable to the reactants to be adsorbed and combine to form products. Thus, the catalytic poison retards or destroys the activity of the catalyst.

SUMMARY

Catalyst and catalysis: A catalyst is the substance which increases the rate of the reaction without undergoing any change and can be recovered as such at the completion of the reaction. The phenomenon of increase in the rate of reaction with the help of a catalyst is known as *catalysis*.

General characteristics of catalyst/catalysis: A catalyst remains unchanged in mass and chemical composition at the end of the reaction, only a small quantity of the catalyst is generally required to catalyze unlimited reaction, the catalyst does not initiate the reaction, the catalyst is generally specific in nature, the catalyst cannot change the position of equilibrium in a reversible reaction.

Types of catalysts/catalysis: Homogeneous catalysis: When the reactants and the catalyst are in the same phase. **Heterogeneous catalysis:** When the reactants and the catalyst are in different phases.

Positive catalysis: When the rate of reaction is accelerated by foreign substances is called *positive catalysis*.

Negative catalysis: The substances which, when added to the reaction mixture, slow down the reaction rate instead of increasing it.

Catalytic poisons: Substances which destroy the activity of the catalyst by their presence.

Example: The activity of iron catalyst is destroyed by the presence of H_2S or CO in the synthesis of ammonia by Haber's process.

Auto-catalysis: In certain reactions, one of the products acts as a catalyst.

Induced catalysis: The phenomenon in which one reaction influences the rate of other reaction, which does not occur under ordinary conditions, is known as induced catalysis.

Theories of catalysis: There are two theories of catalysis. (1) intermediate compound formation theory and (2) Adsorption theory.

Nature of solid catalysts: The solid catalysts may be metals, alloys, metal oxides or metal sulphides. The effectiveness of a catalyst depends upon the two important aspects, i.e. activity and selectivity.

Shape selective catalysis by zeolites: Zeolites are microporous aluminosilicates of the general formula $M \cdot x/n[(AlO_2)_x (SiO_2)_y] \, mH_2O$. These are three-dimensional networks of silicates in which some silicon atoms are replaced by aluminium atoms. The most remarkable feature of zeolite catalysis is the shape selectivity.

Acid–base catalysis: Generally, homogeneous catalysis in solution is carried out by acids or bases or both. A reaction which is catalyzed by proton (H^+ ions) but not by other Bronsted acids (i.e. proton donors) is called *specifically proton-catalyzed reaction*. While the reaction which is catalyzed by OH^- ions is called *specifically base-catalyzed reaction*. On the other hand, the reaction which is catalyzed by Bronsted base is called *general base catalysis*.

Enzyme catalysis: Enzymes are complex nitrogenous organic compounds, i.e. protein molecules of high molecular mass which are produced by living plants and animals. They form colloidal solutions in water and are very effective catalysts. The enzymes are, thus, termed as *bio-chemical catalysts*.

Characteristics of enzyme catalysis: Most efficient catalysts, highly specific nature, temperature dependent, pH dependent, no effect on equilibrium state, colloidal nature, activators or co-enzymes, inhibitors and poisons.

Effect of temperature on enzyme catalysis: The rate of reaction generally increases with rise in temperature. Enzymes are very sensitive to high temperatures as they are protenious in nature and are denatured at high temperatures. Hence we can say that the catalytic activity of enzymes is effective up to a certain temperature limit ($\approx 45°C$).

Michaelis–Menten equation: $\boxed{r = \dfrac{k_2[E_0][S]}{K_m + [S]}}$ K_m is called Michaelis–Menten constant,

$$K_m = \frac{k_1' + k_2}{k_1}.$$

Characteristics of enzyme-catalyzed reaction: Most efficient catalysts, highly specific in nature, temperature dependent, pH dependence, no effect on equilibrium state, colloidal nature, activators or co-enzymes, inhibitors and poisons.

Effect of temperature on enzyme catalysis: The rate of reaction generally increases with rise in temperature but after a certain limit of temperature (60°C) they are denatured.

pH-dependence enzyme-catalyzed reaction: The optimum pH for enzymatic reactions is 5.

EXERCISES

1. Define the following terms:
   (i) Catalyst, (ii) catalysis, (iii) homogeneous catalysis, (iv) heterogeneous catalysis, (v) positive catalyst, (vi) negative catalyst, (vii) auto catalyst, (viii) induced catalyst, (ix) catalytic promoters, (x) catalytic poison, (xi) enzymes.

2. Explain/justify the following:
   (i) Rough surface of a catalyst is more effective than smooth surface.
   (ii) When acidic solution of $KMnO_4$ is added to a warm solution of oxalic acid, the colour is discharged slowly in the beginning but after sometime it disappears rapidly.
   (iii) Chloroform used as anesthetic is kept in dark coloured bottle and mixed with ethanol.
   (iv) Industries prefer to use the catalyst which works at room temperature.
   (v) The activity of heterogeneous catalyst is highly dependent.
   (vi) Phosphoric acid is mixed with hydrogen peroxide.
   (vii) It is necessary to remove arsenic impurity in the manufacture of SO_3 by contact process.
   (viii) It is necessary to remove CO when ammonia is obtained by Haber process.
   (ix) Why all adsorptions are exothermic?
   (x) The ester hydrolysis is slow in the beginning but becomes faster after sometime.
   (xi) Why a catalyst is more effective in finely divided state?
   (xii) Which will adsorb more, a lump of charcoal or charcoal powder?
   (xiii) Solid catalyst works best as fine power.
   (xiv) Catalyst speed up the reactions.

3. Answer the following:
   (i) What is the role of catalyst in establishing equilibrium?
   (ii) Describe at least three characteristics of a catalyst.
   (iii) How does a promoter and catalyst poison work?
   (iv) Can a catalyst increase the yield of there reaction?
   (v) Define homogeneous catalysis with one example?
   (vi) Define heterogeneous catalysis with one example.
   (vii) Name three important applications of adsorption.
   (viii) Give at least three points of difference between physisorption and chemisorptions.
   (ix) What is the sign of H and S when a gas gets adsorbed on a solid?

4. What do you understand by catalyst and catalysis? Explain with examples.
5. How the catalyst increases the rate of reaction? Explain.
6. Explain the general characteristics of catalyst.
7. Explain the different types of catalysis giving suitable examples.
8. Describe the mechanism of homogeneous catalysis.
9. Describe the mechanism of heterogeneous catalysis (contact catalysis).

10. What are enzymes? How the enzymes play active role in biological processes?
11. What are the catalyst poisons? How they retard or stop the reactions?
12. Derive the Mechaelis–Menten equation for the rate of enzyme-catalyzed reactions.
13. What is solid catalyst? Explain the nature of solid catalyst.
14. What do you understand by shape selective catalyst? Explain their functions by giving suitable examples.
15. What do you understand by acid–base catalysis? Explain the kinetics and mechanism of Acid–base catalysis.
16. Describe the characteristics of enzyme-catalyzed reactions.
17. Explain mechanism and kinetics of enzyme-catalyzed reactions.
18. Explain the effect of temperature on enzyme catalysis.
19. What shall be the rate of digestion of food if enzymes are not present in the body of living organisms?
20. Give equations of the reactions in which the following catalysts are used:
    (i) zymase and invertase, (ii) anhydrous $AlCl_3$, (iii) nitric oxide, (iv) cupric chloride, (v) manganese dioxide, (vi) conc. H_2SO_4.

CHAPTER 4

WATER TREATMENT PART I

4.1 INTRODUCTION

Water is one of the most widely used and abundant chemical in the nature. About 80% of the earth's surface is covered by water, out of which only 0.3% can be used for domestic, agriculture and industrial purposes and rest is locked in oceans, polar ice caps, giant glaciers and rock crevices. Water is so essential part of animal and plant life that without which life can not survive.

4.2 SOURCES OF WATER

There are two types of sources of water. These are:

4.2.1 Surface Water

(a) *Rain water:* It is the purest form of all the natural waters. However, it may contain dissolved impurities gases from industries like CO_2, SO_2, NO_2, etc. and suspended solid particles of organic and inorganic origin.

(b) *Lake water:* It contains much lesser amount of dissolved minerals than even well water but quantity of organic matter present in it is very high.

(c) *River water:* Since the rivers flow over the land surface, the river water get the soluble minerals of soil dissolved in it. River water thus contains dissolved minerals of the soil like bicarbonates, chlorides, sulphates, etc. of sodium, calcium, magnesium and iron. This water also contains organic matter and suspended impurities.

(d) *Sea water:* It is the most impure form of all the natural waters. The impurities of dissolved salts are thrown into the sea by the rivers joining the sea. Out of the total dissolved salt (3.5%) present in the sea water, 2.6% is sodium chloride.

Surface water as such is not safe for human consumption as it contains different impurities, pathogenic bacteria and suspended matter.

4.2.2 Underground Water

(a) Spring and well water: This water has clearer appearance but contains more dissolved salts. Underground water has least organic matter therefore contains more hardness.

4.3 IMPURITIES IN WATER

There may be following types of impurities present in the water:
1. *Dissolved impurities:* The following impurities may be present in the water in the dissolved form:
    (a) *Inorganic salts:* cations like Ca^{2+}, Mg^{2+}, Na^+, K^+, Fe^{2+}, Al^{3+} and sometimes traces of Cu^{2+} and Zn^{2+}. Anions like Cl^-, SO_4^{2-}, HCO_3^-, NO_3^- and sometimes F^- and NO_2^-.
    (b) *Organic salts:* like dies, drugs, etc.
    (c) *Dissolved gases:* like O_2, CO_2, N_2, oxides of nitrogen and sulphur, and sometimes NH_3 and H_2S.
    (d) *Toxic elements:* like As, Hg, Cr, etc.
2. *Colloidal impurities:* like clay, silica, $Al(OH)_3$, $Fe(OH)_3$, organic waste products, humic acids, amino acids, complex proteins, etc.
3. *Suspended impurities:* Inorganic matter like clay and sand; organic metter like oils globules, vegetables and animal matter and other forms of animals and vegetable life.
4. *Biological impurities:* Bacteria and other microorganisms like virus, algae, fungi, diatoms, etc.

4.4 PROPERTIES IMPARTED BY IMPURITIES PRESENT IN WATER

The effect of impurities on water quality can be seen as follows:

4.4.1 Colour

Oftenly surface water and sometimes underground water has colour due to colouring compounds, colloidal and other impurities like algae, fungi, diatoms, peat, tannin, humus, iron and manganese salts, etc. and also due to industrial effluents. The colour of natural water ranges from pale yellow to yellowish brown or dark brown. The presence of colour and colouring materials is often objectionable in dyeing, scouring, laundry and pharmaceutical industries. The colour and organic matter is minimized or removed generally by coagulation, absorption, settling, filtration and super-chlorination.

4.4.2 Taste

The taste of water is due to the presence of dissolved minerals and salts. The taste of chlorinated water is due to the compounds formed by the reaction of chlorine on traces of organic matter. The brackish taste of water is due to the presence of excess amount of salts, e.g. NaCl and KCl. Presence of iron, aluminium and manganese sulphates and excess of lime imparts bitter taste to the natural water while CO_2 and nitrates give pleasant taste. Soapy taste of water is due to the presence of large amount of Na_2CO_3. Disagreeable taste is objectionable in beverages, food products, pharmaceutical industries and potable water.

4.4.3 Odour

Unpleasant odour in water may be mainly due to the presence of organic matter, living organisms, decaying vegetation, bacteria, weeds, algae, fungi, etc. In some deep well waters and some industrial effluents give mostly unpleased rotten egg-like odour due to presence of H_2S. Alcohols, phenols, aldehydes, ketones, etc. released from industries into the water bodies causes the offensive odour in water. The disagreeable odour can be removed/minimized by adsorption, oxidation, chlorination and precipitation methods.

4.4.4 Turbidity and Sediments

It is due to the presence of finely divided colloidal, insoluble impurities like inorganic matter, clay, silica, slit, calcium carbonate, ferric hydroxide or organic matter, e.g. micro-organisms, animals matter, oils, fats finely divided vegetable, etc.

4.4.5 Acidity

Acidity of water is imparted due to the presence of free acids, industrial waste, minerals, iron and aluminium salts, dissolved CO_2, SO_2, NH_3, etc.

4.5 HARDNESS OF WATER

"Hardness of water is that characteristic which prevents the formation of lather with soap." It is due to the presence of bicarbonates, chlorides and sulphates of Ca, Mg, Fe, Mn and other heavy metals. When a water sample is treated with soap (sodium or potassium salts of higher fatty acids like oleic, palmitic, stearic acid, etc.), does not form lather because the Ca^{2+} and Mg^{2+}, even Fe^{2+}, Mn^{2+} and Al^{3+} present in water react with soap to form white scum/precipitate of insoluble soaps.

$$2C_{17}H_{35}COONa + Ca(HCO_3)_2 \rightarrow (C_{17}H_{35}COO)_2Ca + 2NaHCO_3$$

$$2C_{17}H_{35}COONa + CaSO_4 \rightarrow (C_{17}H_{35}COO)_2Ca + Na_2SO_4$$

$$2C_{17}H_{35}COONa + MgCl_2 \rightarrow (C_{17}H_{35}COO)_2Mg + 2NaCl$$

4.5.1 Types of Hardness

Hardness of water can be classified into the following two categories:

(a) *Temporary or carbonate hardness:* It is due to the presence of bicarbonates of Ca and Mg, and carbonate of iron. Temporary hardness can be removed easily by boiling the water. On boiling, bicarbonates are decomposed to insoluble carbonates and hydroxides. In other words, it is the soap consuming capacity of the water sample.

$$Ca(HCO_3)_2 \rightarrow CaCO_3 \downarrow + H_2O + CO_2 \uparrow$$

$$Mg(HCO_3)_2 \rightarrow Mg(OH)_2 \downarrow + 2CO_2 \uparrow$$

(b) *Permanent or non-carbonate hardness:* It is due to the presence of chlorides and sulphates of calcium, magnesium, iron and other heavy metals. Permanent hardness can not be removed by boiling the water but can be removed by some chemical methods like lime–soda method, zeolite method and ion exchange methods.

4.5.2 Alkaline and Non-Alkaline Hardness

(a) *Alkaline hardness:* The hardness which is due to the presence of bicarbonates, carbonates, and hydroxides of the hardness producing metals like Cu, Mg, Fe, Al, Mn, is called *alkaline hardness*. It is also known as *carbonate hardness*.

The alkaline hardness in raw water is always associated with the bicarbonates, but in treated water or in boiler water it may also contain hardness due to small quantities of calcium carbonate and magnesium hydroxide.

(b) *Non-alkaline hardness:* The hardness which is due to the presence of chlorides, sulphates, etc. of calcium and magnesium is called *non-alkaline hardness*. It is obtained by subtracting the alkaline hardness from total hardness. It is also known as *non-carbonate hardness*.

4.6 CALCIUM CARBONATE EQUIVALENT (DEGREE OF HARDNESS)

The concentration of hardness and non-hardness producing salts or ions such as bicarbonates, chlorides and sulphates of Ca, Mg, etc. is measured in terms of equivalent amount of calcium carbonate. Thus,

$$\underset{\text{Equivalent hardness}}{CaCO_3} = \frac{\text{Mass of hardness producing subsance} \times \text{Molecular mass of CaCO}_3}{\text{Moecular mass of hardness producing subsance}}$$

or

$$\underset{\text{Equivalent hardness}}{CaCO_3} = \frac{w \times 100}{M}$$

$CaCO_3$ is particularly selected to express the hardness of water because:
1. $CaCO_3$ is the most insoluble salt which can be precipitated during water treatment.
2. Since the molecular mass of $CaCO_3$ is 100 (eq. wt. = 50), the calculation becomes easy. The $CaCO_3$ equivalent hardness of different species can be calculated by using multiplication factor as mentioned in Table 4.1.

Table 4.1 Calculation of calcium carbonate equivalent hardness

S. No.	Salts/ions	Molecular mass	Chemical equivalent	Multiplication factor to convert into CaCO$_3$
1	Ca(HCO$_3$)$_2$	162	81	100/162
2	CaCl$_2$	111	55.5	100/111
3	CaSO$_4$	136	68	100/136
4	CaCO$_3$	100	50	100/100
5	Ca(NO$_3$)$_2$	164	82	100/164
6	Mg(HCO$_3$)$_2$	146	73	100/146
7	MgCl$_2$	95	47.5	100/95
8	MgSO$_4$	120	60	100/120
9	MgCO$_3$	84	42	100/84
10	Mg(NO$_3$)$_2$	148	74	100/148
11	HCO$_3^-$	61	61	100/61
12	CO$_3^{2-}$	60	30	100/60
13	OH$^-$	17	17	100/17
14	CO$_2$	44	22	100/44
15	H$^+$	2	1	100/2
16	Al$_2$(SO$_4$)$_3$	342	57	100/114
17	FeSO$_4 \cdot$ 7H$_2$O	278	139	100/278
18	NaAlO$_2$	82	82	100/164

4.6.1 Degree of Hardness

"The degree of hardness is defined as the number of parts of $CaCO_3$ equivalent hardness present in a particular number of parts of water regarding the unit employed."

4.6.2 Units of Hardness

1. Milligrams per litre (mg l^{-1}): It is the number of milligrams of $CaCO_3$ equivalent hardness present in one litre of water, i.e. 1 mg = 1 mg of $CaCO_3$ equivalent hardness per litre of water.

2. Parts per million (ppm): It is the number of parts of $CaCO_3$ equivalent hardness present per millioin (10^6) parts of water.

   1 ppm = 1 parts of $CaCO_3$ equivalent hardness per 10^6 parts of water.

   Weight of one litre water = 1 kg = 1000 × 1000 mg = 10^6 mg. Therefore, 1 mg l = 1 mg $CaCO_3$ per 10^6 mg of water.

3. Equivalents per million (epm): It is defined as the number of milliequivalents of hardness per million weight units of solution. In dilute solution the density does not differ very much from unity. Therefore, for dilute solutions, 1 ppm = 1 mg equivalent per litre. 1 ppm is conventionally taken as equal to 1 ml of 1 N solution per litre.

   Thus, 1 meq = 50 mg of $CaCO_3$ = 50 ppm.

4. Degree Clarke's or grains per gallon: It is the number (grain) of $CaCO_3$ equivalent hardness per gallon (1 grain = 1/7000 lb) of water. 1°Clarke = 1 grain of $CaCO_3$ equivalent hardness per gallon (70,000 parts) of water 1°Cl = 1 part of $CaCO_3$ equivalent hardness per 70,000 parts of water.

5. Degree French (°Fr): It is the parts of $CaCO_3$ equivalent hardness per 10^5 parts of water. 1°Fr = 1 part of $CaCO_3$ equivalent hardness per 10^5 parts of water.

Relationship between various units

1 mg l^{-1} = 1 ppm = 0.07°Cl = 0.1°Fr = 0.02 Meq
1°Cl = 19.3 ppm = 14.3 mg l^{-1} = 1.43°Fr

4.7 DETERMINATION OF HARDNESS

The hardness of water can be determined by the following methods:

1. O. Hehner's method
2. Soap Titration method
3. EDTA Titration method

Among the above mentioned methods the EDTA titration method is most commonly used because it is the most accurate and easy method. So here only EDTA titration method is mentioned.

4.7.1 EDTA Titration Method

EDTA titration method is based on the fact that when solution of EDTA is added to the water sample at higher pH, it forms stable, soluble complex with Ca and Mg ions present in the hard water. The volume of EDTA solution used is equivalent to the hardness of water. The end point is detected with the help of Eriochrome Black–T (EB–T) indicator.

Ethylenediamine-tetraacetic acid (EDTA) is a very good complexing agent. Its disodium salt ionizes to give the ions as follow:

$$\begin{array}{c} ^-OOC-H_2C \\ HOOC-H_2C \end{array} N-CH_2-CH_2-N \begin{array}{c} CH_2-COOH \\ CH_2-COO^- \end{array}$$

(Disodium salt of EDTA)

which can be represented as H_2Y^{2-}. It forms complexes with Ca^{2+} and Mg^{2+}, and other divalent cations as shown by the following reactions:

$$M^{2+} + H_2Y^{2-} \rightarrow MY^{2-} + 2H^+$$

$$M^{n+} + H_2Y^{2-} \rightarrow MY^{(n-4)} + 2H^+$$

The dissociation of these complexes is governed by the pH of the solution and hardness causing divalent ions which are stable in alkaline medium (pH ≈ 10). The indicator used is a complex organic compound [sodium 1-(1-hydroxy 2-napthylazo) 6-nitro 2-naphthol 4-sulphonate], commonly known as *Eriochrome Black-T*.

Eriochrome Black-T indicator

It has two phenolic hydrogen atoms and for simplicity can be represented as $Na^+H_2In^-$.

$$H_2In^- \underset{5.5}{\overset{7.0}{\rightleftharpoons}} HIn^{2-} \underset{10.0}{\overset{10.5}{\rightleftharpoons}} In^{3-}$$
$$\text{Red} \qquad\qquad \text{Blue} \qquad\qquad \text{Yellowish orange}$$

When a small amount of the indicator solution is added to a hard water sample (pH ≈ 10) the indicator reacts with Mg^{2+} or Ca^{2+} to produce wine red colour complex.

$$\underset{\text{Blue}}{HIn^{2-}} + Ca^{2+}/Mg^{2+} \rightarrow \underset{\text{Wine red}}{Ca/Mg-In^-} + H^+$$

The optimum pH of the sample for the experiment is 10.0 ± 0.1 as the complex formation is effective at higher pH. It is adjusted by adding ammonia buffer ($NH_4OH + NH_4Cl$).

When this solution (water sample) is titrated with EDTA (H_2Y^{2-}), the EDTA first gets complexed with Ca^{2+} ions and then with Mg^{2+} ions to form the stable complexes:

$$Ca^{2+} + H_2Y^{2-} \rightarrow CaY^{2-} + 2H^+$$

$$Mg^{2+} + H_2Y^{2-} \rightarrow MgY^{2-} + 2H^+$$

These complexes Ca/Mg–Y^{2-} are more stable than Ca/Mg–In^- indicator complex. Therefore, when all the hardness causing ions get complexed with EDTA, the indicator is liberated in the free form and appears as blue colour at the end point.

$$\underset{\text{Wine red}}{Ca/Mg-In^-} + H_2Y^2 \rightarrow \underset{\text{Pure blue}}{Ca/MgY^{2-} + H-In^2} + 2H^+$$

Completion of the above reaction makes the end point of the titration.

Procedure

1. *Standardization of EDTA solution with standard hard water (SHW):* A known volume of standard hard water (SHW) is buffered with ammonia solution to maintain its pH around 10,

and four drops of Eriochrome Black-T indicator is added to it. A wine red colour appears. This mixture is titrated against standard EDTA solution till the appearance of blue colour at the end point. The volume of EDTA used up to the end point is recorded as V_1 ml.

2. *Determination of total hardness:* A known volume of water sample is buffered with ammonia solution to maintain its pH around 10 and four drops of Eriochrome Black-T indicator is added to it. A wine red colour appears. This mixture is titrated against standard EDTA solution till the appearance of blue colour. The volume of EDTA used is recorded as V_2 ml which correspondence to the total hardness of water.

3. *Determination of permanent hardness:* The temporary hardness is first removed by boiling and then filtering the hard water sample. Now a known volume of this boiled out water is buffered and titrated in same the way against standard EDTA solution using Eriochrome Black-T indicator. The volume of EDTA used (V_3 ml) corresponds to permanent hardness.

4. *Determination of temporary hardness:* Temporary hardness is obtained by subtracting permanent hardness from the total hardness.

Observations and calculation

Strength of SHW = 1000 PPM

Volume of SHW = x ml

Volume of hard water sample = y ml

Volume of EDTA solution used for its standardization = V_1 ml

Volume of EDTA solution used with hard water sample = V_2 ml

Volume of EDTA solution used with boiled out hard water sample = V_3 ml

$$\text{Strength of EDTA solution} = \frac{\text{Strength of SHW} \times \text{Volume of SHW } (x)}{\text{Volume of EDTA used } (V_1)} = a \text{ ppm}$$

$$\text{Total hardness} = \frac{\text{Strength of EDTA} \times \text{Volume of EDTA used } (V_2)}{\text{Volume of water sample}} = b \text{ ppm}$$

$$\text{Permanent hardness} = \frac{\text{Strength of EDTA} \times \text{Volume of EDTA used } (V_3)}{\text{Volume of water sample}} = c \text{ ppm}$$

Temporary hardness = Total hardness (b) – Permanent hardness (c) = d ppm.

4.8 DISADVANTAGES OF HARD WATER

Hard water has great disadvantages in different fields. It affects human health, crop yields, various industrial applications, etc., which can be explained under the following heads.

4.8.1 Domestic Use

(a) *Drinking water:* Continuous use of hard water may affect the digestive system and may cause constipation, loss of appetite, etc. It may also cause the formation and deposition of calcium oxalate crystal in the urinary track.

(b) *Cooking:* Due to presence of hardness causing salts in hard water, its boiling point increases. So when food is cooked in hard water more heat and time is required, ultimately it causes the wastage of fuel and money. Moreover, the food cooked in hard water has an unpleasant taste.

(c) *Washing:* When hard water is used for washing the clothes it consumes lot of soap. It is because the Ca and Mg ions present in hard water react with sodium soap (soluble) to form insoluble Ca/Mg soap. So, firstly the soap is consumed in removing the hardness (Ca^{2+} and Mg^{2+} ions) and then it starts cleaning the fabric.

$$\underset{\substack{\text{Sodium soap}\\\text{Water soluble}}}{2R-COONa} + \underset{\text{From hard water}}{Ca^{2+}} \rightarrow \underset{\substack{\text{Calcium soap}\\\text{Water insoluble}}}{(R-COO)_2Ca} + 2Na^+$$

$$\underset{\text{Sodium soap}}{2R-COONa} + \underset{\text{From hard water}}{Mg^{2+}} \rightarrow \underset{\text{Magnesium soap}}{(R-COO)_2Mg} + 2Na^+$$

Hence, washing of clothes in hard water causes wastage of soap, spotting of clothes, and sometimes due to presence of iron salts it may cause staining (yellowing) of the fabric.

(d) *Bathing*: When we take the bath with hard water it produces sticky scum on the body and in the bathing tub which sticks on the body and hairs. So, the cleansing of body is not good. Moreover, it causes the wastage of soap.

4.8.2 Industrial Use

(a) *Steam production in the boilers*: The water is used in the boiler for steam generation, should be soft and pure. If the hard water is fed into the boiler, it results in the sludge and scale formation, priming and foaming, and caustic embrittlement.

(b) *Pharmaceutical industries:* To prepare the medicines it requires soft and pure water. If drugs, syrup, injections, ointment, etc. are prepared in hard water, the hardness causing salts may produce undesirable products in them which may be dangerous for the use in the body.

(c) *Sugar industries:* If sugar is crystallized in hard water, the salts like chloride, sulphate, nitrate, etc. may cause difficulty in the crystallization and sugar so produced may be deliquescent.

(d) *Textile industries:* Hard water causes the wastage of lot of soap during washing the fabrics, fibres, etc. The precipitate of Ca and Mg compounds adheres on the fabric which reduces the quality of fabrics. The iron and manganese salts may spoil the beauty by spotting on fabrics. The dissolved Ca, Mg and iron salts present in hard water may react with dyes resulting in the dull shade and spots on the fabric used for dying.

(e) *Paper industries:* If the paper pulp is prepared in the hard water the salts present in it may react with the materials like cellulose, etc. used in paper, resulting in the poor glaze and finishing of paper. Moreover, the paper so obtained may be brittle and deliquescent.

(f) *Laundry:* Use of hard water in laundry causes the wastage of soap, staining and spotting of clothes.

4.9 ALKALINITY OF WATER

"Alkalinity of water may be defined as the concentration of OH^- ions and the concentration of ions or species which increase the concentration of OH^- ions due to dissociation or hydrolysis."

The alkalinity of natural water is generally due to the presence of HCO_3^-, SiO_3^{2-}, $HSiO_3^-$, CO_3^{2-}, and also due to the presence of salts of some weak organic acids (hamates).

4.9.1 Types of Alkalinity

Depending on the anions (HCO_3^-, CO_3^{2-}, OH^-) present in the water the alkalinity may be classified into two categories:

(a) *Bicarbonate alkalinity:* It is caused due to the presence of HCO_3^- ions
(b) *Carbonate alkalinity or hydroxide alkalinity:* It causes due to the presence of OH^- and CO_3^{2-} ions. Highly alkaline water may lead to caustic embrittlement and also the deposition of precipitate and sludge in boiler tubes and pipes. The possible combinations of ions causing alkalinity in water are:

1. Only OH^- ions
2. Only CO_3^{2-} ions
3. Only HCO_3^- ions
4. OH^- and CO_3^{2-} ions
5. CO_3^{2-} and HCO_3^- ions

The possibility of OH^- and HCO_3^- existing together is ruled out because of the fact that they combine together to form CO_3^{2-}

$$OH^- + HCO_3^- \rightarrow CO_3^{2-} + H_2O$$

The knowledge of alkalinity in water is important for controlling the corrosion, conditioning the boiler feed water (internally), calculating the amount of lime and soda needed for water softening, and also in neutralizing the acidic solution produced by the hydrolysis of salts. In boilers, high alkalinity in water may lead caustic embrittlement and also the precipitation of sludge and deposition of scales.

4.9.2 Experimental Determination of Alkalinity

The alkalinity of water is due to the presence of hydroxide ion (OH^-), carbonate ion (CO_3^{2-}) and bicarbonate ion (HCO_3^-) present in the given sample of water. These can be estimated separately by titrating against standard acid and using phenolphthalein and methyl orange as indicators. The chemical reaction involved can be shown by the following equations:

(i) $OH^- + H^+ \rightarrow H_2O$
(ii) $CO_3^{2-} + H^+ \rightarrow HCO_3^-$ $\Big] P \Big] M$
(iii) $HCO_3^- + H^+ \rightarrow H_2O + CO_2$

The titration of the water sample against a standard acid up to phenolphthalein end point shows the completion of reactions (i) and (ii) only. This amount of acid used thus corresponds to hydroxide plus one half of the normal carbonate present. The titration of the water sample against a standard acid to methyl orange end point marks the completion of reactions (i), (ii) and (iii). Hence, the amount of acid used after the phenolphthalein end point corresponds to one half of normal carbonate plus all the bicarbonates, while the total amount of acid used represent the total alkalinity (due to hydroxide, bicarbonate and carbonate ions).

Procedure

A known volume of water sample is taken in a conical flask and titrated against standard acid (N/50) using phenolphthalein indicator until the pink colour caused by phenolphthalein just disappeared. The volume (V_1) of standard acid used is equivalent to the phenolphthalein alkalinity. This water sample is further titrated against standard acid using methyl orange indicator until the light yellow colour changes to red. The total volume ($V_1 + V_2$) of standard acid used is equivalent to the methyl orange alkalinity.

Observation and calculations

Volume of water sample taken for each titration = V ml
Volume of standard acid used to phenolphthalein end point = V_1 ml
Additional volume of standard acid used to methyl orange end point = V_2 ml

$$\text{Phenolphthalein alkalinity} = \frac{\text{Normality}\left(\frac{N}{50}\right) \times \text{Volume of acid used }(V_1)}{\text{Volume of water sample}} \times 50 \times 1000 = a \text{ ppm}$$

$$\text{Methyl orange (total alkalinity)} = \frac{\text{Normality}\left(\frac{N}{50}\right) \times \text{Volume of acid used }(V_1 + V_2)}{\text{Volume of water sample}} \times 50 \times 1000 = b \text{ ppm}$$

4.9.3 Calculation of Types of Alkalinity

Following reactions take place during titration of water sample

(i) $OH^- + H^+ \rightarrow H_2O$
(ii) $CO_3^{2-} + H^+ \rightarrow HCO_3^-$ ⎤ P
(iii) $HCO_3^- + H^+ \rightarrow H_2O + CO_2$ ⎦ M

Here P and M represent the volume of standard acid used during titration of water sample for phenolphthalein and methyl orange indicator end point, respectively. From the above relations we can calculate the types of alkalinity as mentioned in Table 4.2.

Table 4.2 Calculation of types of alkalinity on the basis of P and M

S. No.	Value of P and M	Hydroxide (OH^-)	Carbonate (CO_3^{2-})	Bicarbonate (HCO_3^-)
1	$P = 0$	0	0	M
2	$P = M$	P or M	0	0
3	$P = 1/2M$	0	2P or M	0
4	$P < 1/2M$	0	2P	M − 2P
5	$P > 1/2M$	2P − M	2(M − P)	0

(a) **When $P = 0$:** In this case the alkalinity is due to the presence of only HCO_3^- ions, rest two species, i.e. OH^- and CO_3^{2-} are absent.
(b) **When $P = M$:** In this case the alkalinity is due to the presence of only OH^- ions which is equal to P or M, rest two species, i.e. CO_3^{2-} and HCO_3^- are absent.
(c) **When $P = 1/2M$:** In this case the alkalinity is due to the presence of only CO_3^{2-} ions which is equal to $2P$ or M, rest two species, i.e. OH^- and HCO_3^- are absent.

The phenolphthalein end point indicates half neutralization of CO_3^{2-} (up to HCO_3^-)

$$CO_3^{2-} + H^+ \rightarrow HCO_3^-$$

While the methyl orange end point indicates the complete neutralization of CO_3^{2-}

$$CO_3^{2-} + H^+ \rightarrow HCO_3^-; \quad HCO_3^- + H^+ \rightarrow H_2O + CO_2$$

Hence, the alkalinity due to CO_3^{2-} = 2P or M.

(d) **When P < 1/2M:** In this case, the alkalinity is due to the presence of CO_3^{2-} and HCO_3^- ions. The OH^- ions are absent. The phenolphthalein end point shows the half neutralization of CO_3^{2-} (up to HCO_3^-), hence, for complete neutralization it will require twice of phenolphthalein end point.

$$CO_3^{2-} + H^+ \rightarrow HCO_3^-$$

Thus, the alkalinity due to CO_3^{2-} = 2P.

On the other hand, methyl orange indicator shows the complete neutralization of CO_3^{2-}

$$CO_3^{2-} + H^+ \rightarrow HCO_3^-$$
$$HCO_3^- + H^+ \rightarrow H_2O + CO_2$$

Therefore, the alkalinity due to HCO_3^- = M − 2P.

(e) **When P > 1/2M:** In this case, the alkalinity is due to the presence of OH^- ions in addition to CO_3^{2-} ions. The HCO_3^- ions are absent. The phenolphthalein end point shows the complete neutralization of OH^- and half neutralization of CO_3^{2-} (up to HCO_3^-) which is equal to 2P − M and 2(M − P), respectively.

$$\left.\begin{array}{l} OH^- + H^+ \rightarrow H_2O \\ CO_3^{2-} + H^+ \rightarrow HCO_3^- \end{array}\right\} P$$

On the other hand, the methyl orange indicator shows the complete neutralization of OH^- and CO_3^{2-}.

$$\left.\begin{array}{l} OH^- + H^+ \rightarrow H_2O \\ CO_3^{2-} + H^+ \rightarrow HCO_3^- \\ HCO_3^- + H^+ \rightarrow H_2O + CO_2 \end{array}\right\} M$$

The neutralization of HCO_3^- = M − P and, therefore, the complete neutralization of CO_3^{2-} = 2(M − P) (twice of the neutralization of HCO_3^- ions).

The alkalinity due to OH^- = M − 2(M − P) = M − 2M + 2P = 2P − M

4.10 SLUDGE AND SCALE

Due to the heating and continuous evaporation of water in the boiler the concentration of salts in the water increases. When the ionic products exceed the solubility product, the salts are thrown out in the form of precipitate. The precipitate may appear in two forms, i.e. loose slimy precipitate or hard adhering material. (Figures 4.1(a) and (b)).

4.10.1 Sludge

"Sludge is a soft, loose and slimy precipitate formed within the boiler." It is formed at comparatively colder portion of the boiler and collects in the bends and areas of the system where the flow rate is slow.

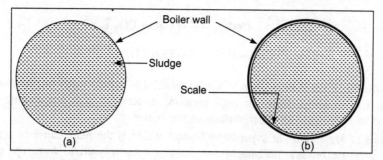

Figure 4.1 (a) Sludge and (b) scale formation in the boiler.

Composition

Sludge is formed by substances which have greater solubility in hot water than in cold water, e.g. $MgCO_3$, $MgCl_2$, $CaCl_2$, $MgSO_4$, etc. Sludge can easily be scrapped off with a wire brush.

Disadvantages of sludge formation

1. Sludge is poor conductor of heat, so it tends to waste a portion of heat.
2. If sludge is formed along with scale then it gets entrapped in the scale and both get deposited as scale.
3. Excessive sludge formation disturbs the working of the boiler.
4. It causes the chocking and blocking of the boiler openings like valves and tubes, and hence causes the trouble in boiler operation.
5. It causes the decay/corrosion of boiler material and hence weakening of the boiler.

Prevention of sludge formation

Sludg formation can be prevented by:
 (i) using well softened water and
 (ii) frequently blowdown operations.

4.10.2 Scales

"Scale is hard adhering deposit which sticks firmly to the inner surfaces of the boiler." Scale is difficult to remove even with the help of hammer and chisel. Scale is the main source of boiler trouble.

Composition

The formation of scale may be due to the presence of impurities like:

$CaCO_3$, $Ca(OH)_2$, $CaSO_4$, $CaSiO_3$, $Mg(OH)_2$, $MgSiO_3$, etc. in water.

Causes of scale formation

(a) *Due to the decomposition of calcium bicarbonate:*

$$Ca(HCO_3)_2 \xrightarrow{\Delta} \underset{Scale}{CaCO_3} \downarrow + H_2O + CO_2 \uparrow$$

In low pressure boilers, $CaCO_3$ causes the scale formation but in high pressure boilers, $CaCO_3$ becomes soluble.

$$CaCO_3 \xrightarrow{\Delta} CaO + CO_2 \uparrow$$

$$CaO + H_2O \xrightarrow{\Delta} Ca(OH)_2$$

(b) *Decomposition of calcium sulphate:* $CaSO_4$ is soluble in cold water but solubility decreases with the rise in temperature. In high pressure boilers (at high temperature) $CaSO_4$ gets deposited as hard scale on inner surface of the boiler.

(c) *Hydrolysis of Mg salts:* In super-heated water which is the main cause of scale formation in high pressure boilers, the magnesium salts get hydrolyzed to form the precipitate of $Mg(OH)_2$ as scale.

$$MgCl_2 + 2H_2O \xrightarrow{\Delta} \underset{\text{Soft scale}}{Mg(OH)_2 \downarrow} + 2HCl$$

(d) *Presence of silica (SiO_2):* Silica if present in water deposited as $CaSiO_3$ and/or $MgSiO_3$. These silicates are sparingly soluble in cold water but almost insoluble in hot water and stick firmly to the inner wall of the boiler.

Disadvantages of scale

(a) *Wastage of fuel:* Due to low thermal conductivity of the scale the heat transfer to the boiler water is poor and hence excessive heating is required which results in the wastage of fuel. This heat loss depends on the thickness of the scale (Table 4.3).

Table 4.3 Thickness of scale and wastage of fuel

Thickness of scale (mm)	0.325	0.625	1.25	2.5	12
Wastage of fuel	10%	15%	50%	80%	150%

(b) *Decrease in efficiency:* As the scale may deposit in the valves and condenser of boiler, and can choke and block openings, which results in decreasing the efficiency of boiler.

(c) *Lowering of boiler safety:* The overheating of the boiler tube makes the boiler material soft and weaker, and causes distortion of boiler tubes and also makes the boiler unsafe to bear the pressure of steam, especially in high pressure boilers.

(d) *Danger of explosion:* When thick scales crack due to uneven expansion, the water comes suddenly in contact with overheated iron plats which causes the formation of large amount of steam and so sudden high pressure is developed which may even cause the explosion of the boiler.

(e) *Bagging:* Because of scale there may be uneven heating at different portions of boiler which may cause the bagging, i.e. distortion of boiler.

Removal of scale

1. By using scraper or piece of wood or wire brush.
2. By giving thermal shocks, i.e. strong heating and then sudden cooling.
3. By blowdown operation.
4. By adding/spraying the chemicals if the scale is adherent and hard. $CaCO_3$ scales can be dissolved in 5–10% HCl. Ca_3PO_4 scale can be removed by using EDTA solution.

Prevention of the scale formation

Scale formation can be prevented by the following methods:

(a) *External treatment:* This involves the softening of water, i.e. removal of hardness causing salts and silica before feeding to the boiler. Softening is done by various methods like lime–soda method, ion exchange method, reverse osmosis, etc. These methods are given in detail in Chapter 5 (Section 5.4).

(b) *Internal treatment (sequestration):* It involve the addition of chemicals directly to the boiler to prevent the scale formation by the salts left during external treatment. This is mainly used as corrective treatment to remove the slight residual hardness in water and sometimes corrosive medium also.

4.10.3 Internal Treatment Methods

(a) *Colloidal conditioning:* In low pressure boilers the scale formation can be minimized or prevented by adding organic compounds such as glue, agar-agar (a gel), kerosene, tannin, etc. which get coated over the scale forming precipitate, thereby yielding non-sticky and loose deposits which can easily be removed by belowdown operation.

(b) *Carbonate conditioning:* In low pressure boilers, scale formation can be avoided by adding sodium carbonate to the boiler water. In this conditioning when sodium carbonate solution is added to boiler water, the concentration of CO_3^{2-} ions increases and when it becomes greater than SO_4^{2-} ions ($K = K_{CaCO_3}/K_{CaSO_4}$), only $CaCO_3$ gets precipitated and $CaSO_4$ remains in solution. Hence, the deposition of scale forming $CaSO_4$ is prevented.

$$CaSO_4 + Na_2CO_3 \xrightarrow{\Delta} CaCO_3 \downarrow + Na_2SO_4$$

Consequently deposition of $CaSO_4$ as scale is prevented and Ca is precipitated as loose sludge of $CaCO_3$ which is removed off by blowdown operation.

(c) *Phosphate conditioning:* In high pressure boilers scale formation can be avoided by adding sodium phosphate. In this conditioning, an excess of soluble phosphate is added to the boiler water to precipitate the residual hardness which reacts with hardness (Ca and Mg) of water forming non-adherent and easily removable soft sludge of Ca and Mg phosphates. This soft sludge can be removed off by blowndown operation.

$$3CaCl_2 + 2Na_3PO_4 \longrightarrow Ca_3(PO_4)_2 \downarrow + 6NaCl$$

The three salts of phosphate, e.g. NaH_2PO_4, Na_2HPO_4 and Na_3PO_4 are used for alkaline, neutral and acidic water, respectively.

(d) *Calgon conditioning:* In this conditioning, scale forming salts are converted into highly soluble complexes by adding sodium hexametaphosphate $Na_2[Na_4P_6O_{18}]$ (calgon) which are not easily precipitated under the boiler conditions. This calgon reacts with hardness causing ions Ca^{2+} and Mg^{2+} to form highly soluble complex of calcium and magnesium hexametaphosphates.

$$Na_2[Na_4P_6O_{18}] \rightleftharpoons 2Na^+ + [Na_4P_6O_{18}]^{2-}$$

$$2Ca^{2+} + [Na_4P_6O_{18}]^{2-} \rightleftharpoons \underset{\text{Highly soluble complex ion}}{[Ca_2P_6O_{18}]^{2-}} + 4Na^+$$

(e) *Conditioning with sodium aluminate (NaAlO₂):* When $NaAlO_2$ is added to water, it gets hydrolyzed to give sodium hydroxide and gelatinous precipitate of aluminium hydroxide.

$$NaAlO_2 + H_2O \rightleftharpoons Al(OH)_3 + NaOH$$

Also the magnesium salt get hydrolyzed to give magnesium hydroxide.

$$MgCl_2 + H_2O \rightleftharpoons Mg(OH)_2 + NaCl$$

Finely suspended impurities including silica and oil drops get entrapped into the flocculants precipitate of $Al(OH)_3$ and $Mg(OH)_2$ which is removed off by blowdown operation. Thus, the scale formation is prevented.

(f) *Conditioning with EDTA:* EDTA is very good complexing agent it forms stable and soluble complex with hardness causing ions (Ca and Mg ions). When 1.5% alkaline solution of EDTA is added to the boiler feed water, scale gets dissolved. Moreover, it prevents wet steam and iron oxide formation in the boiler.

(g) *Radioactive conditioning:* In this method the capsules containing radioactive materials are placed at different points inside the boiler. The radiations emitted by radioactive materials prevent the coagulation, i.e. precipitation of impurities in the form of scale.

4.11 BOILER CORROSION

"Decay or disintegration of boiler material, tubes, pipelines, plates, etc. due to the chemical or electrochemical attack of its environment (acids, alkalies, dissolved gases, Mg salts, soda, etc.) is called boiler corrosion."

4.11.1 Causes of Boiler Corrosion

It is mainly caused due to: (i) the presence of alkalinity, (ii) presence of free mineral acids, (iii) presence of dissolved gases like O_2, CO_2, H_2S, etc. in water, (iv) formation of alkali due to the hydrolysis of Mg salts, (v) formation of sludge and scale in the boiler, (vi) formation of galvanic cell.

4.11.2 Control of Boiler Corrosion

It can be minimized by—(i) neutralizing the free acids or bases in the water before feeding it to the boiler, (ii) removing the dissolved gases from the boiler feed water, (iii) feeding the soft and salt free water and (iv) preventing the formation of sludge and scale in the boiler.

4.12 CAUSTIC EMBRITTLEMENT

It is the specific type of boiler corrosion which is caused due to the feeding highly alkaline water into the boiler. When water is treated by lime–soda method, some Na_2CO_3 left unreacted in the softened water. This water when used in high pressure boilers, the Na_2CO_3 decomposes to give NaOH which makes the water alkaline.

$$Na_2CO_3 + H_2O \rightarrow NaOH + CO_2$$

This alkaline water when circulated in minute cracks and pits, the NaOH attacks on these sensitive areas and dissolves the iron metal to form sodium ferrate. This leads to the corrosion of boiler parts particularly at stressed areas.

The mechanism of caustic embrittlement can be explained by the formation of concentration cell:

⊕ Iron	NaOH	NaOH	⊖ Iron
(At stressed portions)	(Concentrated solution)	(Dilute solution)	(At plane surface)
Anode			Cathode

The iron at concentrated NaOH becomes anodic and gets corroded.

The caustic embrittlement can be prevented by adding sodium sulphate or lignin to the boiler water and by using sodium phosphate as softening agent in place of Na_2CO_3.

4.13 PRIMING AND FOAMING

Due to the rapid heating or presence of oily impurities in the boiler feed water two types of problems may arise, i.e. priming and foaming.

4.13.1 Priming

"It is defined as the process of wet steam formation in the boiler due to excess of dissolved solids, and uneven or sudden heating of water."

Causes of priming

Priming is mainly caused due to: (i) the presence of suspended as well as dissolved impurities in the water, (ii) formation of sludge and scale in the boiler, (iii) uneven and sudden heating of water, (iv) feeding the water above the water level, i.e. excess water feeding, and (v) defective boiler design.

Minimization of priming

Priming can be minimized by: (i) feeding the water free from salt and other impurities and hardness, (ii) using the properly designed boiler which can maintain adequate and uniform heating, (iii) by preventing the scale and sludge formation, and (iv) maintaining constant heat supply.

4.13.2 Foaming

"It is defined as the production of persistent foam or bubbles in the boiler due to the presence of oily impurities in the water."

Causes of foaming

Foaming is mainly caused due to: (i) the presence of oily, soapy, floating as well as dissolved impurities in the water, (ii) feeding the water treated by lime–soda method having unreacted soda (Na_2CO_3), (iii) formation of sludge and scale in the boiler, (iv) uneven and sudden heating of water, and (v) presence of impurities which increase the viscosity and surface tension of water.

Minimization of priming

Priming can be minimized by: (i) feeding the water free from oily, soapy impurities as well as hardness, (ii) preventing the scale and sludge formation, (iii) maintaining the constant heat supply, (iv) removing the foaming agents like oil, soap, soda, silisic acid, aluminium hydroxide, etc. from water, (v) adding the anti-foaming agents like caster oil, polyamide, etc., (vi) removing the concentrated water from boiler, and (vii) connecting the steam purifiers to the boiler.

Both priming as well as foaming occurs together in the boiler.

Disadvantages of priming and foaming

The salts carried out with super-heated steam are deposited over super-heater and turbine blades, and other machine parts which reduce the efficiency and life of boiler and machines.

4.14 CARRY OVER

"The phenomenon of carrying of water by steam along with impurities during steam production in the boiler is called *carry over*."

4.14.1 Causes

Carry over is caused due to strong priming and foaming which is due to the presence of suspended oily, soapy, floating and dissolved impurities, sludge and scale in the boiler feed water and uneven, sudden and strong heating of water.

4.14.2 Minimization of Carry Over

Carry over can be minimized by reducing the causes of priming and foaming, i.e. by feeding the water free from impurities and hardness, using the properly designed boiler, preventing the scale and sludge formation, maintaining constant heat supply.

SOLVED NUMERICAL PROBLEMS

Hardness of water

1. Calculate the hardness of a water sample containing the following impurities per litre:

$$CaSO_4 = 15.2 \text{ mg}, Ca(HCO_3)_2 = 10.2 \text{ mg}, MgCl_2 = 9.0 \text{ mg}.$$

Solution:

$$CaCO_3 \text{ Equivalent hardness} = \frac{\text{Mass of hardness producing substance} \times \text{Moecular mass of } CaCO_3}{\text{Moecular mass of hardness producing substance}}$$

Salts/ions	Mass of hardness producing substance (mg l⁻¹ (ppm))	Molecular mass of hardness producing substance	Hardness in terms of CaCO₃ equivalent
$CaSO_4$	15.2	136	$= \frac{15.2}{136} \times 100 = 11.17$ ppm
$Ca(HCO_3)_2$	10.2	162	$= \frac{10.2}{162} \times 100 = 6.29$ ppm
$MgCl_2$	9.0	95	$= \frac{9.0}{95} \times 100 = 9.47$ ppm

Hence, hardness of water = 11.17 + 6.29 + 9.47 = **26.93 ppm**

2. A sample of water contains the following dissolved salts: $Ca(HCO_3)_2$ = 20.0 ppm, $Mg(HCO_3)_2$ = 17.5.0 ppm, $MgCl_2$ = 12 ppm, $CaCl_2$ = 22.2 ppm and $CaSO_4$ = 28 ppm. Calculate the temporary and permanent hardness of water.

Solution:

Salts/Ions	Mass of hardness producing substance mg l^{-1} (ppm)	Molecular mass of hardness producing substance	Hardness in terms of CaCO$_3$ equivalent
Ca(HCO$_3$)$_2$	20.0	162	$= \dfrac{20.0}{162} \times 100 = 12.34$ ppm
Mg(HCO$_3$)$_2$	17.5	146	$= \dfrac{17.5}{146} \times 100 = 11.98$ ppm
MgCl$_2$	12.0	95	$= \dfrac{12.0}{95} \times 100 = 13.15$ ppm
CaCl$_2$	22.2	111	$= \dfrac{22.2}{111} \times 100 = 20$ ppm
	28.0	136	$= \dfrac{28.0}{136} \times 100 = 20.58$ ppm

Temporary hardness = Ca(HCO$_3$)$_2$ + Mg(HCO$_3$)$_2$ = 12.34 + 11.98 = **24.32 ppm**

Permanent hardness = MgCl$_2$ + CaCl$_2$ + CaSO$_4$ = 13.15 + 20 + 20.58 = **53.73 ppm**.

3. On analysis, a municipal water sample found to contain Ca(HCO$_3$)$_2$ = 26.4 mg l^{-1}, Mg(HCO$_3$)$_2$ = 13.8 mg l^{-1}, CaSO$_4$ = 15.5 mg l^{-1}, MgSO$_4$ = 6.5 mg l^{-1}, MgCl$_2$ = 20.5 mg l^{-1} and CaCl$_2$ = 15.0 mg l^{-1}, NaCl = 25 mg l^{-1}. Calculate temporary, permanent and total hardness.

Salts/Ions	Mass of hardness producing substance mg l^{-1} (ppm)	Molecular mass of hardness producing substance	Hardness in terms of CaCO$_3$ equivalent
Ca(HCO$_3$)$_2$	26.4	162	$= \dfrac{26.4}{162} \times 100 = 16.296$ ppm
Mg(HCO$_3$)$_2$	13.8	146	$= \dfrac{13.8}{146} \times 100 = 9.45$ ppm
CaSO$_4$	15.5	136	$= \dfrac{15.5}{136} \times 100 = 11.39$ ppm
MgSO$_4$	6.5	120	$= \dfrac{6.5}{120} \times 100 = 5.41$ ppm
MgCl$_2$	20.0	95	$= \dfrac{20}{95} \times 100 = 21.05$ ppm
CaCl$_2$	15.0	111	$= \dfrac{15.0}{111} \times 100 = 13.51$ ppm

Temporary hardness = Ca(HCO$_3$)$_2$ + Mg(HCO$_3$)$_2$ = 16.296 + 9.45 = 25.746 ppm

Permanent hardness = CaSO$_4$ + MgSO$_4$ + MgCl$_2$ + CaCl$_2$

= 11.39 + 5.41 + 21.05 + 13.51 = **51.36 ppm.**

Alkalinity of water

4. A water sample is alkaline to only methyl orange but not for phenolphthalein indicator. However, its 100 ml on titration using methyl orange indicator consumed 17.2 ml of N/50 HCl. Calculate the alkalinity present in water sample in terms of CaCO$_3$ equivalent.

Solution:

$$\text{Alkalinity} = \frac{\text{Normality of acid} \times \text{Volume of acid used}}{\text{Volume of water sample}} \times 50 \times 1000$$

$$\text{Phenolphthalein alkalinity} = \frac{N_A \times V_A}{V_W} \times 50 \times 1000 = \frac{17.2}{50 \times 100} \times 50 \times 1000 = \textbf{172 ppm.}$$

5. 100 ml of water sample gave the titer value of 15 ml N/50 HCl for phenolphthalein end point. Another water sample gave the titer value of 30 ml for methyl orange end point. Calculate the amount and types of alkalinity present in water sample.

Solution:

$$\text{Alkalinity} = \frac{\text{Normality of acid} \times \text{Volume of acid used}}{\text{Volume of water sample}} \times 50 \times 1000$$

$$\text{Phenolphthalein alkalinity} = \frac{N_A \times V_A}{V_W} \times 50 \times 1000 = \frac{1 \times 15}{50 \times 100} \times 50 \times 1000 = \textbf{150 ppm}$$

$$\text{Methyl orange alkalinity} = \frac{30}{50 \times 100} \times 50 \times 1000 = \textbf{300 ppm.}$$

Type of alkalinity present

Since, $P = 1/2M$, the alkalinity is present only due to CO_3^{2-} which is equal to $2P$ or M, Hence, CO_3^{2-} alkalinity = 2×150 = **300 ppm.**

6. A water sample is alkaline to both the phenolphthalein and methyl orange indicators. 100 ml of this water sample required 12 ml of N/50 H_2SO_4 up to phenolphthalein end point and on further titration using methyl orange indicators it required 18 ml more N/50 H_2SO_4 for complete neutralization. Calculate the amount and types of alkalinity present in water sample.

Solution:

$$\text{Alkalinity} = \frac{\text{Normality of acid} \times \text{Volume of acid used}}{\text{Volume of water sample}} \times 50 \times 1000$$

$$\text{Phenolphthalein alkalinity} = \frac{N_A \times V_A}{V_W} \times 50 \times 1000 = \frac{20}{50 \times 100} \times 50 \times 1000 = \textbf{120 ppm}$$

$$\text{Methyl orange alkalinity} = \frac{(12+18)}{50 \times 100} \times 50 \times 1000 = \frac{30}{50 \times 100} \times 50 \times 1000 = \textbf{300 ppm.}$$

Type of alkalinity present

Since, $P < 1/2M$, the alkalinity is present due to CO_3^{2-} and HCO_3^- which is equal to $2P$ and $(M - 2P)$, respectively, Hence, CO_3^{2-} alkalinity = 2×120 = **240 ppm** and HCO_3^- alkalinity = $300 - 2 \times 120$ = **60 ppm.**

7. A water sample is alkaline to both the phenolphthalein and methyl orange indicators. 100 ml of this water sample required 20 ml of N/50 H_2SO_4 up to phenolphthalein end point. Another 100 ml of this water sample titration using methyl orange indicators required 20 ml N/50 H_2SO_4. Calculate the amount and type of alkalinity present in water sample.

Solution:

$$\text{Alkalinity} = \frac{\text{Normality of acid} \times \text{Volume of acid used}}{\text{Volume of water sample}} \times 50 \times 1000$$

$$\text{Phenolphthalein alkalinity} = \frac{N_A \times V_A}{V_W} \times 50 \times 1000 = \frac{20}{50 \times 100} \times 50 \times 1000 = \textbf{200 ppm}$$

$$\text{Methyl orange alkalinity} = \frac{20}{50 \times 100} \times 50 \times 1000 = \textbf{200 ppm.}$$

Type of alkalinity present

Since, $P = M$, the alkalinity is present due to only OH^- which is equal to P or M respectively, Hence, OH^- = **200 ppm.**

8. 50 ml of water sample on titration with N/50 H_2SO_4 consumed 6.4 ml of acid up to phenolphthalein end point. Another 100 ml of same water sample consumed 24 ml of N/50 H_2SO_4 up to methyl orange end point. Calculate the amount and type of alkalinity present in water sample.

Solution:

$$\text{Alkalinity} = \frac{\text{Normality of acid} \times \text{Volume of acid used}}{\text{Volume of water sample}} \times 50 \times 1000$$

$$\text{Phenolphthalein alkalinity} = \frac{N_A \times V_A}{V_W} \times 50 \times 1000 = \frac{6.4}{50 \times 50} = \textbf{128 ppm}$$

$$\text{Methyl orange alkalinity} = \frac{24}{50 \times 100} \times 50 \times 1000 = \textbf{240 ppm.}$$

Type of alkalinity present

Since, $P > 1/2\ M$, the alkalinity is present due to CO_3^{2-} and HCO_3^- which is equal to $2P - M$ and $2(M - P)$ respectively,

Hence, $$CO_3^{2-} = 2P - M = 2 \times 128 - 240 = \textbf{16 ppm} \quad \text{and}$$
$$HCO_3^- = 2(M - P) = 2(240 - 128) = \textbf{224 ppm.}$$

9. A water sample is alkaline to both the phenolphthalein and methyl orange indicators. 100 ml of raw water sample required 30 ml of N/50 H_2SO_4 up to phenolphthalein end point and 26 ml up to methyl orange indicators on further titration. Calculate the type of alkalinity present in water sample.

Solution:

$$\text{Alkalinity} = \frac{\text{Normality of acid} \times \text{Volume of acid used}}{\text{Volume of water sample}} \times 50 \times 1000$$

$$\text{Phenolphthalein alkalinity} = \frac{N_A \times V_A}{V_W} \times 50 \times 1000 = \frac{30}{50 \times 100} \times 50 \times 1000 = \textbf{300 ppm}$$

$$\text{Methyl orange alkalinity} = \frac{56}{50 \times 100} \times 50 \times 1000 = \textbf{560 ppm.}$$

Type of alkalinity present

Since, $P > 1/2M$, the alkalinity is present due to the presence of CO_3^{2-} and HCO_3^- which is equal to $2P - M$ and $2(M - P)$ respectively.

Hence, $\qquad CO_3^{2-} = 2P - M = 2 \times 300 - 560 = $ **40 ppm** and

$\qquad HCO_3^- = 2(M - P) = 2(560 - 300) = $ **520 ppm.**

EDTA method

10. 50 ml of standard hard water (1000 ppm) required 40 ml of EDTA solution. 50 ml of hard water sample required 30 ml of EDTA solution. After boiling and filtering 50 ml of this water required 12 ml of EDTA solution. Calculate the temporary and permanent hardness of water in ppm.

Solution:

Strength of standard hard water (SHW) = 1000 ppm

$$\text{Strength of EDTA} = \frac{\text{Strength of SHW} \times \text{Volume of SHW}}{\text{Volume of EDTA solution}}$$

$$= \frac{1000 \times 50}{40} = 1250 \text{ ppm}$$

$$\text{Total hardness} = \frac{\text{Strength of EDTA} \times \text{Volume of EDTA}}{\text{Volume of hard water sample}}$$

$$= \frac{1250 \times 30}{50} = 750 \text{ ppm}$$

$$\text{Permanent hardness} = \frac{\text{Strength of EDTA} \times \text{Volume of EDTA}}{\text{Volume of hard water sample}}$$

$$= \frac{1250 \times 12}{50} = 300 \text{ ppm}$$

Temporary hardness = Total hardness − Permanent hardness

$\qquad\qquad\qquad\qquad = 750 - 300 = $ **450 ppm Ans.**

11. A standard hard water is prepared by dissolving 0.51 g $CaCO_3$ in 400 ml of distilled water. 50 ml of this required 25 ml of EDTA solution. 50 ml of hard water sample required 20 ml of EDTA solution. The same sample after boiling required 14 ml of EDTA solution. Calculate the temporary and permanent hardness of water.

Solution:

Strength of standard hard water (SHW) = 0.51g per 400 ml = $\frac{0.51}{400} \times 1000 = 1275$ ppm

$$\text{Strength of EDTA} = \frac{\text{Strength of SHW} \times \text{Volume of SHW}}{\text{Volume of EDTA solution}}$$

$$= \frac{1275 \times 50}{25} = 2550 \text{ ppm}$$

$$\text{Total hardness} = \frac{\text{Strength of EDTA} \times \text{Volume of EDTA}}{\text{Volume of hard water sample}}$$

$$= \frac{2550 \times 20}{50} = 1020 \text{ ppm}$$

$$\text{Permanent hardness} = \frac{\text{Strength of EDTA} \times \text{Volume of EDTA}}{\text{Volume of hard water sample}}$$

$$= \frac{2550 \times 14}{50} = 714.2 \text{ ppm}$$

Temporary hardness = Total hardness − Permanent hardness

$$= 1020 - 714.2 = \mathbf{305.8 \text{ ppm Ans.}}$$

12. A standard hard water is prepared by dissolving 0.78 g $CaCO_3$ in HCl and the solution was made up to one litre by adding distilled water. 100 ml of this standard hard water required 62 ml of EDTA solution. 100 ml of hard water sample required 52 ml of EDTA solution. After boiling and filtering 100 ml of this sample required 34 ml of EDTA solution. Calculate the total, temporary and permanent hardness of water.

Solution:

Strength of standard hard water (SHW) = 0.78 gl^{-1} = 0.78 × 1000 = 780 mgl^{-1} or ppm

$$\text{Strength of EDTA} = \frac{\text{Strength of SHW} \times \text{Volume of SHW}}{\text{Volume of EDTA solution}}$$

$$= \frac{780 \times 100}{62} = 1258 \text{ ppm}$$

$$\text{Total hardness} = \frac{\text{Strength of EDTA} \times \text{Volume of EDTA}}{\text{Volume of hard water sample}}$$

$$= \frac{1258 \times 52}{100} = 654.16 \text{ ppm}$$

$$\text{Permanent hardness} = \frac{\text{Strength of EDTA} \times \text{Volume of EDTA}}{\text{Volume of hard water sample}}$$

$$= \frac{1258 \times 34}{100} = 427.72 \text{ ppm}$$

Temporary hardness = Total hardness − Permanent hardness

$$= 654.16 - 427.72 = \mathbf{226.44 \text{ ppm Ans.}}$$

SUMMARY

Sources of water: There are two types of sources of water, i.e. surface water (rain water, lake water, river water and sea water), underground water (spring water and well water).

Types of impurities in water: There may be four types of impurities present in water, i.e. dissolved impurities (inorganic salts, organic salts, dissolved gases), colloidal impurities, suspended impurities, and biological impurities. These impurities can impart the properties to water like colour, taste, odour, turbidity and sediments, acidity, etc.

Hardness of water: Hardness of water is that characteristic which prevents the formation of lather with soap. Hardness is of two types, i.e. temporary or carbonate and permanent or non-carbonate hardness.

Calcium carbonate equivalents (degree of hardness): The concentration of hardness and non-hardness producing salts or ions such as bicarbonates, chlorides and sulphates of Ca, Mg, etc. is measured in terms of equivalent amount of calcium carbonate.

Disadvantages of hard water: Hard water has great disadvantages in different fields like drinking, cooking, washing, bathing, steam production in the boilers, pharmaceutical industries, sugar industries, textile industry, paper industry, laundry, etc.

Alkalinity of water: Alkalinity of water may be defined as the concentration of OH^- ions and the concentration of ions which increases the concentration of OH^- ions due to dissociation or hydrolysis. Alkalinity is of two types, i.e. bicarbonate alkalinity and carbonate alkalinity or hydroxide alkalinity.

Sludge and scale: Sludge is a soft, loose and slimy precipitate formed within the boiler and scale is hard adhering deposit which sticks firmly to the inner surfaces of the boiler.

Disadvantages of scale and sludge: Scale and sludge have disadvantages like wastage of fuel, decrease in efficiency, lowering of boiler safety, danger of explosion, bagging, etc.

Prevention of sludge and scale formation: It can be prevented by: external treatment (i.e. softening) and internal treatment, i.e. sequestration or conditioning like: colloidal conditioning, carbonate conditioning, phosphate conditioning, calgon conditioning, conditioning with sodium aluminate, conditioning with EDTA and radioactive conditioning.

Caustic embrittlement: It is defined as the corrosion of boiler materials due to the attack of highly alkaline water.

Priming and foaming: Priming is defined as the process of wet steam formation in the boiler due to excess of dissolved solids, uneven and sudden heating of water, and foaming is defined as the production of persistent foam or bubbles in the boiler due to the presence of oily impurities in the water.

Carry over: The phenomenon of carrying of water by water along with impurities during steam production in the boiler is called *carry over*.

EXERCISES

1. What are the various sources of water? Explain the impurities present in the water obtained from different sources.
2. Explain different types of properties imparted by the impurities present in natural water.
3. What is hardness of water? Explain the various types of hardness of water and its causes.
4. What do you understand by temporary and permanent hardness of water? Distinguish between the same.
5. What do you mean by alkaline and non-alkaline hardness of water?
6. Explain the carbonate and non-carbonate hardness of water.
7. Why do we express hardness of water in terms of calcium carbonate equivalent?
8. What do you understand by degree of hardness? Explain the various units of hardness.
9. Give the relationship among various units of hardness.
10. Distinguish between hard water and soft water.
11. Why does hard water consume lot of soap?

12. Why is water softened before using in boiler?
13. Why should natural water not be fed to boiler?
14. What are the disadvantages of hard water in different fields?
15. Justify that 1ppm is equal to 1 mgl^{-1}.
16. The presence of CO_2 in boiler feed-water should be avoided. Why?
17. What do you understand by alkalinity of water? How is it determined experimentally?
18. How are the types of alkalinity calculated by using the value of P and M.
19. What is hardness of water? How is it determined experimentally by EDTA method?
20. Why EDTA titration method is preferred over soap titration and O. Hehner's method for the determination of hardness of water?
21. Why the pH of water sample is maintained around 10 before titration?
22. Why is NH_3–NH_4Cl buffer solution added during determination of hardness of water EDTA titration?
23. What do you understand by scale and sludge? Explain the disadvantages of scale and sludge formation in the boiler. How these are prevented?
24. What are the causes of scale and sludge formation in the boiler? Explain various methods for the removal of the same.
25. What do you understand by external and internal treatment of boiler water? Explain the different types of conditioning used to prevent the scale formation.
26. Why calgon conditioning is better than phosphate conditioning?
27. What is priming and foaming? Give the causes of priming and foaming and disadvantages in boiler.
28. What do you understand by boiler corrosion? Explain the mechanism of caustic embrittlement.
29. Why can caustic embrittlement be controlled by adding sodium sulphate to boiler-feed water?
30. Why is presence of $NaAlO_2$ in water equivalent to presence of equivalent of $Ca(OH)_2$?
31. How is hardness expressed? Give various units of hardness. How are they related to each other?
32. Why does magnesium bicarbonate require double amount of lime for softening?
33. Explain why is indicator EBT added in estimation of hardness of water by EDTA method?
34. State the zeolite process for the removal of hardness of water. Discuss its merits over lime soda process.
35. Why do we express hardness of water in terms of $CaCO_3$ equivalent?
36. How will you determine the alkalinity of water sample containing hydroxide and carbonate ions?
37. Calculate the hardness of a sample of water containing the following salts per litre: $CaSO_4$ = 16.2, $Mg(HCO_3)_2$ = 1.4 mg; $MgCl_2$ = 9.5 mg.
38. A sample of water contains the following dissolved salts $Mg(HCO_3)_2$ = 20 mg l^{-1}, $MgCl_2$ = 25 mg l^{-1}, $CaCl_2$ = 8.0 mg l^{-1} and $CaSO_4$ = 26 mg l^{-1}. Calculate the temporary and permanent hardness.
39. A water sample contains $Ca(HCO_3)_2$ = 34.4 mg l, $Mg(HCO_3)_2$ = 30.2 mg l and $CaSO_4$ = 13.4 mg l^{-1}. Calculate temporary and permanent hardness of water sample.
40. What is the carbonate and non-carbonate hardness of a sample of water is ppm containing: $Ca(HCO_3)_2$ = 16.2 mg l^{-1}, $Mg(HCO_3)_2$ = 7.3 mg l^{-1}, $MgCl_2$ = 9.5 mg l^{-1}, $CaSO_4$ = 13.6 mg l^{-1}?

41. Standard hard water contains 1.4 g of $CaCO_3$ per litre. 21 ml of this required 28 ml of EDTA solutions. 50 ml of sample of water required 24 ml EDTA solution. The same sample after boiling required 15 ml EDTA solution. Calculate the temporary hardness of the given sample of water, in terms of ppm.
42. 50 ml of a standard hard water containing 1 mg of pure $CaCO_3$ per ml consumed 25 ml of EDTA solution using Eriochrome Black-T indicator. 50 ml of a water sample consumed 28 ml of same EDTA solution. Calculate the total hardness of water sample in ppm.
43. Calculate the hardness of a water sample, whose 50 ml required 25 ml of EDTA solution of 700 ppm.
44. A sample of water was alkaline both to phenolphthalein and methyl orange. 50 ml of this water required 15 ml of N/50 sulphuric acid for phenolphthalein end point and another 10 ml for complete neutralization. Calculate the type and amount of alkalinity in ppm and mg l^{-1}.
45. 100 ml of a sample of water required 20 ml of N/40 H_2SO_4 for complete neutralization. Calculate the temporary hardness.

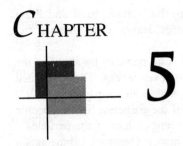

CHAPTER 5

WATER TREATMENT PART II

5.1 INTRODUCTION

Natural water from well, rivers, canals, etc. does not satisfy all the required specifications of drinking water. It may have various types of floating, soluble, suspended, toxic and micro-biological impurities. Water for domestic use should be free from these impurities, i.e. it should be very safe and potable for drinking purposes.

5.2 CHARACTERISTICS OF POTABLE WATER

The municipal water (potable water) should maintain the following requirements:
1. It should be clear and colourless.
2. It should be odourless, i.e. free from unpleasant smell.
3. It should be pleasant to taste.
4. Its turbidity should be less than 10 ppm.
5. Its hardness should not exceed 200 ppm.
6. It should be free from pathogenic (diseases causing) organisms.
7. It should be free from objectionable gases, like F_2, H_2O, H_2S, SO_2, etc. and minerals like Pb, Cr, As, Ni and Mn salts.
8. Total dissolved salts (TDS) should not exceed 500 ppm.
9. Its pH value should be around 7.5–8.0.

5.3 TREATMENT OF WATER FOR DOMESTIC USE

The water for domestic use should be obtained from the source which is least/not contaminated by vegetables, animal matters and industrial effluents. Generally, the treatment of municipal water is carried out in the following steps for removing various type of impurities.

5.3.1 Removal of Floating and Suspended Impurities

The floating and suspended impurities in water can be removed off by using the following methods.

(a) *Screening:* When raw water is passed through the screens (bar screen, band and drum screen and micro-strainers) the floating matter like wood pieces, leaves, dead animals, etc. are removed.

(b) *Sedimentation:* It is the process of removing the suspended solid particles by allowing the water undisturbed in large settlement tanks for a few days or even weeks. The suspended impurities settled down due to force of gravity. The velocity with which a particle in water will fall under the action of gravity depends on: (i) the size of the particles, (ii) the specific gravity of the particles, (iii) the horizontal flow rate and (iv) temperature of the particles. Plain sedimentation removes off only 70–75% of the suspended matter. Therefore, sedimentation with coagulation is preferred.

(c) *Sedimentation with coagulation:* By simple sedimentation process, finely divided silica, clay organic matter and colloidal impurities cannot be removed because these are not settled down easily. Such impurities are generally removed by chemically assisted sedimentation (called coagulation). In this process, the added chemicals produce charged ions that neutralize the positively charged colloidal particles. These particles fall through still water under gravity. The following are the most commonly used coagulants:

1. *Potash alum* $[K_2SO_4 \cdot Al_2(SO_4)_3 \cdot 24H_2O]$*:* It is the most commonly used coagulant. When it is added to water it gets hydrolyzed to form colloidal aluminium hydroxide.

$$Al_2(SO_4)_3 + 6H_2O \longrightarrow 2Al(OH)_3\downarrow + 3H_2SO_4$$
Aluminium hydroxide
(flocculent)

The aluminium hydroxide so formed acts as coagulant as the fine particles get trapped into the precipitate and hence it removes the finely divided colloidal impurities by neutralizing the charge on them.

2. *Sodium aluminates* $(NaAlO_2)$*:* Sodium aluminate acts as a good coagulant. It is generally used in combination with aluminium sulphate. The reactions can be shown as:

$$NaAlO_2 + 2H_2O \longrightarrow NaOH + 2Al(OH)_3\downarrow$$

$$Al_2(SO_4)_3 + 6NaAlO_2 + 12H_2O \longrightarrow 8Al(OH)_3\downarrow + 3Na_2SO_4$$

3. *Ferrous sulphate* $(FeSO_4 \cdot 7H_2O)$*:* It is also a good coagulant and on hydrolysis it gives $Fe(OH)_2$ which is a heavy flock and causes quick sedimentation. To increase the efficiency of coagulation process some coagulants additives like Fuller's earth, bentonite clay and poly-electrolytes are also added.

(d) *Filtration:* Filtration is the process which involves removal of coarse impurities from water by passing it through a porous filtering medium. The common materials used for filtration are quartz, sand (grain size 0.5–1.0 mm, crushed anthracite (particle size 0.8–1.5 mm) and porous clay (particle size 0.8–1.5 mm). Filtration of water takes place due to the difference between the pressures at the filter and that of underneath.

Following types of filters may be used for the filtration of water depending on the flow rate and quality of water:

1. Slow sand filters are used for treating municipal water at flow rate of 2 gallon $ft^{-2}\ h^{-1}$. But these filters are not capable of removing colloidal impurities.
2. Rapid-gravity filter used for treating municipal water at high flow rate of 100 gallon $ft^{-1}\ h^{-1}$.

3. Rapid pressure filter are widely used in industries for continuous rapid filtration at a rate of > 200 gallon ft^{-2} h^{-1}.

The sand filter (Figure 5.1) consists of a large tank having three columns. The central column is fitted with the three-layered filtering medium. The thick top layer of fine sand is placed over coarse sand layer and gravels layer which is supported with iron net. The water is passed through the inlet and percolates evenly on fine sand bed. During passing of water through the sand filter the impurities are held up by the sand. This filtered water is collected in the under drain and send to the other side column for sedimentation from where it is sent to the water outlet.

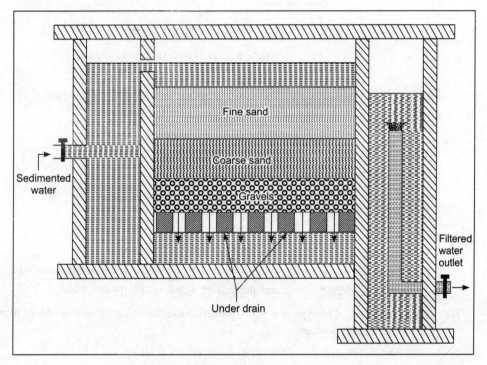

Figure 5.1   Structure of sand filter.

During filtration, the pores of sand get blocked and rate of filtration becomes slow. So, to regenerate the filter, 2–3 cm upper layer is scrapped off and replaced with fresh sand followed by a flush back with clean water to wash away the accumulated impurities.

5.3.2   Removal of Micro-Organism (Disinfection or Sterilization of Water)

The filtered water may contain some percentage of pathogenic (disease causing) micro-organisms like viruses (infectious for hepatitis), intestinal protozoa (*Entamoeba histolytica*), some types of worms (schitosomes), enteric bacteria (salmonella, shigella and vibrio group), etc. Therefore, it requires the elimination of pathogenic germs, bacteria, etc. to make the water fit for drinking purposes.

"The process of removing or killing the pathogenic germs (i.e. bacteria, viruses and micro-organism) in the water and making it fit for use is called *disinfection or sterilization*".

The following methods can be applied for the disinfection of domestic or municipal water:

(a) *By boiling:* When water is boiled for about 15 min, all the pathogenic germs are killed and water becomes safe for drinking. But this process is very costly and cannot be used at large scale, i.e. for the municipal water works. Moreover, water keeping for long may contaminate with micro-organisms.

(b) *By bleaching powder:* Bleaching powder, chemically known as *calcium hypochlorite*, when added to the water it produces hypochlorous acid which is a powerful germ killer.

$$CaOCl_2 + H_2 \longrightarrow Ca(OH)_2 + Cl_2$$
Bleaching powder

$$Cl_2 + H_2O \longrightarrow HOCl + HCl$$

$$HOCl \longrightarrow [O] + HCl$$
Nascent oxygen

This liberated nascent oxygen destroys the germs and bacteria by oxidation. The OCl^- ions destroy the cell membrane of the germs, thus kills the germs

$$HOCl + Germs \longrightarrow Killed\ germs$$

Advantages

1. It is cheap.
2. It is readily available.
3. It is good for treating the acidic water.

Disadvantages

1. Bleaching powder is unstable during storage.
2. It increases the calcium contents in water and hence increases the hardness.
3. Its excess gives bad test and unpleasant smell to the treated water.

(c) *By chlorination:* Chlorine is a good, powerful sterilizing agent as it produces hypochlorous acid while added to water.

$$Cl_2 + H_2O \longrightarrow HOCl + HCl$$

$$HOCl + Germs \longrightarrow Killed\ germs$$

Disinfection by chlorine is due to the chemical reaction of hypochlorous acid (HOCl) with the enzymes in the cell of micro-organisms. Thus, the micro-organisms are killed because of the destruction of enzymes in the cell by HOCl. The chlorine is more effective disinfectant at lower pH (pH < 7).

Advantages of chlorine

1. It is powerful disinfectant.
2. It is economical to use.
3. It requires limited space for storage and transportation is easy.
4. It does not deteriorate on storage even for long duration.
5. It does not introduce any other impurities and/salts in water.
6. It can be used within a wide temperature range.

Disadvantages
1. Excess of chlorine causes irritation to the eyes and mucous membrane.
2. Its excess gives unpleasant taste to the water.
3. It is less effective at higher pH.

Super-chlorination
When an excess quality of chlorine is added to the water, it destroys not only the germs but also the other organic impurities present in the water. This process is called super-chlorination. The super-chlorination ensures rapid and complete disinfection. To remove unused chlorine, ammonia or SO_2 is added to the chlorinated water.

Break point chlorination
It is the process of adding chlorine to the water in a more precisely controlled manner. In this process, firstly sufficient amount of chlorine is added which oxidizes all the organic matter, destroys the bacteria, germs and reacts with ammonia (to form chloramines) leaving a slight excess of free chlorine.

It is clear from Figure 5.2 that the chlorine dose is consumed completely up to point B in destroying the reducing compounds. After point B the quantity of residual chlorine increases with increasing chlorine dose, giving a straight line up to a certain chlorine dose (point C); it is because of the combination of chlorine to form chloro-organic compounds and chloramines. After point C, a sudden decrease in the residual chlorine is noticed and we get the lowest point D. This point is called *break point*. This indicates that the point after which free residual chlorine begins to appear that more or less agrees with the chlorine dose added. This available free chlorine destroys the pathogenic germs and we get the sterilized water. For pure water the curve AP (a straight line) is obtained which shows that pure water does not consume any chlorine dose.

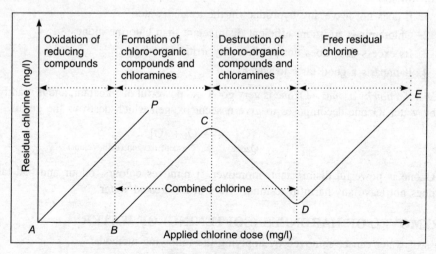

Figure 5.2 Break Point Chlorination.

Thus, the break point chlorination may be defined as "the process of adding the chlorine dose to destroy the germs in addition to the chlorine dose used to oxidize the reducing compounds to form chloro-organic compounds and chloramines to destroy the chloro-organic compounds and chloramines".

Advantages of break point chlorination
1. Up to the break point all the organic compounds, ammonia and other reducing compounds are destroyed.
2. After break point the free residual chlorine is used to destroy completely (100%) all the pathogenic germs.
3. It removes unpleasant smell due to the presence of organic matter in water. It removes the colour in water.
4. The growth of any weed is prevented by chlorine.

De-chlorination

The excess dose of chlorine after break point causes disagreeable taste and smell in water. Therefore, this excess amount of chlorine is removed by filtering the chlorinated water through activated carbon filter and then treating it with SO_2 or Na_2SO_3.

$$SO_2 + Cl_2 + 2H_2O \longrightarrow H_2SO_4 + 2HCl$$
$$Na_2SO_3 + Cl_2 + H_2O \longrightarrow Na_2SO_4 + 2HCl$$

(d) *Disinfection by chloramines ($ClNH_2$):* The chloramine compound is obtained by mixing chlorine with ammonia in 2:1 ratio.

$$Cl_2 + NH_3 \longrightarrow \underset{\text{Chloramine}}{ClNH_2} + HCl$$

This chloramine when added to water, gives HOCl

$$ClNH_2 + H_2O \longrightarrow \underset{\text{Disinfectant}}{HOCl} + NH_3; \quad HOCl + \text{Germs} \rightarrow \text{Killed germs}$$

Advantages
1. It does not leave any byproduct in the treated water.
2. Chloramines are more efficient disinfectant than chlorine alone.
3. Its excess dose does not produce any irritating odour.
4. It imparts a good taste to water.

(e) *Disinfection by ozone:* Ozone is very good and powerful disinfectant, and is readily absorbed by water. Ozone decomposes to give nascent oxygen which destroys the germs.

$$\underset{\text{Ozone}}{O_3} \longrightarrow O_2 + \underset{\text{Nascent oxygen (disinfectant)}}{[O]}$$

Ozone is powerful disinfectant, moreover, it removes colours, odour and bad taste, and it does not leave any harmful residual effect in the treated water.

5.4 REMOVAL OF HARDNESS (SOFTENING) OF WATER

Hardness of water can be removed by applying the following methods:

5.4.1 Lime-soda Method

This method is based on the fact that when calculated amount of lime and soda added to the hard water, all the soluble impurities of temporary as well as permanent hardness are converted into insoluble precipitate form which is removed off by settling and filtration. The following reactions may take place during the treatment as shown in Table 5.1.

Table 5.1 Various reactions involved in lime-soda treatment method

S. No.	Impurities/hardness	Reactions	Requirement of lime/soda
1.	Temporary hardness		
	(i) Ca^{2+}	$Ca(HCO_3)_2 + Ca(OH)_2 \longrightarrow CaCO_3\downarrow + H_2O + CO_2$	L
	(ii) Mg^{2+}	$Mg(HCO_3)_2 + 2Ca(OH)_2 \longrightarrow Mg(OH)_2\downarrow + 2CaCO_3 + 2H_2O$	2L
2.	Permanent hardness:		
	(i) Ca^{2+}	$CaCl_2 + Na_2CO_3 \longrightarrow CaCO_3\downarrow + 2NaCl$	S
	(ii) Mg^{2+}	$MgSO_4 + Ca(OH)_2 \longrightarrow Mg(OH)_2\downarrow + CaSO_4$	L + S
		$CaSO_4 + Na_2CO_3 \longrightarrow CaCO_3\downarrow + Na_2SO_4$	
3.	H^+ (free acids)	$2HCl + Ca(OH)_2 \longrightarrow CaCl_2 + 2H_2O$	L + S
	HCl or H_2SO_4	$CaCl_2 + Na_2CO_3 \longrightarrow CaCO_3\downarrow + 2NaCl$	
4.	CO_2	$CO_2 + Ca(OH)_2 \longrightarrow CaCO_3\downarrow + H_2O$	L
5.	HCO_3^- as $NaHCO_3$	$2NaHCO_3 + Ca(OH)_2 \longrightarrow CaCO_3\downarrow + Na_2CO_3 + H_2O$	L – S
6.	Coagulants		
	(i) Al^{3+} as $Al_2(SO_4)_3$	$Al_2(SO_4)_3 + 3Ca(OH)_2 \longrightarrow 2Al(OH)_3\downarrow + 3CaSO_4$	L + S
		$CaSO_4 + Na_2CO_3 \longrightarrow CaCO_3\downarrow + Na_2SO_4$	
	(ii) Fe^{2+} as $FeSO_4$	$FeSO_4 + Ca(OH)_2 \longrightarrow Fe(OH)_2\downarrow + CaSO_4$	L + S
		$CaSO_4 + Na_2CO_3 \longrightarrow CaCO_3\downarrow + Na_2SO_4$	
	(iii) $NaAlO_2$	$NaAlO_2 + 2H_2O \longrightarrow Al(OH)_3 + NaOH$	–L

On the basis of Table 5.1, we can find the formula for the requirement of lime and requirement of soda:

$$\text{Lime requirement} = \frac{74}{100} \begin{bmatrix} \text{Temporary}\,(Ca^{2+} + 2\times Mg^{2+}) + \text{Permanent} \\ (Mg^{2+} + Fe^{2+} + Al^{3+}) + CO_2 + H^+(HCl\ or\ H_2SO_4) \\ + HCO_3^- - NaAlO_2\ (\text{all in terms of } CaCO_3\ \text{eq.}) \end{bmatrix}$$

$$\text{Soda requirement} = \frac{106}{100} \begin{bmatrix} \text{Permanent}\,(Ca^{2+} + Mg^{2+} + Fe^{2+} + Al^{3+}) \\ + H^+(HCl\ or\ H_2SO_4) - HCO_3^- \\ \text{all in terms of } CaCO_3\ \text{equivalent} \end{bmatrix}$$

Lime–soda method can be operated in two ways: at room temperature (15–30°C) or at higher temperature (90–100°C). So on the basis of operation temperature it is known as cold lime–soda process and hot lime–soda process.

(a) *Cold lime–soda process:* In this process, calculated quantity of lime and soda with coagulants is added to the hard water at room temperature. These chemicals react with hardness causing ions and convert them into insoluble precipitate which is removed off by decantation and then filtration.

The coagulants get hydrolyzed to give the flocculent, gelatinous precipitate and entrap the fine precipitate.

$$NaAlO_2 + 2H_2O \longrightarrow Al(OH)_3 \downarrow + NaOH$$
$$Al_2(SO_4)_3 + 3Ca(HCO_3)_2 \longrightarrow 2Al(OH)_3 \downarrow + 3CaSO_4 + 6CO_2 \uparrow$$
$$\text{Coagulant} \qquad \text{Hardness}$$

Method: The instrument (softener) used in cold lime-soda process is shown in Figure 5.3. It consists of two cylindrical chambers, outer and inner chambers or reaction tank. The outer chamber is fitted with wood fibre filter, soft water outlet and sludge outlet. The inner chamber is fitted with mechanical stirrer. The hard water and calculated quantities of chemical, i.e. lime, soda and coagulants are supplied to the inner chamber where these react to form the precipitate of $CaCO_3$, $Mg(OH)_2$, etc. This precipitate with mud, etc. decanted at the conical bottom of the outer chamber from where it is removed off. When this treated water passed through the wood fibre filter, remaining sludge and impurities are filtered off and we obtain soft water of 50–60 ppm residual hardness from soft water outlet.

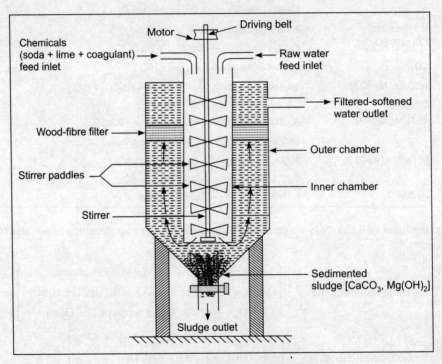

Figure 5.3 Cold lime-soda softener.

(b) *Hot lime–soda process:* The apparatus used in hot lime–soda process is shown in Figure 5.4. It consists of outer chamber or settling chamber, inner chamber or reaction tank and sand filter. The raw water and calculated quantities of chemicals (lime and soda) are supplied to the reaction tank and the hot steam is passed simultaneously, which supplies the heat to the water and allows the mixing of the same, properly. The precipitate is settled down in the outer chamber which is then removed off through sludge outlet. This treated water is then sent to the sand filter where remaining impurities and sludge are removed and we obtain the soft and filter water of about 15–30 ppm residual hardness.

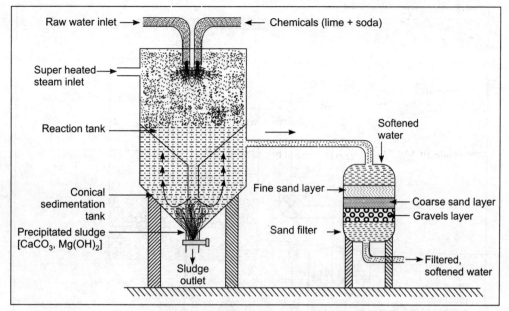

Figure 5.4 Hot lime–soda softener.

5.4.2 Zeolite or Permutit Method

Zeolite (permutite) ($Na_2O \cdot Al_2O_3 \cdot xSiO_2 \cdot yH_2O$) is hydrated sodium aluminosilicate. It is capable of exchanging reversibly its sodium ions with hardness producing ions like Ca^{2+}, Mg^{2+}, etc in water. Zeolites are of two types:

1. *Natural zeolite:* e.g. natrolite—$Na_2O \cdot Al_2O_3 \cdot 4SiO_2 \cdot 2H_2O$ (non-porous).
2. *Synthetic zeolite:* These are obtained by heating china clay with feldspar and soda ash.

Process: The structure of zeolite softener is shown in Figure 5.5. For softening the water by zeolite process, hard water is allowed to permeate through a bed of zeolites kept in a cylinder at a specified rate. The hardness causing ions (Ca^{2+}, Mg^{2+}, etc.) are retained by the zeolite as CaZe and MgZe, respectively while the outgoing water contains sodium salts.

The following reactions take place during the softening process:

$$Na_2Ze + Ca(HCO_3)_2 \longrightarrow CaZe + 2NaHCO_3$$
$$Na_2Ze + Mg(HCO_3)_2 \longrightarrow MgZe + 2NaHCO_3$$
$$Na_2Ze + CaCl_2 \text{ (or } CaSO_4) \longrightarrow CaZe + 2NaCl(Na_2SO_4)$$
$$Na_2Ze + MgCl_2 \text{ (or } MgSO_4) \longrightarrow MgZe + 2NaCl(Na_2SO_4) \text{ (Zeolite) (Hardness)}$$

Advantages

1. It can produce the water of 10 ppm residual hardness.
2. Water treated from zeolite method, if used in boiler, does not cause any sludge formation.
3. It requires less space, less skill and less time for softening of water.

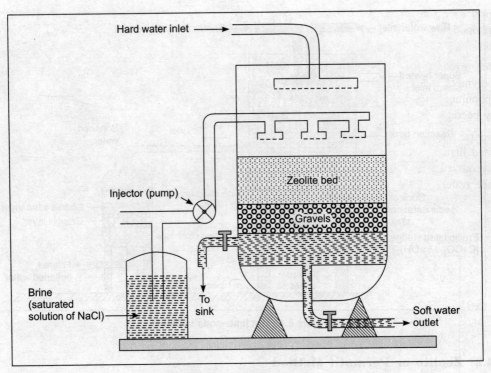

Figure 5.5 Zeolite softener.

Disadvantages

1. This method replaces only Ca^{2+} and Mg^{2+} with Na^+ ions, but all the acidic ions like HCO_3^-, Cl^-, CO_3^{2-}, SO_4^{2-}, etc. are left behind untreated. When this water is used in the boiler, it causes boiler corrosion and caustic embitterment.
2. Highly turbid water cannot be treated efficiently by this method, as the fine impurities get deposited on zeolite bed, creating problems in its working.
3. Water containing excess of acidity or alkalinity may attack the zeolite.
4. Hot water cannot be treated as zeolite tends to dissolve in it.

Regeneration

Using the zeolite continuously, after some time it is completely converted into calcium and magnesium zeolite and it stops softening the water, i.e. it gets exhausted. At this stage the supply of hard water is stopped and the exhausted zeolite is regenerated by treating it with 10% concentrated brine (NaCl) sollution

$$\underset{\text{Exhausted zeolite}}{MgZe} + \underset{\text{Brine}}{2NaCl} \longrightarrow \underset{\text{Regenerated zeolite}}{Na_2Ze} + \underset{\text{Washings}}{MgCl_2}$$

5.4.3 Ion Exchange or Deionization or Demineralization Method

Ion exchange resins are insoluble, cross-linked, long-chain organic polymers with a micro-porous structure (insoluble in water) and the functional groups attached to the chains, which are responsible for the ion exchange properties. The ion exchange resins may be classified as:

Cation exchange resins (R–H): These are mainly styrene divinylbenzene copolymes which on sulphonation or carboxylation become capable to exchange their hydrogen ions with cations in the water.

Anion exchange resins (R–OH): These are styrene divinylbenzene or amine formaldehyde copolymers which contain amino or quaternary ammonium or quaternary phosphonium or tertiary sulphonium groups as an integral part of the resin matrix. After treatment with dilute. NaOH solution they become capable of exchanging their OH⁻ anions with anions present in water.

Process: The apparatus used for deionization of water is shown in Figure 5.6. The hard water is passed first through cation exchange column, where all the cations like Ca^{2+}, Mg^{2+}, Na^+, K^+, etc. are removed off by exchanging with H^+ ions of resin and equivalent amount of H^+ ions is released to the water. Thus,

$$2RH + Ca^{2+} \rightarrow R_2Ca + H^+$$

$$2RH + Mg^{2+} \rightarrow R_2Mg + 2H^+$$

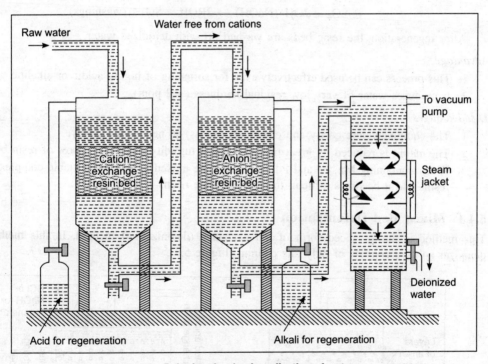

Figure 5.6 Deionizer for demineralization of water.

After cation exchange column, the hard water is passed through anion exchange column where all the anions like HCO_3^-, Cl^-, SO_4^{2-}, etc. present in the water are removed off by exchanging with OH^- ions of resin and equivalent amount of OH^- ions is released to water. Thus,

$$ROH + Cl^- \longrightarrow RCl + OH^-$$

$$ROH + HCO_3^- \longrightarrow RHCO_3 + OH^-$$

$$2ROH + SO_4^{2-} \longrightarrow R_2SO_4 + 2OH^-$$

H^+ and OH^- ions released from cation and anion exchange column, respectively combine to produce water molecule:

$$H^+ + OH^- \longrightarrow H_2O$$

Thus, the water coming out of the exchanger is free from cations as well as anions. This ion free water is known as *deionized* or *demineralized water*.

Regeneration

When capacities of cation and anion exchangers to exchange H^+ and OH^- ions, respectively are vanished, they are then said to be exhausted. The exhausted cation exchanger is regenerated by passing dilute HCl or H_2SO_4 through it.

$$R_2Ca + 2H^+ \text{ (HCl or } H_2SO_4) \rightarrow 2RH + Ca^{2+} \text{ (washings)}$$

The exhausted anion exchange column is regenerated by passing dilute NaOH solution through it.

$$R_2SO_4 + 2OH^-(NaOH) \rightarrow 2ROH + SO_4^{2-} \text{ (washings)}$$

After regeneration, the resin beds are washed off with deionized water.

Advantages

1. This process can be used effectively even for softening of highly acidic or alkaline waters.
2. It produces water of very low residual hardness (≈ 2 ppm).

Disadvantages

1. The equipments are costly and chemicals (resins) so needed are costly.
2. The output is reduced by great turbidity as the turbidity blocks the pores of resin bed.
3. Water containing Mn, Fe and Pd ions cannot be treated as they form stable complexes with resins and it becomes difficult to regenerate the resins.

5.4.4 Mixed Bed Deionization

This method is the advanced form of ion exchange (deionization) method. In this method, the deionizer so used consists of only one column (Figure 5.7).

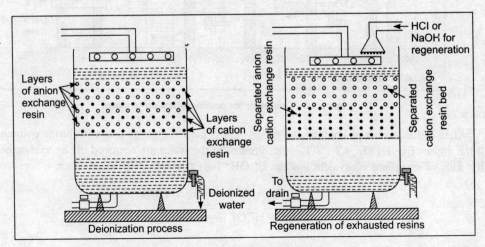

Figure 5.7 Mixed bed deionizer.

The ion exchange bed in the column has different successive layers of cation exchange resin and anion exchange resin. When hard water is passed through the ion exchange bed, all the cations, e.g. Ca^{2+}, Mg^{2+}, Na^+, K^+, Fe^{2+}, Mn^{2+}, etc. get exchanged with the H^+ ions of cation exchange resin and all the anions like Cl^-, SO_4^{2-}, HCO_3^-, etc. get exchanged with the OH^- ions of anion exchange resin. If any of the cations or anions left unexchanged in the first layer, are further removed off when passed through other successive layers. This way we get the water free from all the types of ions, i.e. soft, deionized water.

Advantages of mixed bed deionization over ion exchange method
1. This method is more accurate and better than ion exchange method as it allows the multiphase deionization.
2. This method is more efficient than ion exchange method as we can get the water of even 0 ppm residual hardness.

Disadvantage
The demerit of this method is that it is difficult to regenerate the resins as they are mixed. So for the regeneration firstly they are separated and regenerated separately.

5.5 INTRINSIC OR POLISHED WATER

Water used in the manufacture and washing of transistors, TV tubes and some other types of electronic equipments must be extremely pure. This water is called *intrinsic* or *polished water*. To obtain the intrinsic or polished water the normal water is first passed through the column of cation and then anion exchangers, and is again passed through a bed of mixed exchange resins. This water is prepared as and when required and used immediately to avoid contamination.

5.6 BOILER FEED WATER

When water is converted into steam in a boiler the suspended and dissolved matter may get deposited in the boiler. If these substances form adherent deposits on the walls of the boiler, they are scales which turn sludge or mud into the non-adherent crust/coating on the inner walls of the boiler.

The treatment of the boiler feed water is usually carried out in two stages. In the first stage, the raw water is treated to remove the scale forming, sludge forming and corrosion causing constituents. In the second stage, called internal treatment, chemicals are added to take care of any residual objectionable constituent not removed in the first stage. This way the water is made fit for feeding the boilers.

5.7 DESALINATION OF BRACKISH WATER

"The water which contains excess salt concentration with a peculiar salty or brackish taste is called *brackish water*. The process of removing salts from the saline water is termed as *desalination*."

The salinity of water is expressed in terms of mg l^{-1} or ppm of dissolved solids, Cl^- ion or NaCl. The water quality is usually graded as:

Fresh water—which contains the dissolved solids <1000 ppm.

Brackish water—which contains the dissolved solids >1000 ppm but <3500 ppm.

Sea water—which contains the dissolved solids >3500 ppm.

Brackish water or sea water is totally unfit for drinking purposes. So, for the desalination of brackish water and to make it fit for drinking, following techniques are being used:
1. Electrodialysis and
2. Reverse osmosis.

5.7.1 Electrodialysis

Dialysis is the process of separating the particles of colloids from those of crystalloids by means of diffusion through a suitable membrane. This method is used for the purification of sea water. To increase the rate of purification, the dialysis is carried out by applying electric field which is called *electrodialysis*. Thus, "the process of dialysis applying electric field is called *electrodialysis*."

This method is based on the fact that when the salty water is passed through the ion-selective membrane in the presence of electric field, the ions migrate towards oppositely charged electrode. An electrodialysis cell consists of cation and anion permeable membranes. The cathode is placed near the cation permeable membrane and anode is placed near the anion permeable membrane (Figure 5.8).

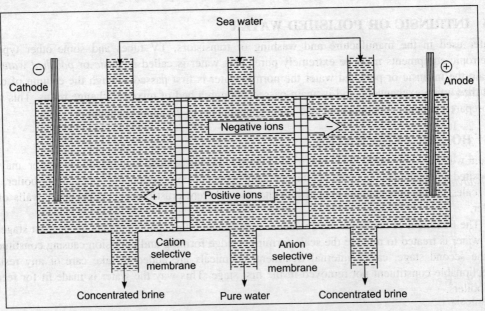

Figure 5.8 Desalination of water by electrodialysis.

Working: The movement of ions takes place through an ion-selective membrane. Impure water is allowed to pass through the middle chamber as well as side chambers. During passing through the middle chamber, the cations (Na^+, K^+, etc.) and anions (Cl^-) move towards oppositely charged electrodes, i.e. catode and anode, respectively. From the middle chamber outlet we get the pure, deminaralized water. Water saturated with ions in side chambers on electrodes is sent to the drain. To increase the efficiency and to use this method at large scale the electrolytic cells are arranged in a series as shown in Figure 5.9.

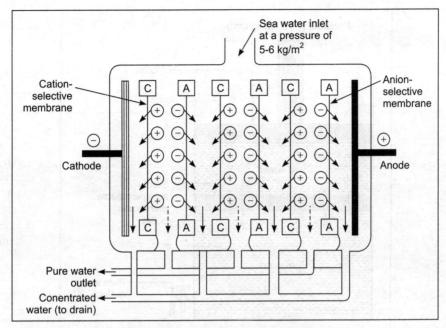

Figure 5.9 Electrodialysis cells in series.

Advantages
1. This method is economical as the installation cost of plant is low.
2. It is continuous process.
3. It can be applied at room temperature.
4. It is the best suited method if the electricity is easily available.

Disadvantages
1. This method can remove only ionic impurities, i.e. Na^+, Cl^-, etc. but not other non-ionic and colloidal impurities.
2. It is not economical where the electricity is costly and rarely available.
3. It cannot be operated without electricity.

5.7.2 Reverse Osmosis

Osmosis is defined as the flow of solvent (water) from a solution of lower concentration to a solution of higher concentration through semipermeable membrane. Osmotic pressure is defined as the excess pressure which must be applied to the solution side to prevent passage of solvent into the solution through a semipermeable membrane, i.e. to prevent the osmosis.

If the pressure applied on the solution side is more than the osmotic pressure, the process of osmosis is reversed and the solvent starts moving from solution towards the pure solvent. Thus, the process of movement of solvent from the solution to the pure solvent through a semipermeable membrane by applying excess pressure on solution side is called *reverse osmosis*, sometimes called *super-filtration* or *hyper-filtration*. This process can be used for the desalination of the sea water or other salty water to get the drinking water.

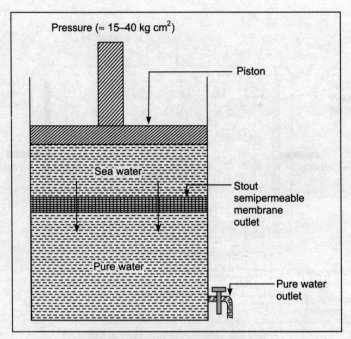

Figure 5.10 Reverse osmosis unit.

Procedure: In this method, the apparatus is used as shown in Figure 5.10. The tank is fitted with a semipermeable membrane (thin cellulose film) and a piston. A pressure (15–40 kg cm^{-2}) greater than osmotic pressure is applied on sea water (brackish water) side. Due to this pressure the osmosis process gets reversed and the water flows from salty water side to pure water side through a semipermeable membrane. Thus, the pure water is obtained which does not contain any undesirable dissolved salts.

Advantages

1. This process consumes extremely low energy.
2. It removes all types of impurities, i.e. ionic, non-ionic, colloidal, organic matter and even colloidal silica which is not removed by demineralization.
3. The maintenance cost is low as it depends only upon the replacement of semipermeable membrane.
4. The semipermeable membrane has its life time quite high. Moreover, it can be replaced within a few minutes.

Disadvantages/demerits

The cellulose acetate membrane (100–150 μm thick) must be capable of withstanding pressures of the order of 20–100 kg cm^{-2}.

SOLVED NUMERICAL PROBLEMS

1. Calculate the amount of lime and soda required to soften 25000 litres of water having following impurities: $Ca(HCO_3)_2$ = 24.0 ppm, $Mg(HCO_3)_2$ = 7.5 ppm, $CaSO_4$ = 6.9 ppm, $MgCl_2$ = 5.0 ppm, $MgSO_4$ = 9.6 ppm, SiO_2 = 3.5 ppm.

Solution:

Salts/ions	Mass of hardness producing substance mg l⁻¹ (ppm)	Molecular mass of hardness producing substance	Hardness in terms of $CaCO_3$ equivalent
$Ca(HCO_3)_2$	24.0	162	$= \dfrac{26.4}{162} \times 100 = 14.81$ ppm
$Mg(HCO_3)_2$	7.5	146	$= \dfrac{7.5}{146} \times 100 = 5.41$ ppm
$CaSO_4$	6.9	136	$= \dfrac{6.9}{136} \times 100 = 5.07$ ppm
$MgSO_4$	9.6	120	$= \dfrac{9.6}{120} \times 100 = 8.0$ ppm
$MgCl_2$	5.0	95	$= \dfrac{5}{95} \times 100 = 5.26$ ppm
$CaCl_2$	15.0	111	$= \dfrac{15.0}{111} \times 100 = 13.51$ ppm

On the basis of above table we can find the formula for the requirement of lime and requirement of soda:

$$\text{Lime requirement} = \frac{74}{100}\begin{bmatrix} \text{Temporary } Ca^{2+} + 2 \times Mg^{2+} + \text{Permanent } (Mg^{2+} + Fe^{2+} + Al^{3+}) \\ + CO_2 + H^+ (HCl \text{ or } H_2SO_4) + HCO_3^- - NaAlO_2 \\ \text{(all in terms of } CaCO_3 \text{ equivalent)} \end{bmatrix}$$

$$\text{Lime requirement} = \frac{74}{100}[\text{Temporary } (14.81 + 2 \times 5.41) + \text{Permanent } (8.0 + 5.26)] \times 25000 \text{ litre}$$

$$= \frac{74}{100}[25.63 + 13.26] \times 25000 = \frac{74}{100} 38.89 \times 25000$$

$$= 719465 \text{ mg}/10^6 = \mathbf{0.719465 \text{ kg}}$$

$$\text{Soda requirement} = \frac{106}{100}\begin{bmatrix} \text{Permanent}(Ca^{2+} + Mg^{2+} + Fe^{2+} + Al^{3+}) + H^+(HCl \text{ or } H_2SO_4) \\ - HCO_3^- \text{ (all in terms of } CaCO_3 \text{ equivalent)} \end{bmatrix}$$

$$\text{Soda requirement} = \frac{106}{100}[\text{Permanent}(8.0 + 5.26 + 13.51)] \times 25000 = \frac{106}{100}[26.77] \times 25000$$

$$= 709405 \text{ mg} = 709405 \text{ mg}/10^6 = \mathbf{0.709405 \text{ kg}}$$

2. Calculate the quantity of lime and soda required for softening 50,000 litres of water containing the following salts per litre:

$Ca(HCO_3)_2$ = 27.5 mg, $Mg(HCO_3)_2$ = 17.6 mg, $CaSO_4$ = 23.6 mg, $MgSO_4$ = 4.2 mg, $MgCl_2$ = 9.7 mg, dissolved CO_2 = 3.0 mg, free acid (HCl) = 4.0 mg, coagulant ($FeSO_4$) added = 5 mg, 10 mg and NaCl = 14.7 and SiO_2 = 15 mg.

Solution:

Salts/ions	Mass of hardness producing substance mg l^{-1} (ppm)	Molecular mass of hardness producing substance	Hardness in terms of CaCO$_3$ equivalent
Ca(HCO$_3$)$_2$	27.5	162	$= \dfrac{27.5}{162} \times 100 = 16.97$ ppm
Mg(HCO$_3$)$_2$	17.6	146	$= \dfrac{17.6}{146} \times 100 = 12.05$ ppm
CaSO$_4$	23.6	136	$= \dfrac{23.6}{136} \times 100 = 17.35$ ppm
MgSO$_4$	4.2	120	$= \dfrac{4.2}{120} \times 100 = 3.5$ ppm
MgCl$_2$	9.7	95	$= \dfrac{9.7}{95} \times 100 = 10.21$ ppm
CO$_2$	3.0	44	$= \dfrac{3.0}{44} \times 100 = 6.81$ ppm
H$^+$	1.2	2	$= \dfrac{1.2}{2} \times 100 = 60$ ppm
FeSO$_4 \cdot$ 7H$_2$O	5.0	278	$= \dfrac{5}{278} \times 100 = 1.79$ ppm

Since NaCl and SiO$_2$ do not impart any hardness to water and do not consume any lime or soda hence these are not considered in the calculation.

$$\text{Lime requirement} = \frac{74}{100}\begin{bmatrix}\text{Temporary } Ca^{2+} + 2 \times Mg^{2+} + \text{Permanent}(Mg^{2+} + Fe^{2+} + Al^{3+}) \\ + CO_2 + H^+(HCl \text{ or } H_2SO_4) + HCO_3^- - NaAlO_2 \\ \text{(all in terms of CaCO}_3 \text{ equivalent)}\end{bmatrix}$$

$$= \frac{74}{100}[\text{Temporary }(16.97 + 2 \times 12.05 + \text{Permanent}$$

$$(3.5 + 10.21 + 1.79)\; 6.81 + 60] \times 50000$$

$$= \frac{74}{100}[41.07 + 15.5 + 6.81 + 60] \times 50000 = 74 \times 123.38 \times 500$$

$$= 4565060 \text{ mg} = 4565060/10^6 = \mathbf{4.565 \text{ kg}}$$

$$\text{Soda requirement} = \frac{106}{100}\begin{bmatrix}\text{Permanent }(Ca^{2+} + Mg^{2+} + Fe^{2+} + Al^{3+}) + H^+ (HCl \text{ or } H_2SO_4) \\ - HCO_3^- \text{ (all in terms of CaCO}_3 \text{ equivalent)}\end{bmatrix}$$

$$= \frac{106}{100}[\text{Permanent }(17.35 + 3.5 + 10.21 + 1.79) + 60] \times 50000 = 106$$

$$\times [92.85] \times 500$$

$$= 4921050 \text{ mg} = 4921050 \text{ mg}/10^6 = \mathbf{4.921 \text{ kg}}$$

3. Calculate the amount of lime (84% pure) and soda (92% pure) required for the treatment of 30,000 litres of water containing the following impurities:

$Ca(HCO_3)_2$ = 35.5 ppm, $Mg(HCO_3)_2$ = 28.2 ppm, $CaSO_4$ = 16.0 ppm, $CaCl_2$ = 37.05 ppm, $MgSO_4$ = 12.0 ppm, HCO_3^- = 6.1 ppm and NaCl =10.0 ppm

Also calculate the temporary and permanent hardness of water sample.

Solution:

Salts/ions	Mass of hardness producing substance mg l⁻¹ (ppm)	Molecular mass of hardness producing substance	Hardness in terms of CaCO₃ equivalent
$Ca(HCO_3)_2$	35.5	162	$= \dfrac{35.5}{162} \times 100 = 21.91$ ppm
$Mg(HCO_3)_2$	28.2	146	$= \dfrac{28.2}{146} \times 100 = 19.31$ ppm
$CaSO_4$	16.0	136	$= \dfrac{16}{136} \times 100 = 11.76$ ppm
$CaCl_2$	37.05	111	$= \dfrac{37.05}{111} \times 100 = 33.37$ ppm
$MgSO_4$	12.0	120	$= \dfrac{12}{120} \times 100 = 10$ ppm
HCO_3^-	6.1	122	$= \dfrac{6.1}{122} \times 100 = 10$ ppm

Since NaCl does not impart any hardness to water and does not consume any lime or soda hence these are not considered in the calculation.

$$\text{Lime requirement} = \frac{74}{100}\begin{bmatrix}\text{Temporary } Ca^{2+} + 2 \times Mg^{2+} + \text{Permanent }(Mg^{2+} + Fe^{2+} + Al^{3+}) + \\ CO_2 + H^+ (HCl \text{ or } H_2SO_4) + HCO_3^- - NaAlO_2 \\ \text{(all in terms of } CaCO_3 \text{ equivalent)}\end{bmatrix}$$

$$\text{Lime requirement} = \frac{74}{100}[\text{Temporary }(21.91 + 2 \times 19.31 + \text{Permanent }(10)$$
$$+ 10] \times \frac{100}{84} \times 30000$$

$$= \frac{74}{84}[70.53] \times 30000 \text{ litre} = 0.8809 \times 70.53 \times 30000$$

$$= 1863896 \text{ mg} = 1863896/10^6 = \mathbf{1.863 \text{ kg}}$$

$$\text{Soda requirement} = \frac{106}{100}\begin{bmatrix}\text{Permanent }(Ca^{2+} + Mg^{2+} + Fe^{2+} + Al^{3+}) + H^+ (HCl \text{ or } H_2SO_4) \\ - HCO_3^- \text{ (all in terms of } CaCO_3 \text{ equivalent)}\end{bmatrix}$$

$$= \frac{106}{100}[\text{Permanent }(11.76 + 33.37 + 10) - 10] \times \frac{100}{92} \times 30000$$

$$= 106 \times [45.13] \times \frac{100}{92} \times 300$$

$$= 1559928 \text{ mg} = 1559928/10^6 = \mathbf{1.56 \text{ kg}}$$

4. A water sample using $FeSO_4 \cdot 7H_2O$ as a coagulant treated at the rate of 278 ppm, gave following data on analysis for raw water: Ca^{2+} = 240 ppm, Mg^{2+} = 96 ppm, CO_2 = 22 ppm, HCO_3^- = 762 ppm. Calculate the lime and soda required to soften 150000 litres of water. Purity of lime is 90% and that of soda is 97%.

Solution:

Salts/ions	Mass of hardness producing substance mg l^{-1} (ppm)	Molecular mass of hardness producing substance	Hardness in terms of $CaCO_3$ equivalent
Ca^{2+}	240	40	$= \dfrac{240}{40} \times 100 = 600$ ppm
Mg^{2+}	96	24	$= \dfrac{96}{24} \times 100 = 400$ ppm
CO_2	22	44	$= \dfrac{22}{44} \times 100 = 50$ ppm
HCO_3^-	968	122	$= \dfrac{395}{122} \times 100 = 327.77$ ppm
$FeSO_4 \cdot 7H_2O$	278	278	$= \dfrac{278}{278} \times 100 = 100$ ppm

$$\text{Lime requirement} = \frac{74}{100} \begin{bmatrix} \text{Temporary } Ca^{2+} + 2 \times Mg^{2+} + \text{Permanent} \\ (Mg^{2+} + Fe^{2+} + Al^{3+}) + CO_2 + H^+(HCl \text{ or } H_2SO_4) + \\ HCO_3^- - NaAlO_2 \text{ (all in terms of } CaCO_3 \text{ equivalent)} \end{bmatrix}$$

$$\text{Lime requirement} = \frac{74}{100}[400 + 100 + 50 + 327.77] \times \frac{100}{90} = \frac{74}{90} \times 877.77$$

$$= 649.54 \text{ mg l}^{-1}$$
$$= 649.54 \times 150000$$
$$= 97432470 \text{ mg}$$
$$= 97.43 \text{ kg}$$

$$\text{Soda requirement} = \frac{106}{100} \begin{bmatrix} \text{Permanent } (Ca^{2+} + Mg^{2+} + Fe^{2+} + Al^{3+}) + \\ H^+(HCl \text{ or } H_2SO_4) - HCO_3^- \\ \text{(all in terms of } CaCO_3 \text{ equivalent)} \end{bmatrix}$$

$$\text{Soda requirement} = \frac{106}{100}[(600 + 400 + 100) - 327.77] \times \frac{100}{97} [772.23] \times \frac{106}{97}$$

$$= 843.880 \text{ mg l}^{-1}$$
$$= 843.88 \times 150000 = 126582032 \text{ mg} = \mathbf{126.582 \text{ kg Ans.}}$$

5. A hard water sample on analysis gave the following results: Ca^{2+} = 162 ppm, Mg^{2+} = 68 ppm, CO_2 = 44 ppm, HCO^-_3 = 236 ppm, free acid = 2.1 ppm, $NaAlO_2$ used as coagulant = 41.0 ppm. The rate of softening of this water was 4000 litres per minute. Calculate the lime and soda required per hour. Purity of lime is 86% and that of soda is 94%.

Solution:

Salts/ions	Mass of hardness producing substance mg l⁻¹ (ppm)	Molecular mass of hardness producing substance	Hardness in terms of CaCO₃ equivalent
Ca^{2+}	160	40	$= \dfrac{160}{40} \times 100 = 400$ ppm
Mg^{2+}	68	24	$= \dfrac{68}{24} \times 100 = 283.3$ ppm
CO_2	44	44	$= \dfrac{44}{44} \times 100 = 100$ ppm
HCO_3^-	236	122	$= \dfrac{236}{122} \times 100 = 193.44$ ppm
$NaAlO_2$	41	82	$= \dfrac{278}{278} \times 100 = 100$ ppm

$$\text{Lime requirement} = \frac{74}{100} \begin{bmatrix} \text{Temporary } Ca^{2+} + 2 \times Mg^{2+} + \text{Permanent} \\ (Mg^{2+} + Fe^{2+} + Al^{3+}) + CO_2 + H^+ (HCl \text{ or } H_2SO_4) \\ + HCO_3^- - NaAlO_2 \text{ (all in terms of } CaCO_3 \text{ equivalent)} \end{bmatrix}$$

$$\text{Lime requirement} = \frac{74}{100}[283.3 + 100 + 193.44 - 100] \times \frac{100}{86} \times \text{rate (l min}^{-1}) \times \frac{60 \text{ min}}{h}$$

$$= \frac{74}{100} \times 476.44 \times \frac{100}{86} \times 4000 \times 60$$

$$= \frac{74}{100} \times 476.44 \times \frac{100}{86} \times 4000 \times 60 = 98390400 \text{ mg/h}$$

$$= 93.39 \text{ kg/h}$$

$$\text{Soda requirement} = \frac{106}{100} \begin{bmatrix} \text{Permanent } (Ca^{2+} + Mg^{2+} + Fe^{2+} + Al^{3+}) + \\ H^+(HCl \text{ or } H_2SO_4) - HCO_3^- \\ \text{(all in terms of } CaCO_3 \text{ equivalent)} \end{bmatrix}$$

$$\text{Soda requirement} = \frac{106}{100}[400 + 283.3 - 193.44] \times \frac{100}{94} \times \text{rate (l min}^{-1}) \times \frac{60 \text{ min}}{h}$$

$$= [489.56] \times \frac{106}{94} \times 4000 \times 60 = 24238365.95 \text{ mg/h}$$

$$= 24.238 \text{ kg/h}$$

6. Calculate the quantity of lime and soda required for the softening of 50,000 litres of water containing the following impurities: $Ca(HCO_3)_2 = 9.7$ mg, $Mg(HCO_3)_2 = 7$ mg, $CaSO_4 = 13.4$ mg, $MgSO_4 = 11$ mg, $MgCl_2 = 2.5$ mg, $NaAlO_2$ (used as coagulant) = 4 mg and $NaCl = 7.8$ mg.

Solution:

Salts/ions	Mass of hardness producing substance mg l⁻¹ (ppm)	Molecular mass of hardness producing substance	Hardness in terms of CaCO₃ equivalent
$Ca(HCO_3)_2$	9.7	162	$= \dfrac{9.7}{162} \times 100 = 5.98$ ppm
$Mg(HCO_3)_2$	7.0	146	$= \dfrac{7.0}{146} \times 100 = 4.79$ ppm
$CaSO_4$	13.4	136	$= \dfrac{13.4}{136} \times 100 = 9.85$ ppm
$MgSO_4$	11.0	111	$= \dfrac{11.0}{120} \times 100 = 9.16$ ppm
$MgCl_2$	2.5	95	$= \dfrac{2.5}{95} \times 100 = 2.63$ ppm
$NaAlO_2$	4	82	$= \dfrac{4.0}{82} \times 100 = 4.87$ ppm

Since NaCl does not impart any hardness to water and does not consume any lime or soda hence these are not considered in the calculation:

$$\text{Lime requirement} = \frac{74}{100}\begin{bmatrix}\text{Temporary } Ca^{2+} + 2 \times Mg^{2+} + \text{Permanent} \\ (Mg^{2+} + Fe^{2+} + Al^{3+}) + CO_2 + H^+ (HCl \text{ or } H_2SO_4) \\ + HCO_3^- - NaAlO_2 \text{ (all in terms of } CaCO_3 \text{ eq.)}\end{bmatrix}$$

$$\text{Lime requirement} = \frac{74}{100} [\text{Temporary }(5.98 + 2 \times 4.79) + \text{Permanent }(9.16 + 2.63) - 4.87]$$

$$\times\ 50000 \text{ litre}$$

$$= 74 \times [27.35 - 4.87] \times 500 \text{ litre} = 0.8809 \times 70.53 \times 500$$

$$= 831760 \text{ mg} = 831760 = \mathbf{0.8317\ kg/50000\ litres}$$

$$\text{Soda requirement} = \frac{106}{100}\begin{bmatrix}\text{Permanent }(Ca^{2+} + Mg^{2+} + Fe^{2+} + Al^{3+}) \\ + H^+ (HCl \text{ or } H_2SO_4) - HCO_3^- \\ \text{(all in terms of } CaCO_3 \text{ equivalent)}\end{bmatrix}$$

$$\text{Soda requirement} = \frac{106}{100} [\text{Permanent }(9.85 + 9.16 + 2.63)] \times 50000 = 106 \times [21.64] \times 500$$

$$= 1146920 \text{ mg} = 1146920 = \mathbf{1.1469\ kg/50000\ litres}$$

7. The hardness of 500 litre of water was completely removed using zeolite softener. After getting exhausted the zeolite bed required 25 litres of NaCl solution containing 2000 mg l⁻¹ of NaCl for regeneration. Calculate the hardness of water sample.

Solution:

Since, 25 litres of NaCl solution contains, 25 × 2000 mg = 50,000 mg = 50 g of NaCl

∴ $CaCO_3$ equivalent hardness = $\dfrac{50}{58.5} \times 50$ (eq. wt. of $CaCO_3$) = 42.73

This hardness is present in 500 litres of water.

∴ Hardness of water sample = $42.73 \times \dfrac{1000}{500}$ = **85.46 ppm**

8. 117 litres of NaCl solution containing 200 gl^{-1} of NaCl was required for regeneration of exhausted zeolite softener (bed). How many litres of hard water sample having 450 ppm hardness can be softened by using this zeolite softener?

Solution:

Amount of NaCl present in 117 litres of solution = 117 × 200 = 23400 g of NaCl

∴ $CaCO_3$ equivalent hardness = $\dfrac{23400}{58.5} \times 50$ = 20000 ppm

∵ One litre of hard water contains 500 ppm hardness = 500 mg of $CaCO_3$ = 0.5 g of $CaCO_3$

∴ Amount of hard water that can be softened by this softener

$$= \dfrac{23400}{0.5 \times 58.5} \times 50 = 40,000 \text{ litres}$$

SUMMARY

Potable water: The water which is safe for drinking is called potable water. It should maintain the following requirements—It should be clear, colourless, odourless, pleasant to taste, having turbidity less than 10 ppm, TDS (total dissolved salts) limit 500 ppm, hardness limit 200 ppm, pH ≈ 7.5–8.0, free from pathogenic germs, toxic matter and objectionable gases.

Treatment of water for domestic use: It is carried out by using the steps—Removal of floating/suspended impurities by screening, sedimentation, filtration and removal of micro-organism by boiling, bleaching powder, chlorination, chloramines and ozone.

Lime-soda method: This method is based on the fact that when calculated amount of lime and soda is added to the hard water; all the soluble impurities of temporary as well as permanent hardness are converted into insoluble precipitate form which is removed off by settling and filtration.

Cold lime–soda process: In this process, calculated quantity of lime and soda with coagulants is added to the hard water at room temperature. These chemicals react with hardness and converts in the form of insoluble precipitate which is removed off by decantation and then filtration.

Hot lime–soda process: It consists of outer chamber or settling chamber, inner chamber or reaction tank and sand filter. The raw water and calculated quantities of chemicals (lime and soda) are supplied to the reaction tank and hot steam is passed simultaneously, which supplies the heat to the water and allows the mixing of the same properly. This treated waters then sent to the sand filter where remaining impurities and sludge are removed and we obtain the soft and filtered water of about 15–30 ppm residual hardness. This process is operated at high temperature, i.e. 90–100°C.

Zeolite or permutite process: Zeolite or permutite–$Na_2O \cdot Al_2O_3 \cdot xSiO_2 \cdot yH_2O$, is hydrated sodium aluminosilicate. It is capable of exchanging reversibly its sodium ions with hardness producing ions like Ca^{2+}, Mg^{2+}, etc in water.

Ion exchange or deionization or demineralization process: Ion exchange resins are insoluble cross-linked, long chain organic polymers with a micro-porous structure (insoluble in water) and the functional groups attached to the chains are responsible for the ion exchange properties. The ion exchange resins may be classified as cation exchange resin and anion exchange resin.

Intrinsic or polished water: Water used for washing in the manufacture of transistors, TV tubes and some other types of electronic equipments must be extremely pure. This water is called intrinsic or polished water.

Boiler feed water: The water which is used to produce steam in the boilers is called boiler feed water. When this water is converted into steam in boiler, the suspended and the dissolved matter may get deposited in the boiler. If these substances form adherent deposits on the walls of the container, they are called scales, which combine with sludge or mud to convert into the non-adherent crust/coating on the inner walls of the boiler.

Desalination of brackish water: The water which contains excess salt concentration with a peculiar salty or brackish taste is called *brackish water*. The process of removing salts from the saline water is termed as *desalination*.

Electrodialysis: The process of dialysis applying electric field is called *electrodialysis*. This method is based on the fact that when the salty water is passed through the ion-selective membrane in presence of electric field, the ions migrate towards oppositely charged electrodes.

Reverse osmosis: The process of movement of solvent from the solution to the pure solvent through a semipermeable membrane by applying excess pressure on solution side is called *reverse osmosis*, sometimes called *super-filtration* or *hyper-filtration*. This process can be used for the desalination of the sea water to get the drinking water.

EXERCISES

1. What are the general characteristics of potable water?
2. Describe the methods for the treatment of municipal water.
3. Explain the various methods used for the sterilization of potable water.
4. What is the principle of lime–soda process? Explain the various reactions take place in lime–soda treatment.
5. Explain the cold lime–soda process for the removal of hardness of water and give the difference between cold and hot lime–soda process.
6. What do you understand by zeolite or permutite? How zeolite plays the role in softening the hard water?
7. Describe the zeolite or permutite method for the softening the hard water. Give its merits and demerits.
8. What do you understand by ion exchange process? Describe the role of ion exchange resins in the softening of water.
9. Describe the ion exchange or deionization or demineralization method of softening the hard water. Give its merits and demerits.
10. A sample of permanent hard water contains micro-organisms. Name one method for each to remove hardness and micro-organisms.

11. How is exhausted ion exchange resins regenerated?
12. Describe the zeolite process for the removal of hardness of water. Discuss its merits over lime–soda process.
13. Why is hot lime–soda process better than cold lime–soda process? Explain in detail.
14. What do you understand by free chlorine? How it works in the disinfection of water?
15. What do you understand by sterilization of potable water? How it is carried out by using the chlorine as disinfectant?
16. Explain the zeolite process for removal of hardness of water. Give its limitations and merits over lime–soda process.
17. What do you understand by conditioning of boiler feed water? Explain different types of conditioning?
18. Give the advantages and limitations of zeolite process and demineralization process.
19. What is disinfectant? What are its main characteristics? Give its uses in the disinfection of municipals water.
20. Explain mixed bed deionization of the removal of hardness of water. Why it is better than simple demineralization method?
21. What do you understand by brackish water? Explain any one suitable method for desalination of brackish water.
22. Explain the reverse osmosis method for the purification of brackish water. Give its advantage over electro dialysis method.
23. Write short notes on the following:
    (i) Sedimentation with coagulation
    (ii) Filtration of municipal water
    (iii) Disinfection or sterilization
    (iv) Break point chlorination
    (v) Deionization or demineralization
    (vi) Brackish water
    (vii) Desalination
    (viii) Electro dialysis
    (ix) Reverse osmosis
    (x) Intrinsic or polished water
    (xi) Boiler feed water
    (xii) Super-chlorination
    (xiii) De-chlorination
24. Differentiate between:
    (i) Soda–lime method and permutite method
    (ii) Cold lime–soda and hot lime-soda method
    (iii) Zeolite method and ion exchange method
    (iv) Ion exchange and mixed bed deionization
    (v) Zeolite method and deionization method
    (vi) Electrodialysis and reverse osmosis:
25. Explain the role of coagulants in the purification of water.

26. Give reasons for the following:
    (i) Why is the water softened by zeolite process unfit for use in boilers?
    (ii) Why is demineralization process preferred over zeolite process for softening of water for use in boilers?
    (iii) The presence of carbon dioxide in boiler feed water should be avoided. Why?
    (iv) Why chloramines are preferable to bleaching powder or chlorine for sterilization of drinking water?
    (v) In the deionization process, water is usually first passed through the cation exchanger and then through the anion exchanger. Give reason.
    (vi) Water containing high turbidity cannot be purified by ion exchange method.
    (vii) Water containing iron and manganese salt cannot be treated by ion exchange method.
    (viii) Why in the cold lime–soda process some quantity of coagulants is added with lime and soda to the water?
    (ix) Why does hard water consume lot of soap?
    (x) Why is water softened before using in boiler?
    (xi) Why should natural water not be feed to boiler?
    (xii) Why are coagulants not used in hot lime–soda process?
    (xiii) Why can caustic embrittlement be controlled by adding sodium sulphate to boiler feed water?
    (xiv) Why can water softened by lime–soda process cause boiler–troubles?
    (xv) Why is boiled water not always 100% safe for drinking purposes?
    (xvi) Why does $Mg(HCO_3)_2$ require double amount of lime for softening?
    (xvii) Why is presence of $NaAlO_2$ in water equivalent of $Ca(OH)_2$?

NUMERICAL PROBLEMS

1. Calculate the amount of lime and soda required to soften 25,000 litres of water having following analysis: $Ca(HCO_3)_2$ = 6.86 ppm, $Mg(HCO_3)_2$ = 8.3 ppm, $CaSO_4$ = 5.8 ppm, $MgCl_2$ = 5.7 ppm, $MgSO_4$ = 10.0 mg, SiO_2 = 3.8 ppm.

2. Calculate the quantity of lime and soda required for softening 50,000 litres of water containing the following salts per litre—$Ca(HCO_3)_2$ = 8.1 mg, $Mg(HCO_3)_2$ = 7.5 mg, $CaSO_4$ = 13.6 mg, $MgSO_4$ = 12.0 mg, $MgCl_2$ = 2.0 mg and $NaCl$ = 4.7 mg.

3. Calculate the amount of lime (84% pure) and soda (92% pure) required for treatment of 20,000 litres of water whose analysis is as follows: $Ca(HCO_3)_2$ = 42.5 ppm, $Mg(HCO_3)_2$ = 37.5 ppm, $CaSO_4$ = 30.0 ppm, $CaCl_2$ = 29.75 ppm, and $NaCl$ = 15.00 ppm. Also calculate the temporary and permanent hardness of water sample.

4. Calculate the amounts lime and soda needed for softening 120000 litres of water containing HCl = 8.3 mg l^{-1}, $MgCl_2$ = 12.5 mg l^{-1}, $NaCl$ = 29.25 mg l^{-1}.

5. A water sample using $FeSO_4 \cdot 7H_2O$ as a coagulant at the rate of 278 ppm, gave the following data on analysis of raw water: Ca^{2+} = 210 ppm, Mg^{2+} = 96 ppm, CO_2 = 44 ppm, HCO_3^- = 732 ppm. Calculate the lime and soda required to soften 50,000 litres of water.

6. Calculate the quantity of lime and soda required for softening 40,000 litres of water using 18 ppm of sodium aluminate as a coagulant. Impurities in water are as follows: Ca^{2+} = 120 ppm, Mg^{2+} = 72 ppm, HCO_3^- = 366 ppm dissolved CO_2 = 24 ppm.

7. Calculate the amount of lime and soda required to soften 10,000 litres of water containing the following ions per litre: Mg^{2+} = 14.7 mg, Ca^{2+} = 30.0 mg, HCO_3^- = 273.2 mg.
8. Calculate the quantities of lime ($Ca(OH)_2$) and soda (anhydrous Na_2CO_3) required for cold softening of 120000 l of water with the following impurities using 16 ppm of sodium aluminate as coagulant. Analysis of raw water: Ca^{2+} = 75 ppm, Mg^{2+} = 46 ppm, CO_2 = 66 ppm, HCO_3^- = 264 ppm, H^+ = 2 ppm. Analysis of treated water: CO_3^{2-} = 45 ppm, OH^- = 34 ppm. Write the chemical equations involved.
9. A zeolite softener was completely exhausted and was generated by passing 100 litres of sodium chloride solution containing 160 gl^{-1} of NaCl. How many litres of a sample of water of hardness 500 ppm can be softened by this softener?
10. The total hardness of 1000 litres of water was completely removed by a zeolite softener. The zeolite softener requires 30 litres of sodium chloride solution, containing 12 gl^{-1} of NaCl for regeneration. Calculate the hardness of water.
11. Calculate the amount of lime and soda required to soften 75,000 litres of water sample having: calcium hardness = 360 mg l^{-1}, magnesium = 150 mg l^{-1}, and total alkalinity = 400 mg l^{-1}.
12. Calculate amount of lime and soda required for softening 90,000 litres of water containing the following salts per liter: $Ca(HCO_3)_2$ = 162 mg, $CaSO_4$ = 136 mg and NaCl = 56.1 mg. Purity of lime is 92% and that of soda is 99%.
13. Calculate the amount of lime and soda required for softening 30,000 litres of water containing the following salts per litre: $Ca(HCO_3)$ = 8.1 mg, $Mg(HCO_3)_2$ = 7.3 mg, $CaSO_4$ = 0.64 mg l^{-1} $MgCO_3$ = 0.82 mg l^{-1}, $MgCl_2$ = 0.56 mg l^{-1}, $MgSO_4$ = 0.70 mg l^{-1}, NaCl = 2.34 mg l^{-1}, SiO_2 = 5.32 mg l^{-1}. The purity of lime is 88.3% and of soda is 99.2%.

CHAPTER 6

CORROSION AND ITS PREVENTION

6.1 INTRODUCTION

A huge amount of money is wasted in the maintenance and replacement of machine and its parts because of the corrosion. So, it needs some preventive and corrective measures to save the same.

Corrosion may be defined as the "slow destruction or deterioration of metal/metallic material on its surface due to unwanted chemical or electrochemical attack of its environment (atmospheric gases, moisture, etc.)". Examples: rusting of iron, corrosion of copper in contact of moist air and CO_2.

6.2 CAUSES OF CORROSION

As we know each material has a tendency to go in a lower energy state. Most of the metals (except noble metals Au, Ag, Pt, etc.) in the nature are found in the form of minerals and ores (oxides, halides, carbonates, sulphates, etc.), i.e. in the lower energy state. Therefore, the metal in the combined form are thermodynamically stable. On the other hand, pure metal is in the higher energy state and tends to go in the lower energy state by reacting with its environment chemically or electrochemically, i.e. the metal is corroded to get thermodynamic stability.

$$\underset{\text{(Higher energy)}}{\text{Metal}} \underset{\text{Reduction/metallurgy}}{\overset{\text{Oxidation/corrosion}}{\rightleftharpoons}} \underset{\text{(Lower energy)}}{\text{Metallic compound + Energy}}$$

The above equation shows that energy is required for the extraction of metals. So, it justifies that the metal in the corrosion process passes from a thermodynamically less stable state (high energy) to a more stable (less energy) state by releasing energy.

6.3 CONSEQUENCES OF CORROSION

Following are the social and economic consequences of corrosion:
1. Health hazards due to escaping of corrosion products into the environment.
2. High machine maintenance cost.
3. Plants' shut down due to its failure.

4. Necessity for overdesign to prevent the corrosion.
5. Decrease in efficiency of machines.
6. Contamination or loss of products.

6.4 ELECTROCHEMICAL CELL/GALVANIC CELL

"A galvanic cell is a device in which the chemical energy is converted into electrical energy on account of some chemical or physical changes within the cell." The driving force in the galvanic cell which causes physical or chemical changes is the decrease in free energy.

Basically, the electrochemical cells are of two types, is galvanic cell and electrolytic cell.

6.4.1 Chemical or Formation Cell

In this type of galvanic cell, the electrical energy is produced due to chemical change occurring within the cell. In other words, the electrochemical cell converts chemical energy into electrical energy. The electrochemical cell is also known as voltaic cell or galvanic cell.

An electrochemical cell consists of two electrodes or metallic conductors in contact with an electrolyte as shown in Figure 6.1. An electrode and its electrolyte comprises an electrode compartment. The two electrodes may share the same compartment.

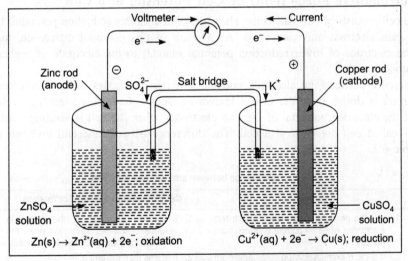

Figure 6.1 Electrochemical cell (Daniel cell).

Daniel cell is an example of electrochemical cell. It consists of zinc electrode dipped in zinc sulphate ($ZnSO_4$) solution. Zinc electrode acts as an anode where oxidation takes place. The copper electrode which acts as a cathode, dipped in copper sulphate ($CuSO_4$) solution where reduction takes place.

The cell reactions may be given as:

At anode (–ve) :	$Zn(s) \longrightarrow Zn^{2+}(aq) + 2e^-$ (Oxidation)
At cathode (+ve) :	$Cu^{2+}(aq) + 2e^- \longrightarrow Cu(s)$ (Reduction)
Cell reaction :	$Zn(s) + Cu^{2+} \longrightarrow Zn^{2+}(aq) + Cu(s)$

The difference between galvanic cell and electrolytic cell is given in Table 6.1.

Table 6.1 Difference between galvanic cell and electrolytic cell

Galvanic cell	Electrolytic cell
1. It converts the chemical energy into electrical energy.	1. In this cell the electrical energy from an external source is used to bring out chemical changes.
2. Anode and cathode are placed in different containers.	2. Both electrodes are placed in the same compartment.
3. Anode is negative and cathode is positive.	3. Anode is positive and cathode is negative.
4. Reaction taking place is spontaneous.	4. Reaction taking place is non-spontaneous.
5. Oxidation takes place at anode and reduction at cathode.	5. Oxidation takes place at cathode and reduction at anode
6. Salt bridge or porous pot is required.	6. No salt bridge or porous pot is required.
7. $\Delta G = -ve$	7. $\Delta G = +ve$
8. Examples—Dry cell or lead storage battery.	8. Examples—Charging of battery, electrolytic purification of metals.

6.4.2 Electromotive Force (emf) or Cell Potential of a Cell

The galvanic cell consists of two half cells. The electrodes at higher reduction potential, have higher tendency to gain electrons and vice versa. As a result of this potential difference, the electrons flow from the electrode of lower reduction potential (anode) to the electrode of higher reduction potential (cathode).

The difference between the electrode potentials of two electrodes of the electrochemical cell when no current is drawn from the cell is known as *electromotive force* (emf) of cell while the difference of the electrode potential of the two electrodes when the cell is sending current through the circuit is called *cell potential* of a cell. The difference between the emf and cell potential is given in Table 6.2.

Table 6.2 Difference between emf and cell potential

emf	Cell potential or potential difference
1. It is the potential difference between two electrodes when no current is drawn from the cell.	1. It is the difference of the electrode potential of the two electrodes when the cell is sending current through the circuit.
2. It is the maximum voltage obtainable from the cell.	2. It is less than the emf of the cell.
3. The work calculated from the emf is the maximum.	3. The work calculated from the potential difference is less than the maximum work obtainable from the cell.
4. emf is responsible for the flow of steady current in the cell.	4. It is not responsible for the flow of steady current in the cell.

This emf acts as driving force for the cell reactions. It is expressed in volts.

emf = Reduction potential of cathode − Reduction potential of anode

or
$$E_{cell} = E_{cathode} - E_{anode} \tag{6.1}$$
or
$$E_{cell} = E_{right} - E_{left}$$

As per cell convention, cathode is always written on right-hand side and anode on the left-hand side. The resulting voltage of Zn–Cu cell may be given as

$$E_0(\text{Cu}) - E_0(\text{Zn}) = 0.345 - (-0.762) = 1.107 \text{ V} \tag{6.2}$$

For concentration, other than 1 gram ions per litre, the emf of cell may be calculated from the Nernst's equation for the cell:

$$\text{Zn(s)}|\text{Zn(aq)}\|\text{Cu(aq)}|\text{Cu(s)}$$

emf of the cell, $E_{\text{cell}} = E_{\text{(cathode)}} - E_{\text{anode}}$

$$E_{\text{cell}} = E^0_{\text{cell}} - \frac{2.303 \, RT}{nF} \log \frac{[\text{Zn}^{2+}(\text{aq})]}{[\text{Cu}^{2+}(\text{aq})]} \tag{6.3}$$

$$E_{\text{cell}} = E^0_{\text{cell}} - \frac{0.0591}{2} \log \frac{[\text{Zn}^{2+}(\text{aq})]}{[\text{Cu}^{2+}(\text{aq})]} \tag{6.4}$$

$(\therefore \ F = 96500 \text{ Coulomb}, R = 8.314 \text{ J K}^{-1}\text{mol}^{-1}, T = 298 \text{ K})$

6.4.3 Concentration Cell

"The cell in which two electrodes of same metal are immersed in solutions containing different concentrations of its ions and connected with a wire or salt bridge is known as *concentration cell*."

In this type of cells, the electrical energy is produced due to the physical change, i.e. transfer of matter from one electrode to another. The potential difference set up between two electrodes is due to the difference in concentrations of the ions in solution. In a concentration cell, the metal which is immersed into dilute solution has a tendency to go to the solution and acts as anode. On the other hand, the metal which is immersed into concentrated solution has a tendency to deposit the ions (released by anode) on it and acts as cathode. Concentration cells are of two types:

(a) *Electrode concentration cell:* In this cell the electrodes of same materials yielding the ions of different concentration are immersed in the same solution.

Example: *Amalgam electrode concentration cell*

Anode	Electrolytic	Cathode
Ag (Hg)	Ag NO$_3$ (soln)	Ag (Hg) ($a_1 > a_2$)
($a_{\text{Ag}^+} = a_1$)		($a_{\text{Ag}} = a_2$)

$$E_{\text{cell}} = E^0_{\text{cell}} - \frac{0.0591}{2} \log \frac{a_1}{a_2} \tag{6.5}$$

Gas electrode concentration cell

Anode	Electrolytic	Cathode	
Pt, H$_2$(g)	HCl(soln)	H$_2$(g), Pt	(P$_1$ > P$_2$)
(P$_{\text{H}_2}$ = P$_1$)	(a_{H^+} = a)	(P$_{\text{H}_2}$ = P$_2$)	

$$E_{\text{cell}} = E^0_{\text{cell}} - \frac{0.0591}{2} \log \frac{P_1}{P_2} \tag{6.6}$$

(b) *Electrolytic concentration cell:* In this cell both electrodes are of same materials, but dipped into solutions of different concentrations of same reversible ions. Example:

1. *Concentration cell with transference/liquid junction*:

   Pt, $H_2(g)$(1 atm) | HCl(soln) a_1, AgCl(s) | Ag || Ag | AgCl(s), HCl(soln) a_2 | $H_2(g)$ (1 atm), Pt |

$$E_{cell} = E^0_{cell} - \frac{0.0591}{2} \log \frac{a_2}{a_1}, \quad (a_2 > a_1) \tag{6.7}$$

2. *Concentration cell without transference/liquid junction*:

   Pt, $H_2(g)$(1 atm) | HCl(soln) a_1, HCl(soln) a_2 | $H_2(g)$(1 atm), Pt |

The emf of this cell can be calculated by the equation:

$$E_{cell} = E^0_{cell} - \frac{0.0591\,t}{2} \log \frac{a_2}{a_1}, \text{ (where } a_2 > a_1) \tag{6.8}$$

These concentration cells play very important role in corrosion.

6.5 ELECTROCHEMICAL SERIES (ACTIVITY SERIES)

The standard electrode potential of a large number of electrodes has been measured using standard hydrogen electrode as the reference electrode. These electrodes can be arranged in increasing or decreasing order of their reduction potential. "The arrangement of elements in order of increasing reduction potential values is called *electrochemical series*." The values of reduction potential of different elements are given in Table 6.3.

Table 6.3 Standard reduction potential of some electrodes at 298 K

Electrode	Electrode reaction (reduction)	E^0
Li	$Li^+(aq)+e^- \rightarrow Li(s)$	−3.05
K	$K^+(aq)+e^- \rightarrow K(s)$	−2.93
Ba	$Ba^{2+}(aq)+2e^- \rightarrow Ba(s)$	−2.90
Ca	$Ca^{2+}(aq)+2e^- \rightarrow Ca(s)$	−2.87
Na	$Na^+(aq)+e^- \rightarrow Na(s)$	−2.71
Mg	$Mg^{2+}(aq)+2e^- \rightarrow Mg(s)$	−2.37
Al	$Al^{3+}(aq)+3e^- \rightarrow Al(s)$	−1.66
Zn	$Zn^{2+}(aq)+2e^- \rightarrow Zn(s)$	−0.76
Cr	$Cr^{3+}(aq)+3e^- \rightarrow Cr(s)$	−0.74
Fe	$Fe^{2+}(aq)+2e^- \rightarrow Fe(s)$	−0.44
Cd	$Cd^{2+}(aq)+2e^- \rightarrow Cd(s)$	−0.40
Co	$Co^{2+}(aq)+2e^- \rightarrow Co(s)$	−0.28
Ni	$Ni^{2+}(aq)+2e^- \rightarrow Ni(s)$	−0.25
Sn	$Sn^{2+}(aq)+2e^- \rightarrow Sn(s)$	−0.14
Pb	$Pb^{2+}(aq)+2e^- \rightarrow Pb(s)$	−0.13
Pt	$2H^+(aq)+2e^- \rightarrow H_2(g)$	0.00
Cu	$Cu^{2+}(aq)+2e^- \rightarrow Cu(s)$	+0.34
I_2	$I_2(aq)+2e^- \rightarrow 2I^-(aq)$	+0.54

Left side: Increasing → (i) Tendency of oxidation (ii) Power as reducing agent

Right side: Increasing → (i) Tendency of reduction (ii) Power as oxidising agent

(Contd.)

Electrode	Electrode reaction (reduction)	E^0
Fe	$Fe^{3+}(aq) + 3e^- \rightarrow Fe(s)$	+0.77
Hg	$Hg_2^{2+}(aq) + 2e^- \rightarrow 2Hg(s)$	+0.79
Ag	$Ag^+(aq) + e^- \rightarrow Ag(s)$	+0.80
Br_2	$Br_2(aq) + 2e^- \rightarrow 2Br^-(aq)$	+1.08
Cl_2	$Cl_2(aq) + 2e^- \rightarrow 2Cl^-(aq)$	+1.36
Au	$Au^{3+}(aq) + 3e^- \rightarrow Au(s)$	+1.42
Mn	$MnO_4(aq) + 8H_3O^+ + 5e^- \rightarrow Mn^{2+}(aq) + 12H_2O(l)$	+1.51
F_2	$F_2(aq) + e^- \rightarrow 2F^-(aq)$	+2.87

Applications

1. To predict the relative strength of oxidizing and reducing agents.
2. To calculate the emf of the cell.
3. To predict the feasibility of the redox reactions.
4. To predict whether a metal can liberate hydrogen from an acid or not.

6.6 TYPES OF CORROSION

Corrosion can be classified into two categories:

1. Dry or chemical corrosion
2. Wet or electrochemical corrosion.

6.6.1 Dry or Chemical Corrosion

This type of corrosion occurs due to direct chemical action of environment/atmospheric gases such as O_2, H_2, Cl_2, N_2, SO_2, CO_2, H_2S, etc. or anhydrous inorganic liquids with metal surfaces in immediate proximity. There are three sub-types of chemical corrosion:

1. Oxidation corrosion
2. Liquid metal corrosion
3. Corrosion by other gases

(a) *Oxidation corrosion:* The corrosion which takes place by the direct chemical action of oxygen on metal at low or high temperature usually in the absence of moisture. For example, alkali metals (Li, Na, K, etc.) and alkaline earth metals (Be, Mg, Ca, etc.) undergo corrosion by this process at ordinary temperature but other metals are oxidized only at high temperatures.

Mechanism: When a metal is exposed to air it gets oxidized at the surface.

$$2M \rightarrow 2M^{n+} + 2ne^- \text{ (Oxidation of metal—anodic half reactions)}$$

$$2ne^- + \tfrac{1}{2}nO_2 \rightarrow nO^{2-} \text{ (Reduction of oxygen—cathodic half reaction)}$$
Oxide ions

$$2M + \tfrac{1}{2}nO_2 \rightarrow M_2O_n \text{ (Formation of metal oxide—Corrosion reactions)}$$

The metallic oxide scale so formed on the metal surface acts as a barrier to resist further oxidation (corrosion) of inside metal. The nature of oxide film (<300 Å thickness) plays an important role in the oxidation corrosion. The oxide film may be of the following types:

1. *Porous (non-protective):* Porous oxide film allows further diffusion of oxygen into the metal surface and therefore, the metal is easily and continuously corroded even at low temperature. Alkali metals (Li, Na, K, etc.) and alkaline earth metals (Be, Mg, Ca, Sr, etc.) form porous (non-protective) layers and undergo rapid continuous corrosion. The porous nature of oxide film may be explained with the help of Pilling Bedworth's rule.

   Pilling Bedworth's rule: It states that an oxide layer is protective or non-porous if the volume of oxide is greater than the volume of metal from which it is formed. On the other hand, if the volume of oxide is less than the volume of metal, the oxide layer is porous (non-protective), i.e.

   $$\text{Specific volume ratio} = \frac{\text{Volume of metal oxide}}{\text{Volume of metal}}$$

   Hence we may conclude that:
   (a) If the specific volume ratio is greater than 1, i.e. volume of oxide is greater than volume of metal, the oxide film will be non-porous (protective) and no further corrosion will take place. For example, Al, Ga, Tl, etc. form non-porous layer of their oxides.
   (b) If the specific volume ratio is less than 1, i.e. volume of oxide is less than the volume of metal, the oxide film will be porous (non-protective) and it will allow further diffusion of O_2 and hence further corrosion.

2. *Non-porous (protective):* If the oxide film is continuous and impervious it acts as protective coating and prevent the metal from further corrosion. For example Al, In, Tl, Pb etc form non-porous (protective) layer.

3. *Stable:* If the oxide film is stable, i.e. fine grained, impervious and adhering to the metal surface it may cut off the penetration or diffusion of oxygen into the underlying parental metal surface. This layer acts as protective coating/shield for the metal surface. Examples—Al, Pb, Cu, Sn, etc.

4. *Unstable:* The oxide film may be unstable, i.e. it decomposes back to give metal and oxygen;

   $$\text{Metal oxide} \longrightarrow \text{Metal} + \text{Oxygen}$$

   Hence, there is no resultant corrosion. The oxidation corrosion is not possible in such cases. For example, metals like Ag, Au, Pt, etc. do not undergo oxidation corrosion.

5. *Volatile:* If the oxide film so formed is volatile then it escapes at high temperature as soon as it is formed. Therefore, the underlying metal surface is exposed to further attack of oxygen which results in the further corrosion. For example, molybdenum (Mo) and tungsten (W) form volatile oxide layer and corroded rapidly.

(b) *Corrosion by other gases:* Some gases like CO_2, SO_2, Cl_2, F_2, etc. attack on certain metals and form protective or non-protective layer on metal surface. The extent of corrosion depends on the chemical affinity between gas and metal, and the nature of the oxide film formed.

(c) *Liquid metal corrosion:* This type of corrosion takes place when liquid metal is allowed to flow over solid metal at a high temperature. It results in the weakening of solid metal due to the dissolution of solid metal into liquid metal.

For example: In the nuclear reactor, the liquid sodium metal leads to the corrosion of cadmium.

6.6.2 Electrochemical or Wet Corrosion

"The corrosion which takes place under wet or moist conditions due to the formation of short circuited electrochemical cells is called *electrochemical corrosion*."

The electrons flow from metal surface (anodic area) towards impure metallic portions (cathodic area) through a conducting medium. Example: rusting of iron.

Conditions for wet corrosion

1. Two dissimilar metals or alloys should be in contact with each other.
2. Conducting medium, e.g. moisture, conducting solution, etc. should be present on the metal surface.
3. Formation of separate anodic and cathodic area/parts between which the current flow through a conducting medium.
4. Metal surface should be heterogeneous.
5. Formation of corrosion product somewhere between anode and cathode.

6.6.3 Electrochemical Theory of Corrosion

According to this theory, the chemically non-uniform surfaces of metals behave as tiny electrochemical cells in the presence of an electrolytic conducting medium (water containing dissolved O_2 and CO_2). Thus, wet corrosion of metal in an aqueous medium is an electrochemical phenomenon in which the electron current flows between the anodic and cathodic areas.

The anodic reaction involves the dissolution of metals in the form of metal ions with the liberation of electrons.

At anode:
$$M \longrightarrow M^{n+} + ne^- \text{ (oxidation)}$$

The cathodic reactions may take place in two different ways, i.e. (a) evolution of hydrogen type and (b) absorption of oxygen.

(a) *Evolution of hydrogen* (*in acidic medium*)

The electrons so released at anode due to the oxidation of metal flow through the metal from anode to cathode. These electrons are absorbed by H^+ ions (in acidic medium) at cathode.

In acidic medium: $2H^+ + 2e^- \longrightarrow H_2$ (Reduction)

In neutral or weakly alkaline medium: $2H_2O + 2e^- \longrightarrow 2OH^- + H_2 \rightarrow$ (Reduction)

(b) *Absorption of oxygen* (*in neutral medium*)

In neutral medium, the oxygen absorbs electrons in the presence of moisture to give OH^- ions:
$$H_2O + \tfrac{1}{2}O_2 + 2e^- \longrightarrow 2OH^-$$

The M^{n+} ions from anode move towards cathode through a conducting medium and combine with OH^- ions at cathode to form corrosion product $M(OH)_n$.

Mechanism of rusting of iron

Rusting of iron is the most common example of wet corrosion. "It may be defined as the deterioration of iron on its surface by the action of moist air, resulting in the deposition of reddish brown coatings on metal surface." The overall reaction of rusting of iron is given as:

$$4Fe(s) + 3O_2(g) + 2xH_2O(g) \longrightarrow 2Fe_2O_3 \cdot xH_2O(s)$$
Iron — Air — Water in air — Brown rust

This mechanism can be explained with the help of electrochemical theory of wet corrosion Figure 6.2.

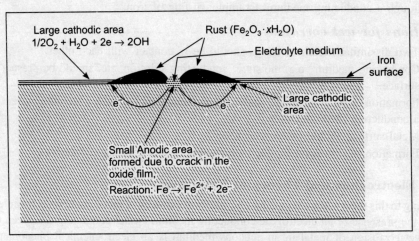

Figure 6.2 Mechanism of wet corrosion (rusting of iron).

1. The non-uniform surface of metal or impurities present in iron or iron lattice has imperfections, behaves like small electric cells or corrosion couples in the presence of water containing dissolved oxygen or carbon dioxide. A film of moisture with dissolved oxygen and CO_2 constitutes electrolytic solution. In each small cell, pure iron part acts as an anode and impure part acts as a cathode.

2. At anode Fe atoms get oxidized to Fe^{2+} ions loosing electrons

$$Fe \longrightarrow Fe^{2+} + 2e^- \quad \text{(Oxidation)}$$

3. These electrons are consumed by water in cathodic reaction

$$H_2O + \tfrac{1}{2}O_2 + 2e^- \longrightarrow 2OH^-$$

4. Ferrous ions move away from anodic region and meet hydroxide ion (OH^-) to form the corrosion product $Fe(OH)_2$ which get deposited at cathode

$$Fe^{2+} + 2OH^- \longrightarrow Fe(OH)_2$$

5. This ferrous hydroxide on drying is converted into black rust

$$Fe(OH)_2 \xrightarrow{\Delta} Fe_3O_4 \cdot xH_2O \text{ (black rust)}$$
Anhydrous magnetite (Fe_3O_4)

If the oxygen supply is sufficient then the ferrous hydroxide further oxidizes into ferric hydroxide which on drying gives yellow rust (Table 6.4).

$$Fe(OH)_2 + O_2 + H_2O \rightarrow Fe(OH)_3 \xrightarrow{\Delta} Fe_2O_3 \cdot xH_2O$$
Ferric oxide (yellow rust)

Table 6.4 Difference between dry (chemical) and wet (electrochemical) corrosion

Dry or chemical corrosion	Wet or electrochemical corrosion
1. It occurs in dry condition, i.e. in the absence of moisture.	1. It occurs in wet condition, i.e. in the presence of moisture.
2. It involves the direct chemical attack of an environment on the metal surface.	2. It involves the formation of a large number of electrochemical cells on the metal surface.
3. It is uniform all along the surface.	3. It is non-uniform and depends on the relative anodic and cathodic areas.
4. It occurs on homogeneous as well as on heterogeneous surface.	4. It occurs only on heterogeneous surface.
5. Corrosion product is deposited at the same place where the corrosion occurs.	5. Corrosion occurs always at anode, but corrosion product is deposited near cathode.
6. It is a slow process.	6. It is comparatively a fast process.

6.7 OTHER TYPES OF ELECTROCHEMICAL CORROSION

6.7.1 Galvanic Corrosion

"The corrosion which takes place when different metals are in contact either directly or through an electrical conductor and jointly exposed to corrosive environment is called *galvanic corrosion*."

In this type of corrosion, a galvanic cell is formed due to the contact of two dissimilar metals. The metal which is at higher position in the electrochemical series with more +ve electrode potential, acts as an anode and undergoes corrosion (oxidation) while other metal which is at lower position in the electrochemical series acts as a cathode.

For example, when zinc (Zn) and copper (Cu) are electrically connected and exposed to an electrolyte, the Zn metal which is more electropositive and at higher position in electrochemical series, constitutes an anode while Cu which is at lower position in the electrochemical series, constitutes a cathode. Moisture or aqueous solution acts as an electrolytic medium (Figure 6.3).

Anode : $Zn \longrightarrow Zn^{2+} + 2e^-$ (Oxidation)

Cathode : $½O_2 + H_2O + 2e^- \longrightarrow 2OH^-$ (Reduction)

Complete reaction: $Zn + O_2 + H_2O \longrightarrow Zn(OH)_2$ (Corrosion)

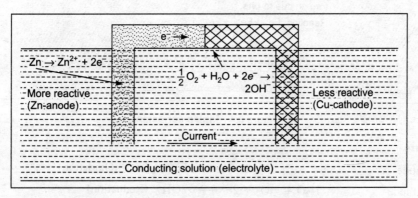

Figure 6.3 Galvanic corrosion of Zn in contact of copper.

Other examples are: Pb–Sb solder around Cu wire, steel screw in a brass marine hardware and steel pipes connected to Cu plumbing.

Factors that affect galvanic corrosion
1. *Potential difference between the two metals*: Greater the potential difference greater will be the rate of corrosion.
2. *Suitable conducting medium*: Greater the conductivity of medium greater will be the rate of corrosion.
3. *Surface area of metals*: Corrosion is greater at rough and uneven surfaces.

Prevention
Galvanic corrosion can be controlled by:
1. avoiding the presence of corroding medium.
2. avoiding the formation of galvanic couple.
3. providing insulating materials between the two metals.

6.7.2 Pitting Corrosion

"The corrosion which takes place due to the breakdown or cracking of a metal surface, i.e. formation of pits, cavities, crevices or pin holes on metal surface is called *pitting corrosion*". It is localized and an accelerated attack of an environment in the presence of extraneous impurities (sand, dust, scale, etc.) deposited on the metal surface (Figure 6.4).

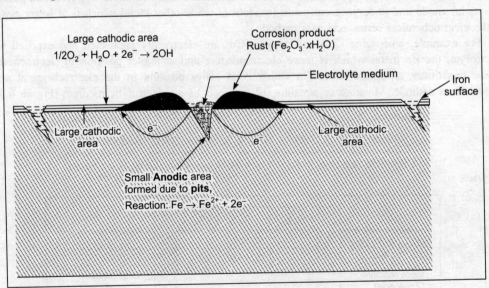

Figure 6.4 Pitting corrosion.

Reactions

At anode   : $Fe \longrightarrow Fe^{2+} + 2e^-$ (Oxidation)

At cathode   : $H_2O + \frac{1}{2}O_2 + 2e^- \longrightarrow 2OH^-$ (Reduction)

Overall reaction : $Fe + H_2O + \frac{1}{2}O_2 \longrightarrow Fe(OH)_2$ (Corrosion)

Factors affecting pitting corrosion
1. Non-uniform and non-homogeneous metal surface.
2. Presence of sharp corners, bents, cavities on the metal surface.
3. Relative anodic and cathodic areas.

Prevention and control
Pitting corrosion may be prevented by:
1. purifying the metal.
2. avoiding inhomogeneities both in the metal and in the corrosive environment.
3. proper designing of the metal/metal article.
4. using protective coating on metal surface.

6.7.3 Stress Corrosion or Stress Cracking

"The Corrosion which occurs due to the combined effect of tensile stress and the corrosive environment on the metal or an alloy is called stress corrosion."

The stresses are developed during the manufacturing, quenching, bending, annealing, welding, etc. and during heavy working, i.e. rolling, drawing, etc. (Figure 6.5). It is highly localized attack. The stress on metal produces strains that results in the localized zones of higher electrode potential.

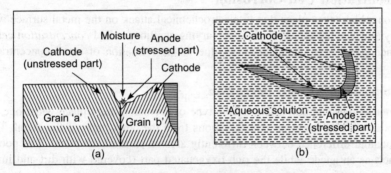

Figure 6.5 (a) and (b) Stress corrosion.

Types of stress corrosion
It is of two types: (a) season cracking, (b) caustic embrittlement (boiler corrosion).

(a) *Season cracking:* This type of corrosion generally takes place in copper alloy, like brass in an atmosphere of ammonia (NH_3) or amine ($R-NH_2$). When brass (alloy of Cu and Zn) exposed to an ammonia environment, both the metals, i.e. Cu and Zn form stable complex ions $[Cu(NH_3)_4]^{2+}$ and $[Zn(NH_3)_4]^{2+}$, respectively. This leads to the dissolution of brass and ultimately the formation of cracks in the presence of stress.

Anode: $Cu + 4NH_3 \longrightarrow [Cu(NH_3)_4]^{2+} + 2e^-$ (Oxidation)

$Zn + 4NH_3 \longrightarrow [Zn(NH_3)_4]^{2+} + 2e^-$

Cathode: $O_2 + 4H^+(aq) + ne^- \longrightarrow 2H_2O$ (Reduction)

(b) *Caustic embrittlement:* This type of corrosion takes place in the high pressure boilers due to the use of highly alkaline water. Water softened by lime–soda process still may have some residual Na_2CO_3. In high pressure boilers, some residual Na_2CO_3 decomposes to give sodium hydroxide (NaOH and CO_2) which makes the water caustic. On evaporation the concentration of caustic soda increases on bents, joints, rivets, etc. and causes corrosion of boiler. This may sometimes cause failure of boiler.

$$2NaOH + O_2 + Fe \longrightarrow Na_2FeO_2 + H_2O$$
$$3Na_2FeO_2 + 4H_2O \longrightarrow 6NaOH + Fe_3O_4 + H_2$$
$$6Na_2FeO_2 + 6H_2O + O_2 \longrightarrow 12NaOH + 2Fe_3O_4$$

NaOH is regenerated and magnetite (Fe_3O_4) is precipitated, hence enhancing further dissolution of iron. Caustic embrittlement results due to the formation of concentration cells.

Prevention

It can be prevented by:

1. avoiding the use of water softened by lime–soda method.
2. using the water softened by sodium phosphate in place of sodium carbonate in external treatment.
3. adjusting the pH of water to 8–9.
4. adding sodium sulphate (Na_2SO_4) or tannin to the boiler water, which blocks the cracks in the boiler wall and prevents the attack of NaOH in such areas.

6.7.4 Concentration Cell Corrosion

"The Corrosion which takes place due to electrochemical attack on the metal surface when exposed to an electrolyte of varying concentrations or varying aerations called *concentration cell corrosion*." The metal undergoes electrochemical attack due to the formation of small concentration cells on its surface.

Types of concentration cell corrosion

(a) *Differential aeration corrosion:* This type of corrosion takes place when one part of metal is exposed to a different air concentrations from the other parts of the metal. This results in the potential difference between differently aerated parts of the metal. The poor oxygenated part acts as an anode while the rich oxygenated part (covered with dirt and impurities) acts as a cathode (Figure 6.6).

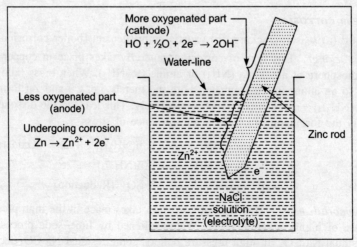

Figure 6.6 Differential aeration (concentration cell) corrosion.

It causes the flow of differential current between anode and cathode and results in the corrosion of anodic part of the metal.

At anode: $\quad\quad\quad\quad\quad\quad\quad\quad Zn \longrightarrow Zn^{2+} + 2e^-$

At cathode: $\quad\quad\quad\quad H_2O + \frac{1}{2}O_2 + 2e^- \longrightarrow 2OH^-$

$$\overline{Zn + H_2O + \frac{1}{2}O_2 \longrightarrow Zn(OH)_2}$$

Differential aeration corrosion is accelerated by the deposition of dirt, scales on metal, presence of salts, impurities in electrolytic medium, cracks, crevices pits, etc. These parts are less oxygenated and act as anode. Similarly, the iron corrodes under drop of water or salt solution.

(b) *Water-line corrosion:* This is the specific type of corrosion which takes place due to the different air concentrations on metal parts below water line.

"The corrosion which takes place by the formation of concentration cell near water line due to the different concentrations of air (O_2) at different portions of metal below water line".

The less oxygenated metal part acts as an anode while the more oxygenated metal part acts as a cathode. (Figure 6.7)

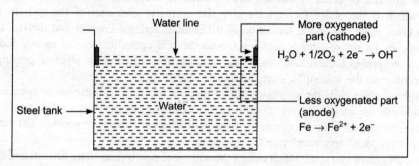

Figure 6.7 Water-line corrosion.

When water stored in a steel tank which remains stagnant for long duration, results in the corrosion. The iron at the water line is more oxygenated and acts as a cathode while the iron below the water is less oxygenated and acts as an anode.

At anode the iron gets oxidized to give ferrous ions (Fe^{2+}):

$$Fe \longrightarrow Fe^{2+} + 2e^- \quad (Oxidation)$$

These electrons released during oxidation are absorbed by oxygen atoms at cathode to give OH^- ions:

$$H_2O + \frac{1}{2}O_2 + 2e^- \longrightarrow 2OH^- \quad (Reduction)$$

The overall corrosion reaction is as follows:

$$Fe + H_2O + \frac{1}{2}O_2 \longrightarrow Fe(OH)_2$$

In the presence of sufficient amount of oxygen this ferrous hydroxide gets further oxidized to ferric hydroxide which on drying gives yellow rust.

$$Fe(OH)_2 + H_2O + \frac{1}{2}O_2 \longrightarrow Fe(OH)_3 \xrightarrow{\Delta} Fe_2O_3 \cdot x H_2O \quad (Yellow\ rust)$$

The water line corrosion is enhanced by the presence of salts, turbidity, suspended impurities and micro-organism. This corrosion can be retarded by the presence of anodic inhibitors like phosphates, chromates, bicarbonates, silicates, etc.

6.7.5 Soil Corrosion

The underground pipeline structures (for oil, gas and water) which remain in the constant contact of soil, moisture and salts get corroded. Thus, "the corrosion of the metal articles, pipelines, etc. which takes place due to the electrochemical attack of moisture and salts present in the soil is called *soil corrosion*".

Following are the important factors which are responsible for the soil corrosion:

1. moisture content
2. acidity of soil
3. content of organic matter
4. content of electrolytes
5. micro-organisms present
6. physical properties of the soils.

In the absence of acids in the soil the conductivity depends on the moisture and electrolyte contents which is the major factor governing the corrosive character of the soil. In such a soil, corrosion takes place due to the formation of differential aeration couples and the rate of corrosion mainly depends on the resistance between the anodic and cathodic areas of metals and the rate of infusion of oxygen to the cathodic area. Thus, the portions of cable or pipeline passing below the paving become anodic and suffer corrosion.

In highly acidic soil, the corrosion takes place is of hydrogen-evolution type in which the anodic and cathodic areas are very close to each other. The conductivity in this case has not much significance. In case of organic matter, the formation of soluble metal complexes and the peptization of corrosion products may accelerate the corrosion.

In case of gravel and sand which makes the soil more porous and more aerated and the corrosion of metal articles depends mostly on the moisture and salt content (Figure 6.8). Water logged soils may generate anaerobic bacteria which may generate the conditions for microbiological corrosion in buried pipelines or cables passing from one type of soil to another, e.g. from less aerated clay to more aerated cinders (having air pockets). These pipelines or cables may get corroded due to differential aeration. Other factors of soil corrosion are the differences in pH and the presence of unburnt carbon in the cinders.

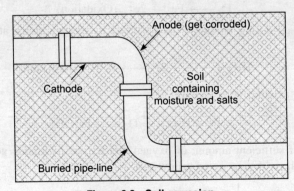

Figure 6.8 Soil corrosion.

6.7.6 Microbiological Corrosion

There are certain bacteria which influence the corrosion process by interacting with corrosion environment.

"The corrosion which takes place due to the attack of bacteria present in water or soil on the metal surface is called *micro-biological corrosion*."

For example, oxygen consuming bacteria present in water or soil decrease the concentration of oxygen in the medium (acidic or basic) in contact with a metal structure. If the metal structure is only partially exposed to the oxygen-depleted environment, then the bacterial action increases the corrosion due to differential aeration.

Anaerobic bacteria like *Microspora vibrio desulfuricus* reduces the sulphates to sulphur which is used as nutrient to prepare their protoplasm. During this process the depolarization effect takes place by the conversion of O_2 from the sulphates into such a form that allows the corrosion of iron take place. After the death of these bacteria, sulfur is liberated in the form of H_2S that converts a fraction of the corrosion product into FeS which is less effective to suppress the corrosion than the oxide or hydroxide. The presence of sulphates and organic matter under anaerobic conditions results in strong and localized corrosion of the cast iron.

Sometimes, barrier of medium created by the growth of low-form algae is capable of digesting the iron. Some film-forming bacteria also cause the metal corrosion.

Some other corrosive bacteria are:

1. Film forming bacteria like flavobacterium, mucoids, aerobacter and pseudomonas lead to corrosion in cooling tower systems.
2. The presence of iron bacteria is sometimes connected with pitting or blockage of pipes due to corrosion products.
3. Iron and manganese bacteria which thrive on iron and manganese compounds, in the presence of oxygen deposit insoluble hydrated oxides on these metal surfaces.
4. Iron and manganese bacteria may promote the rate of corrosion by forming differential aeration couples.

6.8 PASSIVITY OR PASSIVATION

"The phenomenon by which a metal or an alloy shows the greater corrosion resistance than expected from its position in the galvanic series due to the formation of highly protective, very thin ($\approx$ 0.0004 mm) and invisible film on the surface is called *passivity* or *passivation*."

For example, active metals like Al, Tl, Cr, chrome steel, etc. in presence of oxygen tend to form the passive layer (oxide film) on their surface and show outstanding corrosion resistance. This layer is highly protective and has self-healing property, i.e. if it gets damaged, it is automatically repaired in oxidizing environment. Although, in reducing environment the metal becomes chemically active and the corrosion is rapid.

Chrome steel is not corroded even in concentrated HNO_3 as the Cr present in it forms protective oxide film while iron gets readily corroded even in dilute HNO_3.

6.9 FACTORS AFFECTING RATE OF CORROSION

The rate and extent of corrosion depends on the following factors:

(a) *Nature of metal*
   1. Purity of metal
   2. Physical state of metal

3. Passive character of metal
4. Position of the metal in the galvanic series
5. Relative anodic and cathodic areas
6. Nature of corrosion film/layer
7. Solubility of corrosion product

(b) *Nature of environment*
1. Temperature
2. Humidity
3. pH of the medium
4. Presence of impurities in the atmosphere
5. Presence of suspended particles in the atmosphere
6. Conductance of corroding medium
7. Nature of electrolyte
8. Formation of oxygen concentration cell

6.9.1 Nature of Metal

(a) *Purity of Metal:* Pure metal is not corroded but the presence of impurities in the metal results in the heterogeneity and hence formation of tiny electrochemical cells. The anodic part gets corroded and the rate of corrosion depends on the extent of impurities present in the metal, i.e. greater the impurities present in the metal greater will be the rate of corrosion. For example, 99.999% pure Zn corrodes at the rate of one atom s^{-1} while 99.99% pure metal corrodes at 2650 atoms^{-1} and 99.8% at 7200 atoms s^{-1}.

(b) *Physical state of metal:* The rate of corrosion is affected by the physical state of metal like grain size, surface structure, orientation of crystals, stress, etc. Larger the grain size, lesser will be the solubility of metal and hence lesser will be the corrosion. Area under stress tends to be anodic and hence stressed area is corroded. For example, in the boiler the caustic embrittlement takes place at the bends, joints, rivets, etc.

(c) *Passive character of metal:* The metals which form the passive layer, i.e. highly protective thin film ($\approx$ 0.0004 mm) of oxide on its surface, exhibit much higher corrosion resistance than expected from its position in the galvanic series. For example, metals like Al, Gl, Tl, Cr, Ni, etc, show high corrosion resistance due to the formation of passive oxide film on their surfaces.

(d) *Position in the galvanic series:* In the galvanic series, the metals/alloys are arranged on the basis of their oxidation potential and corrosion behaviour. The greater the oxidation potential, i.e. higher up the position in the galvanic series greater will be the tendency of metal to be anodic and hence greater will be the corrosion. If the two metals are in contact, the metal which is at higher position in the galvanic series behaves as an anode and corroded first. Moreover, greater is the difference in the position of metals in the galvanic series faster is the corrosion. Order of oxidation potential of different elements and alloys are shown in Table 6.5. For example, if zinc metal is in contact with iron, the zinc behaves as an anode and corroded as it is at higher position in the galvanic series.

Table 6.5 Galvanic series based on the relative oxidation potential in sea water

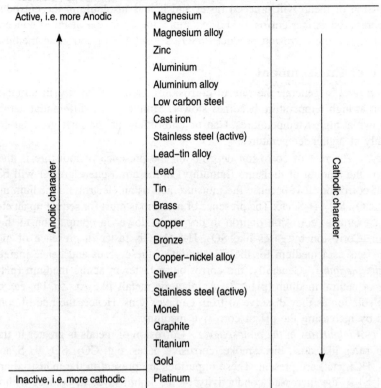

(e) *Relative anodic and cathodic areas on metal surface:* In general, the rate of corrosion of metal is directly proportional to the ratio of cathodic and anodic areas. The corrosion is very rapid and highly localized in case of smaller anodic area. Due to large cathodic area the demand of electrons is high which causes the increased rate of dissolution (oxidation) of metal at anodic region. For example, an iron pipe fitted in a large copper tank corrodes rapidly and severely.

(f) *Nature of corrosion film/layer:* In actual practice, all the metals undergo corrosion in atmosphere or other corrosive environment up to different extents. The corroded metal/metal part is covered by a very thin film of oxide. The corrosion is rapid and severe, if the oxide film is porous or volatile and on the other hand, if the oxide film is non-porous and unstable then the corrosion is very less. The porous or non-porous nature of oxide film depends on the specific volume ratio.

$$\text{Specific volume ratio} = \frac{\text{Volume of metal oxide}}{\text{Volume of metal}}$$

So, if the specific volume ratio is greater than one, the oxide film will be non-porous and hence protective and vice versa. Hence, the rate of corrosion depends on the specific volume ratio. For example, the specific volume ratio of Mg, Cu, Zn, etc. is less than one, therefore, these metals get corroded rapidly.

(g) *Solubility of corrosion product:* If the corrosion product is soluble in the corroding medium, the corrosion of metal will proceed faster and if it is insoluble in the corroding medium it will suppress the further corrosion. For example, the corrosion of Pb in concentrated H_2SO_4 is slow because the corrosion product $PbSO_4$ is insoluble in corroding medium (H_2SO_4).

6.9.2 Nature of Environment

(a) *Temperature:* In general, the rate of corrosion increases with rise in temperature but wet corrosion at high temperature becomes slower. Similarly, the differential aeration corrosion slows down at higher temperatures. Caustic embrittlement and inter-granular corrosion takes place only at higher temperatures.

(b) *Humidity:* The rate of corrosion depends on the presence of moisture in the atmosphere. Greater is the amount of moisture (humidity) in the atmosphere greater will be the rate and extent of corrosion. It is because the moisture acts as an electrolytic medium and solvent for gases like O_2, SO_2, H_2S, etc. The presence of medium is must for setting up an electrochemical cell. For example, corrosion of iron in dry air is slower in comparison to the corrosion in moist air. Corrosion by gases like SO_2, H_2S, etc. is faster in presence of moisture as the moisture acts as a medium for the dissolution of these gases and hence increases acidity.

(c) *pH of the medium:* Generally, the corrosion is faster in acidic medium (pH < 7) than in alkaline or neutral medium (pH ≥ 7). Amphoteric metals like Al, Zn, Pb, etc. corrode in the alkaline solution as they dissolve to from complex ions. Hence, the rate of corrosion can be reduced by increasing the pH of corrosive medium.

(d) *Presence of impurities in the atmosphere:* Corrosion of metals is greater in the areas where the impurities like dust, dirt, smoke, corrosive gases, e.g. CO_2, SO_2, H_2S and acid fumes (H_2SO_4, HCl, etc.) are present. These impurities are present in the industrial areas and cities which lead to the increased conductivity of the liquid layer in contact with metal surface and hence increase the rate of corrosion.

(e) *Presence of suspended particles:* Presence of suspended particles in the atmosphere enhances the rate of corrosion as they absorb moisture, gases and act as strong electrolytes. Suspended particles are of two types: (i) active particles—NaCl, $(NH_4)_2SO_4$, etc. and (ii) Inactive particles—charcoal, etc.

(f) *Conductance of corroding medium:* Rate of wet corrosion depends on the conductance of corroding medium. Greater the conductance of corroding medium greater will be the rate of corrosion. For example, the underground or soil corrosion will be greater in the presence of salts and moisture in the earth, i.e. if the conductance of soil is more the underground corrosion of metal will be more severe.

(g) *Nature of electrolyte:* The presence of ions like Cl^-, SO_4^{2-}, NH_4^+, etc. (electrolyte) increases the rate of corrosion. For example, the chloride ions present in the electrolyte destroy the protective film and metal surface is exposed to further corrosion, e.g. Al metal corroded rapidly in the sea water. But if the electrolyte consists of silicate ions, they form an insoluble silicate layer and prevents the metal from further corrosion.

(h) *Formation of oxygen concentration cell:* The rate of wet corrosion increases with the increase in the supply of oxygen/air to the moist metal surface. This happens due to the formation of oxygen concentration cell (differential aeration). Less oxygenated part acts as an anode while more oxygenated part acts as a cathode and the corrosion takes place at the anode. For example, pitting corrosion and water line corrosion take place due to the formation of oxygen concentration cells.

6.10 CORROSION CONTROL–PROTECTION FROM CORROSION

Corrosion of metal is undesirable and hence it must be prevented. It can be controlled/prevented by various methods. Some important methods are discussed as follows.

6.10.1 Material Selection

While designing any metal article/machine the right type of material should be selected to protect it from corrosion. Following points should be considered while selecting the metal:

1. *Chemical properties and its interaction with environment:* The noble metals are highly immune to corrosion but cannot be used for general purposes because of economic reason.
2. *Purity of metal chosen:* Metal should be as much pure as possible because pure metal does not corrode.
3. It should be free from internal stresses.
4. Contact of two dissimilar metals should be avoided.
5. The use of metal alloys should be preferred.

6.10.2 Proper Designing

Following points should be taken into consideration while designing metal articles or machine parts (Figure 6.9):

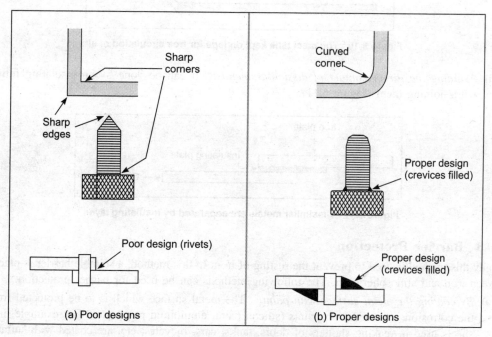

Figure 6.9 Design of articles.

(a) *Avoiding sharp corners and edges*: Accumulation of dirt, dust, etc. can be prevented by avoiding sharp corners and *edges*, and hence the corrosion can be controlled.

(b) *Localized stresses should be avoided:* The metal/metal articles should be designed in such a way that it should have no or minimum localized stressed crevices. The stresses produce stagnant areas with scales and cause the formation of concentration cells hence result in the metal corrosion.

(c) *Arrangement for free passage of air:* The metal articles should be designed/arranged in such a way that there is a free circulation of air to prevent the deposition of dirt, dust scales, etc. This can be done by supporting the machine/tank, etc. on beams or legs (Figure 6.10).

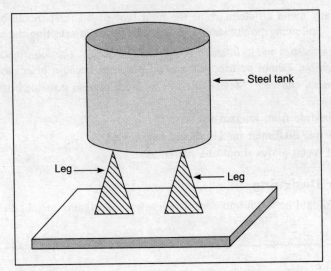

Figure 6.10 The steel tank kept on legs for free circulation of air.

(d) *Avoiding the direct contact of dissimilar metals:* It can be done by an insulating fitting while joining them (Figure 6.11).

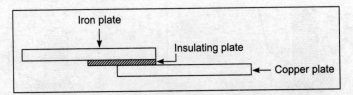

Figure 6.11 Dissimilar metals are separated by insulating layer.

6.10.3 Barrier Protection

Usually this method is used to prevent the rusting of iron. In this method, a suitable barrier is placed between iron and atmospheric air. The following methods can be used for barrier protection:

(a) *By coating the metal surface with paint:* The metal surface which is to be protected from the corrosion is coated with paints (silicon paint, aluminium paint, etc.). For example, iron sheets used in making shutters of doors, tanks, cars, bicycles, etc. are coated with suitable paints.

(b) *By applying a thin film of oil or grease on metal surface:* This method is used to protect the iron tools and machine parts from corrosion by covering the metal surface.

(c) *By coating the metal surface with plastic layer:* This method is used to protect the costly components/instruments like electronic instruments, handle of chairs, iron sheets used in the inner lining of the cars, buses, etc.

(d) *By coating the metal with certain chemicals (anti-rust solutions):* Some anti-rust chemicals like solutions of FePO$_4$, alkaline phosphates and alkaline chromates, etc. are used to protect the iron metal from corrosion.

6.10.4 Metallic Coatings

The metallic coating is usually done by the following methods:
  (i) *Hot dipping:* The metal which is to be coated, is dipped in the molten bath of the coating metal for a sufficient period and then is taken out with coated film on it. The coating of zinc metal on iron is called *galvanization*.
  (ii) *Metal spraying:* The molten metal is sprayed on the base metal with the help of spraying gun. This method is good as: (1) coating can be applied to the finished base metal, (2) coating can be applied to any desired spot, (3) thickness of coating film can be controlled and hence uniform coating can be obtained. This method is widely used for coating of Cu, Ni, Sn, Pb, Brass, etc.
  (iii) *Electroplating/electrodeposition:* "*It is the process of deposition of coating metal on the base metal by passing direct current through an electrolytic solution of soluble salt of a coating metal.*" Electroplating increases the resistance of metal to the corrosion and chemical attack. Moreover, it improves hardness and physical appearance.

The metal to be electroplated is first cleaned and then made cathode of an electrolytic cell. The anode is taken of coating metal or an inert material like graphite. Salt solution of metal (to be deposited) is taken as an electrolyte in the electroplating tank. Both electrodes (anode and cathode) are dipped in electrolytic solution in the tank, on passing direct current the ions of coating metal migrate to cathode and get deposited as thin film/layer on the metal article (cathode). (Figure 6.12):

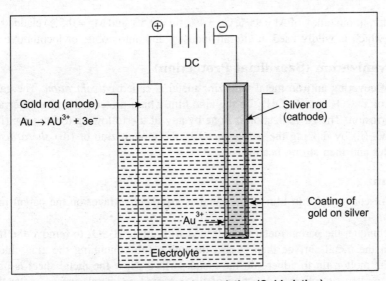

Figure 6.12 Process of electroplating (Gold plating).

The electroplating may be of the following types on the basis of coating metal
Gold plating: It is the process of coating of gold on Ag, Cu, etc.
Chrome plating: It is the process of coating of chromium on iron, brass, bronze, etc.

Tinning: It is the process of coating of tin on steel sheets, tanks, etc.

Plate-forming: It is the process of coating of platinum on steel blades, knife, etc.

6.10.5 Metal Cladding

It is the process of making sandwich of base metal (metal to be protected) between two layers of cladding materials (usually pure Al metal) and then passing this combination through rollers under the action of heat and pressure to form a clad sheet. Cladded material other than Al may be Ni, Cu, Pt, Ag or Pb and even alloys like stainless steel, etc. show the sheets between two Rollers (Figure 6.13).

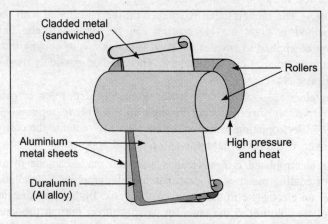

Figure 6.13 Cladding of duralumin alloy with aluminium metal.

For example: duralumin (alloy of Al = 95%, Cu = 4%, tin = 0.5% and Mg = 0.5%) cladded on both sides with pure Al, which is wildly used in aircraft industry and automobile or locomotive industry.

6.10.6 Galvanization (Sacrificial Protection)

The process of covering an iron metal with zinc metal is called *galvanization*. The galvanized iron forms a layer of $ZnCO_3$ and $Zn(OH)_2$ on the zinc film which protects the zinc layer and even iron metal from corrosion. The galvanization is done by any of the following techniques: (i) by spraying the molten zinc, (ii) by dipping the parent metal in the molten zinc or (iii) *sherardising* (spraying the zinc powder and then strong heating).

6.10.7 Tinning

"The process of coating of thin film of tin metal as a protective layer on the parent metal like iron is called *tinning*".

In this technique, the parent metal first cleaned with dilute H_2SO_4 to remove the impurities like oxide layer on the metal surface then passed through a tub containing the zinc chloride solution which helps the molten tin to adhere on the metal surface. Now the metal sheet is passed through a tank having molten tin where a thin film of tin is coated on metal surface. After this coating it is passed through a series of rollers sprayed with palm oil. Roller removes any excess of tin left uncoated and palm oil prevents the coated tin from oxidation. Thus, a continuous, protective layer of tins deposited on the parent metal, e.g. iron. This tin layer prevents the metal from the corrosion. Moreover, tin is a non-toxic metal and, therefore, it is used in the coating of steel used in cooking utensils and refrigerator components.

6.10.8 Cathodic Protection or Electrical Protection or Sacrificial protection

This method is mainly used to protect the underground pipelines and tanks. This is of two types:

(a) *Sacrificial anodic protection:* In this method, the iron pipes or articles are connected with more active metals like magnesium or zinc through a wire. The more active metal has lower reduction potential and hence acts as sacrificial anode. This oxidizes in preference to iron and protects the iron from rusting. The anode gradually disappears due to the oxidation process and is replaced from time to time (Figure 6.14).

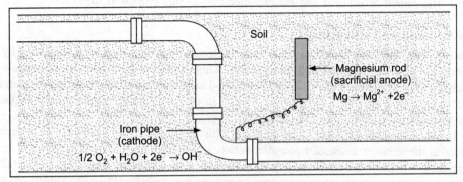

Figure 6.14 Sacrificial anodic protection

(b) *Impressed current cathodic protection:* In this method, the iron article (pipeline, etc.) is connected to the negative terminal of the battery and opposite terminal (positive) is connected to inert anode (graphite or high silica iron). The anode is usually buried in a sufficient strength. Impressed current is applied to the iron pipe which reverses the direction of corrosion to the cathode instead of anode. Hence, the metal articles are protected from corrosion (Figure 6.15).

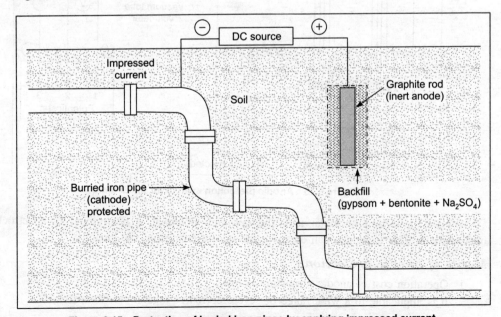

Figure 6.15 Protection of buried iron pipes by applying impressed current.

This method is useful to protect the underground water pipelines, oil pipelines, water tanks, transmission towers, etc. from corrosion.

Limitations of cathodic protection

1. It protects the iron pipelines effectively but simultaneously it may increase the corrosion of adjacent pipelines because of the stray current.
2. If the hydrogen is produced during cathodic reaction, it may have the blistering effect on metal itself.
3. The capital investment and maintenance cost of this process is high.

6.10.9 Anodic Protection

In this method, the metal article which is to be protected from corrosion is made passive by applying current in a direction that makes it more anodic. This method is applicable to only metals or alloys which shows *active–passive behaviour*. This method has been applied to protect the steel and stainless steel, and to some extent Fe, Al and Cr. Similar to that of cathodic protection this method is also used in electrolytic corrosive environment.

In this technique, a potentiostat is fixed with the tank. The potentiostat is a device which maintains the metal at a constant potential between reference electrode and metal tank or article to be protected. It has three terminals—first one of which is connected to the tank or metal article to be protected, second terminal is connected to an auxiliary cathode (Pt or Pt cladded electrode) and third terminal is connected to a reference electrode (usually calomel electrode) (Figure 6.16).

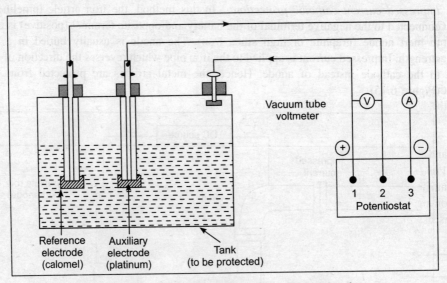

Figure 6.16 Anodic protection system for metal tank.

For the protection of metal tank the optimum potential is determined by electrochemical measurement. This technique can decrease the rate of corrosion, significantly.

Merits of anodic protection

1. Operation cost of anodic protection is low.
2. It is applicable to a wide range of severe corrosive medium/environment.

3. It needs only few auxiliary electrodes.
4. It can protect even complex structures.
5. Corrosion rate can be measured by measuring corrosion current.
6. Feasibility of process can be measured in the laboratory.

Disadvantages of anodic protection
1. It requires high starting current for protection of metal from corrosion.
2. It is suitable for only those metals/alloys which show the active–passive behaviour.
3. Installation cost is high.
4. In uncontrolled condition, severe corrosion may occur.

6.11 INHIBITERS

"Corrosion inhibiter is a substance which reduces the corrosion of metal when added in small quantity to the corrosive environment." Inhibiters are of three types:

(a) *Anodic inhibiters:* These inhibiters react with metal ions and form insoluble precipitate on anodic area of the metal (iron) surface and decrease the rate of corrosion. These inhibitors are phosphates, chromates, tungstate, molybdate, etc.

For example, when an iron article is dipped into boiling solution of alkaline phosphate, the H^+ ions (necessary for the oxidation of Fe to Fe^{2+}, $H_2O \longrightarrow OH^- + H^+$) get neutralized. Thus, the oxidation (rusting) of iron is prevented.

The anodic inhibitors are used to protect different iron parts of engine and car radiator.

(b) *Cathodic inhibiters:* These are the organic substances like amines, substituted urea, thiourea, mercaptanes, heavy metal soaps, etc. which when added to the environment of metallic article get absorbed over the cathodic layer and hence decreases the rate of corrosion.

Some inorganic oxides (arsenic and antimony oxide) are used as cathodic inhibitors as they form adherent layer of As and Sb at the cathodic surface. This metallic film increases the hydrogen over voltage for hydrogen evolution:

$$2H^+_{(aq)} + 2e^- \longrightarrow H_2(g)$$

and thus, prevents the metal from further corrosion.

(c) *Vapour phase inhibiters:* These are the organic compounds (e.g. dicyclohexyl ammonium nitrite) which directly changes into vapour phase, i.e. sublimate to form a protective film on the metal surface. Thus, these inhibitors protect the metal from corrosion. This method is used to protect the parts of sophisticated instruments and equipments from corrosion.

SUMMARY

Corrosion: Corrosion may be defined as the slow destruction or deterioration of metal/metallic material on its surface due to unwanted chemical or electrochemical attack of its environment (atmospheric gases, moisture, etc.). Example—rusting of iron.

Electrochemical cell/galvanic cell: A galvanic cell is a device in which the chemical energy is converted into electrical energy on account of some chemical or physical change within the cell.

Concentration cell: "The cell in which two electrodes of same metal are immersed in solutions containing different concentrations of its ions and connected with a wire or salt bridge is known as *concentration cell*".

Electrochemical series (activity series): The arrangement of elements in order of increasing reduction potential values is called as *electrochemical series*.

Dry or chemical corrosion: The corrosion which occurs due to direct chemical action of environment/atmosphere gases such as O_2, H_2, Cl_2, N_2, SO_2, H_2S, etc. or anhydrous inorganic liquid with metal surfaces in immediate proximity is called *dry corrosion*.

Pilling Bedworth's rule: It states that an oxide layer is protective or non-porous if the volume of oxide is greater than the volume of metal from which it is formed. On the other hand, if the volume of oxide is less than the volume of metal the oxide layer is porous (non-protective), i.e.

$$\text{Specific volume ratio} = \frac{\text{Volume of metal oxide}}{\text{Volume of metal}}$$

Electrochemical or wet corrosion: "The corrosion which takes place under wet or moist conditions due to the formation of short circuited electrochemical cell is called *electrochemical corrosion*. Example—rusting of iron.

Galvanic corrosion: The corrosion which takes place when different metals are in contact (either directly or through an electrical conductor) and jointly exposed to corrosive environment is called as *galvanic corrosion*.

Pitting corrosion: The corrosion which takes place due to the breakdown or cracking, i.e. formation of pits, cavities, crevices or pin holes on metal surface is called *pitting corrosion*.

Stress corrosion or stress cracking: The corrosion which occurs due to the combined effect of tensile stresses and the corrosive environment on metal/alloy is called *stress corrosion*.

Caustic embrittlement: The boiler corrosion which takes place in the high pressure boiler due to the use of highly alkaline water softened by lime–soda process.

Concentration cell corrosion: The corrosion which takes place due to an electrochemical attack on metal surface when exposed to an electrolyte of varying concentration or varying aeration called *concentration cell corrosion*.

Differential aeration corrosion: This types of corrosion takes place when one part of metal is exposed to a different air concentration from the other parts of the metals.

Water line corrosion: The corrosion which takes place due to the formation of concentration cell near water line due to the different concentration of air (O_2) at different portions of metal blow water line.

Soil corrosion: The corrosion of the metal articles, pipelines, etc. takes place due to the electrochemical attack of moisture and salts present in the soil is called *soil corrosion*.

Microbiological corrosion: The corrosion which takes place due to the attack of bacteria present in water or soil on metal surface is called *microbiological corrosion*.

Factors affecting rate of corrosion: (a) *Nature of metal*: Purity of metal, physical state of metal, passive character of metal, position in the galvanic series, relative anodic and cathodic area, nature of corrosion film/layer and solubility of corrosion product. (b) *Nature of environment*: Temperature, humidity, pH of the medium, presence of impurities in the atmosphere, presence of suspended particles in the atmosphere, conductance of corroding medium, nature of electrolyte, formation of oxygen concentration cell.

Barrier protection: Usually this method is used to prevent the rusting of iron. In this method, a suitable barrier is placed between iron and atmospheric air.

Metallic coatings: It is carried out by: hot dipping, metal spraying, electroplating/electrodepositing.

Metal cladding: It is the process of making sandwich of base metal (metal to be protected) between two layers of cladding materials (usually pure Al metal) and then passing this combination through rollers under the action of heat and pressure to form a clad sheet.

Cathodic protection or electrical protection or sacrificial protection: This method is mainly used to protect the underground pipelines and tanks. It is of two types: sacrificial anodic protection and impressed current cathodic protection.

Anodic protection: In this method, the metal article which is to be protected from corrosion is made passive by applying current in a direction that makes it more anodic. This method is applicable to only the metals or alloys which show *active–passive behaviour*.

Inhibiters: A substance which reduces the corrosion of metal when added in small quantity to the corrosive environment. Inhibiters are of three types: anodic, cathodic and vapour phase inhibiters.

EXERCISES

1. What is corrosion of metals? Explain the causes of corrosion and its consequences.
2. Explain the various types of corrosion in detail by giving suitable examples.
3. What is dry corrosion? Explain the role of oxide film in the dry corrosion.
4. What do you understand by chemical corrosion? Explain its mechanism.
5. What is wet corrosion? Describe the mechanism of electrochemical corrosion by:
    (i) hydrogen evolution and
    (ii) oxygen absorption.
6. What is electrochemical corrosion? Describe the mechanism of electrochemical corrosion. Discuss the role of oxygen in corrosion cells.
7. Discuss the principle of electrochemical corrosion by the example of rusting of iron.
8. Write brief notes on the following:
    (i) galvanic corrosion
    (ii) pitting corrosion
    (iii) stress corrosion
    (iv) differential aeration corrosion
    (v) water line corrosion
    (vi) soil corrosion
    (vii) boiler corrosion
    (viii) microbiological corrosion
    (ix) concentration cell corrosion
    (x) crevices corrosion.
9. Discuss the various factors which influence the rate of corrosion.
10. Explain the following factors influencing the corrosion rate:
    (i) nature of corrosion product
    (ii) the ratio of anodic to cathodic area.
11. Discuss the effect of the following factors on the rate of corrosion:
    (i) position of metal in the electrochemical series
    (ii) physical state of metal
    (iii) nature of corrosion products

12. What is corrosion? Briefly discuss the various methods employed for protection of metals from corrosion.
13. Explain the following techniques for the protection of metal from corrosion:
    (i) proper design of metal articles
    (ii) cathodic protection
    (iii) anodic protection
    (iv) electroplating
    (v) tinning
    (vi) galvanization
14. Explain the term cathodic protection. Indicate how metal coatings can effectively prevent corrosion. How is galvanization different from catholic protection?
15. Distinguish between the following:
    (i) dry corrosion and wet corrosion
    (ii) electrochemical series and galvanic series
    (iii) corrosion and erosion
    (iv) cathodic protection and anodic protection
    (v) galvanization and electroplating
    (vi) cathodic inhibitors and anodic inhibitors
16. How does iron corrodes in neutral or alkaline medium?
17. What is soil corrosion? Explain the conditions for soil corrosion and its preventive measures.
18. What is microbiological corrosion? Give the names of different micro-organism which cause the microbiological corrosion.
19. Write short notes on:
    (i) concentration cell corrosion
    (ii) pitting corrosion
    (iii) galvanic corrosion and its control
    (iv) gasket corrosion
20. What are the advantages and limitations of anodic protection?
21. Write in brief about electrochemical series and galvanic series.
22. What are cathodic and anodic protections for controlling corrosion? Discuss their merits and demerits.
23. Explain:
    (i) boiler corrosion
    (ii) intergranular corrosion
24. How are metals protected against corrosion by modifying the environment?
25. Write about the steps and paths for cathodic evolution of hydrogen.
26. What is pitting corrosion? Explain the preventive measures for it.
27. What are inhibitors? Explain its types with examples.
28. What is stress corrosion? Give two examples. How can it be controlled?
29. What are corrosion inhibitors? How anodic and cathodic inhibitors provide protection against corrosion? Explain with examples.
30. Distinguish between anodic and cathodic inhibitors.
31. Discuss any two factors which influence the corrosion rate.

32. Explain the cathodic protection by impressed emf method or sacrificial anode method.
33. Discuss the effect of:
    (i) temperature and
    (ii) nature of corrosion product on the rate of corrosion of metals.
34. Define corrosion. Explain losses due to corrosion. Give the classification of corrosion according to environment or surroundings.
35. What are corrosion inhibitors? Classify different types of inhibitors with examples.
36. What is corrosion? Discuss the corrosion caused due to combination of metals of different electrode potential.
37. Discuss the importance of:
    (i) design and
    (ii) selections of materials in controlling corrosion.
38. Mention the major corrosion causing substances and how are they eliminated?
39. What is galvanic series? Discuss its applications.
40. Discuss the factors influencing atmospheric corrosion? How design and material selection help to control metallic corrosion?
41. What are the conditions for dry and wet corrosion?
42. Name two metals which are noble with respect to corrosion.
43. What is Pilling–Bedworth's rule? Explain its role in the protection of metal from corrosion.
44. Give the mechanism of dry corrosion. Explain the role of oxide film in dry corrosion and classify them.
45. Write short note on "atmospheric corrosion" under different conditions.
46. Explain the rusting of iron with the help of electrochemical theory of corrosion.
47. What is meant by differential aeration corrosion? Illustrate with suitable examples.
48. What is cathodic protection? How is it done by using impressed current and sacrificial anode? Explain with suitable examples.
49. How do the following factors influence the rate of corrosion:
    (i) Polarization and
    (ii) Electrode potential?
50. Explain the following methods of corrosion control:
    (i) Corrosion protection
    (ii) Galvanization
    (iii) Cladding
51. What is sacrificial anode? Mention its role in corrosion control.
52. Explain the differential aeration corrosion. Illustrate with an example.
53. What is sacrificial anode? How does it protects a submerged pipeline?
54. Write notes on:
    (i) Galvanic series
    (ii) Design and material selection in minimizing corrosion
    (iii) Corrosion inhibitors
55. Explain how the corrosion can be controlled by sacrificial anode and impressed emf methods.
56. State the two conditions for wet corrosion to take place. Comment on the use of aluminium in place of zinc for cathodic protection of iron from rusting.

57. Write shorts notes on factors influencing atmospheric corrosion.
58. Explain the different factors affecting the rate of corrosion.
59. Write short notes on (or explain):
    (i) Anodic and cathodic inhibitors (or corrosion inhibitors)
    (ii) Anodic protection
60. What is cathodic and anodic protection for controlling corrosion? Discuss their merits and demerits.
61. Give a brief account of cathodic protection method of preventing corrosion or corrosion control.
62. How are metals protected against corrosion by modifying the environment?
63. Give reason of the following:
    (i) Iron sheets riveted with copper corroded faster.
    (ii) A zinc article is under strain.
    (iii) Zinc plate fixed below the ship.
    (iv) Iron corrodes faster than aluminium, even though iron is placed below aluminium in the galvanic series.
    (v) Wire mesh corrodes faster at the joints.
    (vi) Rusting of iron is quicker in saline water than in ordinary water.
    (vii) Corrosion occurs in steel pipe connected to copper plumbing.
    (viii) Bolt and nut made of the same metal is preferred in practice.
    (ix) Part of a nail inside the wood undergoes corrosion easily.
    (x) Nickel plated steel articles should be free from pores and pin holes.
    (xi) Deposition of extraneous matter on metal surface for a long period is undesirable.
    (xii) Corrosion can be considered as the reverse of the process of metal extraction.
    (xiii) Impure metals are more susceptible to corrosion than pure metals.
    (xiv) Chromium is used for coating iron.
    (xv) Silver and copper metals do not undergo much corrosion like iron in moist atmosphere.
    (xvi) Electrochemical series gives a basis for predicting a process of corrosion of a metal.
    (xvii) Though aluminium has high standard oxidation potential than iron, yet aluminium corrodes to a much small extent.
    (xviii) The oxidation resistance of iron can be much improved by alloying it with chromium in different proportions.
    (xix) Galvanized utensils not used for cooking.
    (xx) Brass utensils are usually tinned.
    (xxi) Galvanization of iron articles is preferred to tinning.
    (xxii) Coating of zinc on iron is called sacrificial anode.
    (xxiii) Stainless steel is not used to build a sea-going ship.
    (xxiv) Corrosion of water filled steel tanks occurs below the water line.
    (xxv) Rusting of iron is quicker in saline water than in ordinary water.
    (xxvi) Silver and copper do not undergo much corrosion like iron in moist atmosphere.
    (xxvii) Metal shows tendency to undergo corrosion.

CHAPTER 7

LUBRICATION AND LUBRICANTS

7.1 INTRODUCTION

The usefulness of lubricants was known from the ancient time. Chariot drivers of 1400 BC were aware about the lubrication. Leonardo da Vinci (1452–1519) revealed the fundamental principal of friction and lubrication, and described the effect of lubrication on the coefficient of friction between two rubbing surfaces. The frequent use of lubricants at large scale in automobile and machinery was started since 1947.

7.2 FRICTION

The relative motion of two moving solid surfaces causes the friction and wear. This frictional resistant is due to the roughness and unevenness of moving surfaces. The roughness of surface may be due to the presence of asperities and cavities. The frictional resistance between moving surfaces may arise due to mechanical interlocking, molecular attraction or electrostatic forces of attraction.

7.3 LUBRICANTS

"Any substance which is introduced between two moving/sliding surfaces to reduce the frictional resistance between them is known as lubricant."

"The process of reducing frictional resistance between moving/sliding surfaces by the use of lubricants in between them is called lubrication."

7.3.1 Functions of Lubricant

1. Lubricant reduces surface deformation, wear and tear as it avoids the direct contact between the surfaces.
2. It minimizes the loss of energy in the form of heat, i.e. it acts as a coolant.
3. It enhances the efficiency of machine as it prevents the wastage of energy.
4. It decreases the expansion of metal by local frictional heat.
5. It avoids the seizure (self-welding) of moving surfaces.
6. It reduces the maintenance and running cost of a machine.

7. It sometimes also acts as a seal. For example, between piston and cylinder of an IC engine.

7.4 MECHANISM OF LUBRICATION

There are three types of mechanism by which lubrication is done:
1. Fluid film or thick film or hydrodynamic lubrication.
2. Thin film or boundary line lubrication.
3. Extreme-pressure lubrication.

7.4.1 Fluid Film or Thick Film or Hydrodynamic Lubrication

In this lubrication, the moving/sliding surfaces are separated from each other by a thick film of lubricant by at least 1000 Å thickness. This thick film of lubricant covers the pits, asperities, etc. and avoids the direct contact of moving surfaces by maintaining them smooth as well as avoids the welding of the joint. The resistance to movement is only due to the internal friction of lubricant (Figure 7.1).

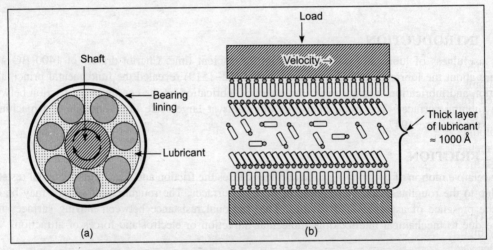

Figure 7.1 Fluid–film lubrication mechanism (a) Journal bearing and (b) Moving surfaces.

This lubrication mechanism is required in case of a shaft running at fair speed and with moderate load. In such cases, the coefficient of friction is very low (0.001–0.03). Delicate instruments/light machines like watches, clocks, gun, sewing machines, scientific instruments, etc. are provided with this type of lubrication.

Hydrocarbon oils and their blends with selected long chain polymers are suitable and satisfactory lubricants for this mechanism.

7.4.2 Thin Film or Boundary Line Lubrication

This type of lubrication is required when a continuous fluid film of lubricant cannot persist and direct metal to metal contact is possible in case:
1. shaft starts moving from rest and experience jerk.
2. at very low speed.

3. at very high load.
4. viscosity of oil is too low.

Under these conditions the lubrication is maintained by a boundary film in which a thin layer (<1000 Å) of oil of moderately high viscosity is adsorbed by physical or chemical forces on the metal surfaces, that avoids the direct contact of moving surfaces. The coefficient of friction in such cases is usually 0.05–0.15 (Figure 7.2).

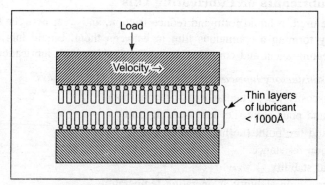

Figure 7.2 Boundary line lubrication mechanism.

The lubricants like polymers of long hydrocarbon chains, possessing polar and active groups or atoms with high viscosity index, good oiliness and high heat and oxidation resistance are best suited for boundary line mechanism.

Vegetable oils, animal oils and their soaps or their blends with graphite or molybdenum disulphide or solid lubricants alone are used for boundary line lubrication as they are adsorbed on the metal surface either physically or chemically.

This mechanism is used in the ball bearings with high load, gears of tractors, rollers, etc.

7.4.3 Extreme-Pressure Lubrication

This type of lubrication is used under the conditions where the moving/sliding surfaces are under very high pressure and speed, excessive heat is generated and a high local temperature is attained. Under such conditions the ordinary fluid film or boundary mechanism is not successful to prevent metal to metal contact as liquid lubricants fail to stick and may decompose and even vaporize. Hence, to fulfill these requirements, special additives are added to mineral oils. These are called *extreme pressure additives*. These additives react with metal to form more durable layer on metal surface which is able to bear very high load and high temperature, moreover, it has self-healing tendency.

Some organic compounds having active radicals or groups such as chlorine as chlorinated esters, sulphur as sulphurized oils and phosphorous as tricrecyle phosphate are very important and commonly used extreme pressure additives. These compounds react with metal surfaces at existing high temperatures to form metallic chlorides, sulphides or phosphides, respectively. These compounds possess high melting point ($FeCl_3$ = 650 °C, FeS = 1100 °C) and serve as good lubricants under extreme pressure and temperature conditions.

This mechanism is used in wire drawing machine, lathe machine, car rear axle, etc.

7.5 CLASSIFICATION OF LUBRICANTS

Lubricants can be classified into following categories on the basis of physical state:
1. Liquid lubricants or lubricating oils
2. Semi-solid lubricants or greases
3. Solid lubricants

7.5.1 Liquid Lubricants or Lubricating Oils

These lubricants are used in liquid form and reduce friction, and wear between two moving/sliding metallic surfaces by forming a continuous film in between them. Liquid lubricants also serve as cooling medium, sealing agent and corrosion preventor together with lubrication.

To be a good and satisfactory lubricant the lubricating oil must possess:
1. Adequate viscosity for particular working conditions.
2. Low cloud and pour point (i.e. low freezing point).
3. High flash and fire point (i.e high boiling point).
4. High oxidation resistance.
5. High thermal stability.
6. High decomposition stability at operating temperature.

Liquid lubricant can further be classified as:
1. Animal and vegetable oils
2. Mineral or petroleum oils
3. Blended oils

(a) *Animal and vegetable oils:* These oils were the most commonly used lubricants in the ancient time before the discovery of petroleum as they possess good oiliness. But they are not suitable for direct use because, they: undergo oxidation easily, form gummy and acidic products when come in contact with air and dust and have some tendency to hydrolyze by moisture/water.
Examples: Animal oils—lard oil, neat foot oil, tallow oil, fish oil, etc.
Vegetable oils—caster oil, palm oil, olive oil, rape seed oil, etc.

(b) *Mineral or petroleum oils:* These oils are basically low molecular weight hydrocarbons (12–50 carbon atoms). These are obtained by fractional distillation of petroleum. These are most widely used lubricants, because, they are: available in abundance, cheap and quite stable under service conditions in comparison to animal and vegetable oils. These mineral oils possess low oiliness, so some high molecular weight compounds like oleic acid and stearic acid, etc. are mixed with petroleum oils to increase oiliness.
Purification: Crude petroleum oils contain impurities like wax, naphthenic asphalt, etc. therefore; it is required to purify them before use. These impurities are removed off by de-waxing, acid refining and solvent refining.

(c) *Blended oils:* No any oil alone can fulfill all the requirements to serve as most satisfactory lubricant for the most of the machines. Therefore, some specific additives are added to oils to improve their properties. The oils so obtained are called *blended oils*. They meet all the requirements to be satisfactory lubricants.

Additives to the lubricants

An additive is a material which imparts a new or desired property to the lubricating oils. Additives are mainly of two types: chemically active additives and chemically inert additives. These additives are summarized in Table 7.1.

Table 7.1 Various additives and their effects on properties of lubricants

S. No.	Additive type	Functions	Typical examples
I.	Performance improver additives		
(i)	Oiliness carriers	Increase sticking tendency to the metal surfaces and prevent metal to metal contact.	Vegetable oils (coconut oil, caster oil) and fatty acids (palmitic and stearic acids).
(ii)	Viscosity index improver	These materials expand with increase in temperature and decrease the rate of viscosity change with temperature.	Polymers and co-polymers of olefins, alkylated styrenes, methacrylates and butadiene.
(iii)	Pour point depressant	Reduce interlocking by modifying wax crystal formation and enable lubricants to flow at low temperature.	Alkylated naphthalene, poly-methacrylates and phenolic polymers.
(iv)	Swelling agents	Cause swelling of elastomers by chemical reactions and act as seal.	Aromatic hydrocarbons and organic phosphates.
II.	Lubricant protective additives		
(i)	Anti-oxidant	Retard the oxidative decomposition by terminating the free radical chain reaction and decomposing peroxides.	Aromatic amine, phenols, sulphides, etc.
(ii)	Anti-foaming agents	Prevent the foam formation by lubricants by reducing the surface tension of lubricant.	Silicon polymers.
(iii)	Metal deactivator	Inactivate the metal by forming an inert layer of metal complex on metal surface and decrease the catalytic effect of metal on oxidation rate.	Amines, sulphides and phosphides.
III.	Surface protective additives		
(i)	Anti-wear and extreme pressure agents	Reduce friction, wear and tear, prevent scaring and seizure by forming a film of chemicals to prevent metal to metal contact.	Chlorinated esters, organic phosphate, sulphurized oils and zinc dithiophosphates.
(ii)	Friction modifier	Change coefficient of friction by adsorbing on metal surface.	Organic phosphorous and phosphoric acid esters of high molecular mass.
(iii)	Rust and corrosion inhibitors	Prevent rusting and corrosion of metal parts in contact with lubricant by neutralizing the corrosive acids, polar constituent, etc. on metal surface.	Metal phosphates, basic metal sulphonates, fatty acids and amines.

7.5.2 Semisolid Lubricants or Greases

"Semisolid lubricant or grease may be defined as a semisolid combination of thickening (gelling) agent like soap dispersed throughout liquid lubricating oil." The liquid lubricant may be petroleum oil, synthetic oil or even its blends depending on the specific requirements.

Greases are essentially thixotropic gels in which the metallic soap is gelling agent and the liquid phase is the lubricating oil (dispersion medium). The commonly used gelling agents are soaps of Ca, Li, Na, Al, Ba, Mg and K as well as fatty acids.

Preparation

Most of the common greases are prepared by saponification of oil/fat with alkali (lime, Na_2CO_3, etc.) followed by adding the hot lubricating oil and soap thickeners with constant stirring. The greases have gel-like structure and soaps are gelling agents. Consistency of greases may vary from a heavy viscous liquid to a stiff solid mass.

Properties

Properties of greases depend upon the nature and amount of thickener used, characteristics of base oil and way of preparation of the grease.

1. They have higher resistance to shear (cut) and rubbing than oils and therefore, can hold much heavier load even at lower speed.
2. They show higher coefficient of friction than oil.
3. On storage they tend to separate into oil and soap.
4. On continuous use oil get vaporized off leaving behind the hard stuff.

Uses

Greases are used in the following situations:

1. Where oils cannot remain in place due to high load, low speed, intermittent operation, sudden jerks, etc. (e.g. rail axle boxes).
2. In bearing and gears working at high temperature and pressure.
3. Where bearing needs to be sealed against entry of dust, dirt, grit or moisture.
4. Where dripping, splitting or spurting of oil is undesirable, e.g. machines preparing paper, textiles, edible articles, etc.
5. Where oil cannot be maintained in position due to bad seal or intermittent operation.

Classification of greases

Greases are classified on the basis of soap used in their manufacture. Important greases are:

(a) *Calcium soap-based greases or cup greases:* These are the emulsions of petroleum oils with calcium soap (10–30%) and are prepared by adding required quantity of calcium hydroxide to hot oil under agitation. They can be prepared over a wide range of consistency, i.e. soft paste to hard smooth solid by changing the composition of the lime soap.

These are the cheapest and most widely used greases. They can be used satisfactorily at low temperatures (below 80°C). These are water resistant and suitable for the lubrication of water pumps, tractors, etc.

(b) *Sodium soap or soda-based greases:* These are prepared by thickening the petroleum oils by mixing sodium soaps. As the sodium soap is water soluble, these cannot be used in contact of water or moisture. They possess high temperature properties and can be used up to 175°C temperature. They are used for lubrication of ball bearings which get heated due to friction.

(c) *Lithium soap greases:* These are prepared by thickening the petroleum oils with lithium soaps. They possess good water resistance, high mechanical stability and oxidation resistance. They are suitable for use at low temperatures only (15°C). Due to high cost they are used in special applications only.

(d) *Aluminium-based greases:* These are prepared by dispersing aluminum soap into hydrocarbon oil. They possess excellent stringiness, adhesion and good clarity. They are relatively water proof. They can be used only up to 90°C temperature. They are used where the adhesiveness is of the prime importance.

(e) *Rosin soap greases or axel greases:* These are prepared by thickening the rosin oil with slurry of slacked lime at 58°C. The resulting grease called cold set grease. They are mainly used as axel grease for farm wagon and low speed machinery. They are very cheap resin greases as they are mixed with cheap fillers like talc, mica and sand to increase load bearing capacity. They are water resistant and applicable under high load and low speeds conditions, e.g. in street railway tracks.

(f) *Non-soap greases:* Prepared by using non-soap thickeners such as carbon black, silica gel, modified clay, organic dyes, etc. They can be used at high temperature and pressure.

(g) *Silicon greases:* Prepared from synthetic lubricating oil of silicon polymers. They possess high V.I., high oxidation resistance, high heat resistance and good water resistance.

7.5.3 Solid Lubricants

The lubricants which are used either in the dry powder form or mixed with oil or water are called *solid lubricants*. These lubricants reduce friction by separating two moving surfaces under boundary conditions. These are used where:

1. Machine working with very high load and low speed.
2. Lubricating oils or greases cannot maintain a lubricating film.
3. Contamination of lubricating oils or greases by dirt, dust, grit, etc. is unacceptable like in commutator bushes of electric generators and motors.
4. The operating temperature or load is too high, even for greases to remain in position.
5. The parts of machine to be lubricated are not easily accessible.
6. Combustible lubricants are avoided.

There are three types of solid lubricants: (i) chemically active, e.g. phosphate, chromate and oxidizing agents, (ii) structural, e.g. graphite, MoS_2, talk and mica, (iii) mechanical, e.g. metals and plastics, (iv) soaps.

Most common solid lubricants are: (a) graphite and (b) molybdenum disulphide. These materials have laminar structure.

(a) *Graphite:* It possesses meshwork of hexagonal carbon rings separated from upper layer of unit crystal cell by a sufficient distance of about 6.79 Å and are held by weak van der Waal's forces. These hexagonal layers can readily slide over each other (Figure 7.3(a)). It has low coefficient of friction. Therefore, graphite is soft and acts as lubricant. It can be used up to very high temperature (375°C) in absence of air.

(b) *Molybdenum disulphide (MoS_2):* It has sandwich like structure in which a layer of molybdenum atoms lie between two layers of sulphur atoms which are 6.26 Å apart from each other. These layers can slide over each other like graphite so this is soapy in touch and therefore used as a lubricant. It can be used satisfactorily up to 400°C temperature (Figure 7.3(b)).

Uses

1. These are used in air compressors, lathes and other machine shop operations and equipments used for food processing.
2. Graphite and MoS_2, both are used as dry powder or as an aerosol, aquadag (dispersion in water) or oildag (dispersion in oil).
3. These are particularly valuable at high temperatures and extreme pressure conditions.
4. These are also used to prepare the oilless bearings.
5. The mixture of graphite (7%) and molybdenum disulphide (70%) bonded with silicates (23%) are useful as lubricants in space vehicles.

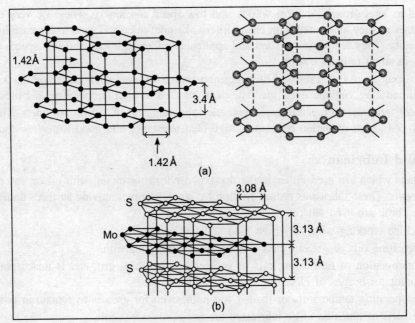

Figure 7.3 (a) Layer structure of graphite and (b) Sandwich-like structure of MoS_2.

Teflon is also a good solid lubricant and used in gasoline gear pumps and underwater machines.

7.6 EMULSIONS

"An emulsion is a mixture of two or more immiscible liquids which is obtained by thoroughly shaking these liquids together to get homogeneous mixture."

Or

"An emulsion is a two-phase system of dispersed phase (fine droplets) and dispersion medium obtained by thoroughly shaking them in the presence of an emulsifier."

In various mechanical operations, e.g. threading, turning, milling and boring the machine parts/tools get heated to a very high temperature. At the cutting edge the pressure is very high ≈ 100000 psi and huge amount of heat is generated which may lead to the oxidation and rusting. In such cases, the overheating and consequent injury of the tool can be prevented by efficient cooling and simultaneous lubrication. This is usually done by using emulsion of oil in water.

Emulsions have a tendency to absorb dust, dirt, grit, etc. and result in the abrasion, wear and tear of lubricated machine parts.

7.6.1 Types of Emulsions

There are two types of emulsions which are used in lubrications. These are:
- (a) *Oil in water type or cutting emulsions:* These are prepared by mixing together an oil and water in the presence of emulsifying agents (3–20%, e.g. soap of alkyl sulphate, alkyl and aryl sulphonate, etc.). These emulsions are used as coolant cum lubricant for cutting tools and diesel motor pistons, etc.

Characteristics of cutting emulsions
1. Provide efficient lubrication as well as cooling.
2. Absorb the heat so as to prevent the wear and damage of the metal.
3. Wash away the fragment of the metal.
4. Reduce the working cost by achieving possible higher cutting speeds.

(b) *Water in oil type or cooling liquid emulsions:* These are prepared by mixing together water and oil in presence of emulsifying agents (1–10%, e.g. alkaline earth metal soap). These lubricating emulsions are used successfully for the lubrication of steam turbine and in compressors handling fuel gases.

Characteristics of cutting emulations:
1. They have higher viscosity than parent oils.
2. They provide efficient lubrication.
3. They reduce the maintenance cost of machines.
4. They can bear very high pressures.

7.7 SYNTHETIC LUBRICANTS

"The lubricants which are prepared or synthesized in the laboratory are known as synthetic lubricants." Synthetic lubricants are much better than ordinary lubricants as they can meet the most drastic and severe conditions such as in the aircraft engines at –50°C to 250°C temperature.

7.7.1 Characteristics of Synthetic Lubricants
1. They can be used over a wide temperature range from –50°C to 250°C.
2. They have high thermal stability and resistance to hydrolysis.
3. They have high VI, high flash point and fire point.
4. They are non-inflammable.
5. They have high corrosion and oxidation resistance.
6. They have high chemical stability.

7.7.2 Important Synthetic Lubricants

(a) *Polymerized hydrocarbons:* These lubricants are prepared by the polymerization of olefin hydrocarbons. They possess high thermal and chemical stability and can withstand at very high temperature. They are less susceptible to oxidation. They are used as lubricant in ships and submarines. *Examples:* poly-ethylene, poly-propylene, poly-butylenes (medium weight range 500–50,000).

(b) *Polyalkylene glycols (PAG):* These are thermally stable materials, possess high shear resistance, high viscosity index, and are free from corrosive action. Hence, they can be used in rubber bearings and joints, compressor pumps, aircraft turbines at high temperatures but cannot be used in contact of water as they are water soluble. Examples: Polyethylene glycol and polypropylene glycol.

(c) *Polyglycidyl ethers:* These are cheap and widely used synthetic lubricants. They are water soluble and have high viscosity index, but they decompose at very high temperatures and undergo subsequent forming the jelly structures. They are used in pumps, gears, compressors, glass manufacturing machines, etc.

(d) *Organic amines and amides:* They are very good lubricants as they possess low pour point and high Viscosity Index (VI). They can be used within a wide temperature range of −50°C to 250°C.

(e) *Silicones:* These are very good synthetic lubricants because they possess high VI and are not oxidized up to 200°C. They are chemically more inert than other synthetic lubricants. These are low temperature lubricants and used in gears, bushes, bearings, clocks, etc. The most commonly synthetic silicone lubricants are dimethyl silicone and methyl phenyl silicone polymers.

(f) *Fluorinated and chlorinated hydrocarbons:* The chlorinated and fluorinated hydrocarbons and their compounds act as very good lubricants. They have high VI, high load bearing capacity, high thermal and decomposition stability, low inflammability, and are chemically inert. Chlorinated diphenyl compounds have extreme pressure lubricating property. Other examples are fluorinated tertiary amines, fluorinated ethers, esters, etc. They are used in boundary line mechanism in ships and submarine.

7.7.3 Uses of Synthetic Lubricants

These are used under the following conditions:

1. In hot running bearings and hot rolling mills.
2. In metal forming processes, e.g. die castings.
3. In air craft turbines where low viscosity and high temperature conditions are required.
4. In reactive environments.

7.8 BIODEGRADABLE LUBRICANTS

In recent years, there has been an increasing demand for developing such lubricants which after disposal should not harm the environment, i.e. they should be environmentally friendly. They should be biodegradable under a variety of environmental conditions (e.g. aerobic, anaerobic, freshwater, marine) and should be nontoxic. This has been driven by the need to comply with environmental legislation emerged from the concern over the accumulation of hydrocarbon compounds in the sediments of the Prodensee lake in Switzerland and to meet the requirements of eco-labelling schemes, standards and customer specifications for 'environmentally acceptable' lubricants.

"The lubricants which degrade or decompose into simpler molecules by the biological actions like eating, digestion, hydrolysis, enzymatic action after use when comes in contact with microorganisms and which are non-toxic, are called *biodegradable lubricants*.

While, the process of degradation or decomposition of lubricants by biological action is called biodegradation of lubricants.

These lubricants basically composed of the following biodegradable polymers: *n*-alkanes up to C-44, branched alkanes, low molecular weight aromatics, cyclic alkanes, mineral oils, alkylated benzenes, PIB, PAO's polyalkylene glycols, vegetable oils (triglycerides), diesters, polyglycol ethers, polyethers and some additives.

7.8.1 Conditions of Biodegradation

Following are the essential requirements for biodegradation of lubricants:

1. The lubricant should contain biodegradable materials (like biodegradable polymer of low molecular mass with heteroatom or functional group, e.g. polylactides (PLA), polyanhydride, polyglycolide (PGL), polycaprolactone (PCL), etc.

2. The lubricant must be in contact with micro-organisms for biodegradation processes.
3. The lubricant must be susceptible to microbial attack.
4. The favourable environmental conditions e.g. amount of moisture, oxygen, concentration of salts and light, and optimum temperature, etc.

7.8.2 Mechanism of Biodegradation of Lubricants

Biodegradations of lubricants can take place in the following two steps:

(a) *Primary biodegradation:* In this step, micro-organisms present in the water, soil, solid waste or sewage encounter the material (polymer) as a source of energy and breakdown them into the simpler chemicals (oligopolymers, dimers and monomers) that they can be digested.

(b) *Final biodegradation:* In this step, the original substance is completely degraded under aerobic or anaerobic conditions into CO_2, H_2O, new microbial mass and non-degradable inorganic material, e.g. metal compounds.

Biodegradation occurs by a series of enzyme-catalyzed oxidation reactions. A terminal alkyl group is oxidized to an aldehyde, then to a long chain carboxylic acid. After repeated degradation through β-oxidation in the fatty acid cycle to acetic acid and a lower carbon number carboxylic acid. The acetic acid is oxidized in the Kreb's cycle to citric acid and carbon dioxide and water (Figure 7.4).

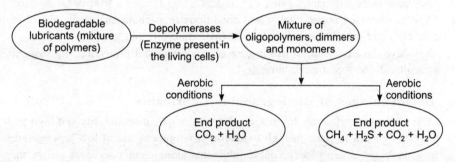

Figure 7.4 Reaction pathway of lubricant degradation.

The ideal biodegradable lubricants are those which can provide all the required performance characteristics. These characteristics are: (i) low temperature fluidity, (ii) thermal stability, (iii) oxidation stability, (iv) load bearing capacity, (v) non-emulsibility, (vi) non-inflammability, (vii) anti-wear and anti-corrosion, (viii) water repellant and, (ix) antifoaming ability.

Although there is no such lubricant available in the market which can fulfill all the above requirements. The hydrocarbon-based lubricants which are biodegradable, but cannot withstand extreme physical and chemical conditions.

7.8.3 Types of Biodegradable Lubricants

There are two types of biodegradable lubricants, i.e. natural and synthetic lubricants which are decomposed by the bacterial and environmental actions.

(a) *Synthetic biodegradable lubricants:* These are prepared from synthetic esters, poly-ethers and poly-hydroxyalkanoates like PHB–PHV.

(b) *Natural biodegradable lubricants:* These are prepared from vegetable and animal oils (e.g. lard oil, fish oil, olive oil, castor oil, etc.) and polymers particularly modified polysaccharides, e.g. starch, cellulose, and chitosan.

On the basis of their application, biodegradable lubricants can be classified into following categories:
  (i) *Hydraulic lubricants:* They are used in hydraulic systems, pumps, marine machinery, etc.
  (ii) *Gear lubricants:* They are used in high pressure gears, etc.
  (iii) *Anti-wear lubricants:* They are used in turbine, pumps, compressors, etc.
  (iv) *Wire line lubricants:* They are used in wire and cable machinery.

7.8.4 Advantages of Biodegradable Lubricants
1. They are biodegradable and do not leave any residue which can harm the environment.
2. They are less toxic to the environment.
3. They possess good oiliness, high Viscosity Index (VI) and can serve in high pressure conditions.
4. They have high flash and fire point, and are suitable in high temperature conditions.
5. They possess good vaporization stability.
6. They increase equipment service life by protecting the parts from wear and tear.
7. They can be used under total waste conditions (e.g. chainsaw, chain oils, 2-stroke engine oils, etc.) and possible accidental leaks (hydraulic oils and greases).
8. They have the advantage of use in case of environmental protection, e.g. aquatic, mountain, agricultural and forest environments.

7.8.5 Disadvantages of Biodegradable Lubricants
1. Most of the biodegradable lubricants possess low oxidation stability and high pour point and therefore cannot be used in high oxidation environment and at low temperatures.
2. In some cases where biodegradable lubricants alone cannot serve as satisfactory lubricants and require the mixing of additives. The additives are poorly biodegradable which reduce the overall degradability of the lubricants. Moreover, the additives increase the overall toxicity to the environment.
3. All conventional, chemically modified or genetically modified oils (tested) also have shown the same levels of biodegradability.

   In future, due to the development of enhanced and naturally more stable oilseeds, reducing prices and other benefits, they will become more prevalent in applications where environmental and safety concerns are high.

7.9 PROPERTIES OF LUBRICATING OILS

7.9.1 Viscosity and Viscosity Index (VI)
Viscosity is one of the most important properties of any lubricating oil because it determines the performance characteristics like oiliness, load bearing capacity, mechanical stability, flash and fire point, cloud and pour point, etc. of the lubricating oils.

"It may be defined as the resistance to the flow of liquid." It is the measure of the internal resistance to the flow of liquid which is due to the forces of cohesion between the liquid molecules.

Absolute viscosity: It may be defined as the tangential force per unit area required to maintain a unit velocity gradient between two parallel surfaces in fluid at unit distance. This force (F) is directly proportional to the area (A) and velocity gradient $\dfrac{dv}{dx}$
Mathematically,

Force,
$$F \propto A \cdot \frac{dv}{dx} = F = \eta \cdot A \cdot \frac{dv}{dx}$$

where dv is the difference in velocities and dx is the distance between two consecutive layers. The proportionality constant η is called *coefficient of viscosity*.

If velocity difference, $dv = 1$ cm s^{-1} and distance between two layers $dx = 1$ cm,

Then,
$$F = \eta$$

Hence, "the coefficient of viscosity may be defined as the force of attraction between the two layers moving at velocity difference of 1 cm s^{-1}, 1 cm apart from each other".

Unit of viscosity is poise or centipoises or dynes per seconds. The ratio of absolute viscosity to density for any liquid is known *kinematics viscosity*. Its unit is stokes or centistokes.

Absolute kinematics viscosity,
$$v = \frac{\eta \,(\text{absolute dynamic viscosity})}{\rho \,(\text{density})}$$

Effect of temperature on viscosity (viscosity index, VI)

Viscosity of liquids decreases with increase in temperature. The rate at which the viscosity of oil changes with temperature is known as *viscosity index (VI)*

$$\text{VI} = \frac{L-U}{L-H} \times 100$$

where
- U = viscosity of oil under test at 100°F.
- L = viscosity at 100°F of the low viscosity standard oil (Gulf oil) having a VI of 0 and also having the same viscosity at 210°F.
- H = viscosity of high viscosity standard oil (Pennsylvanian oil) at 100°F having a VI of 100 and also having the same viscosity at 210°F.

An oil whose viscosity changes rapidly with the change in temperature has low viscosity index. On the other hand, the oil whose viscosity changes only slightly with change in temperature has high viscosity index. The VI of oil can be increased by adding the linear polymers.

Experimental determination

Viscosity of lubricating oils is determined by Redwood viscometer (No. 1 and No. 2).

The Redwood viscometer (Figure 7.5) consists of the following parts: (i) oil cup, (ii) heating bath, (iii) stirrer, (iv) spirit level, (v) levelling screw and (vi) Kohlrausch flask.

A standard cylindrical oil cup made up of brass, has 90 mm height and 46.5 mm diameter. It is fitted with an agate jet in the base. The diameter of agate jet of Redwood No. 1 is 1.62 mm and length 10 mm and for No. 2 the diameter is 3.8 mm and length is 15 mm. The hole of agate jet is covered with valve ball. Inside the cup there is one pointer which indicates the oil level in the cup. The oil cup is surrounded with a stirrer and fitted in a heating bath. Thermometer is provided in the oil cup as well as in heating bath to measure the temperature of oil and water, respectively. The entire apparatus rests on a tripod stand provided with levelling screws (Figure 7.5).

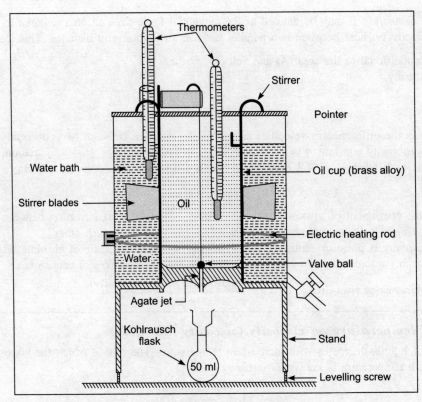

Figure 7.5 Redwood viscometer.

Working: The instrument is levelled with the help of levelling screws. The water is filled in water bath and oil cup is filled with oil up to the pointer. The Kouhlrausch flask of 50 ml capacity is placed below the jet. The oil and water are kept stirred and their respective temperatures are noted. The valve ball is turned from agate jet and oil is allowed to flow into the flask. Flow time for 50 ml of oil is noted with the help of stop watch and valve ball is then turned back to the original position to prevent the overflow of the oil. The time in seconds is equivalent to the Redwood viscosity.

Conversion of redwood viscosity into absolute unit

The Redwood (relative) viscosity can be converted to absolute viscosity (centistokes) with the help of the following equation

$$v = Ct - \frac{\beta}{t}$$

where v = kinematic viscosity in centistokes, t = time of flow in seconds and C and β are constants. The values taken for the constants are given in Table 7.2.

Table 7.2 Values of C and β constants

Instrument	Value of 'C'	Value of 'β'
Redwood No. 1	0.25	172
Redwood No. 2	2.72	1120
Saybolt Universal	0.22	180

7.9.2 Flash and Fire Point

"Flash point is the lowest temperature at which the lubricating oil gives off enough vapours that ignited to produce a flash for a moment when a tiny flame is brought near it."

"While the fire point is that lowest temperature at which the lubricating oil gives off enough vapours which burn (catch fire) continuously for at least five seconds when a tiny flame is brought near it."

Usually, the fire points are 5 to 40°C higher than the flash points.

To be a satisfactory lubricant the oil should have flash point at least above the operation temperature. This ensures the safety against fire hazards during the storage, transportation, handling and use of lubricating oil. Moreover, flash point of oil is used for the identification and detection of contamination in the lubricating oils. The minimum closed up flash point required for the turbine oil is 165°C and that for insulating oil is 146°C.

Experimental determination

The flash and fire points are determined by Pensky–Marten's apparatus (Figure 7.6). The apparatus consists of the following parts: (i) an oil cup, (ii) shutter, (iii) flame exposure device, (iv) air bath, (v) pilot burner.

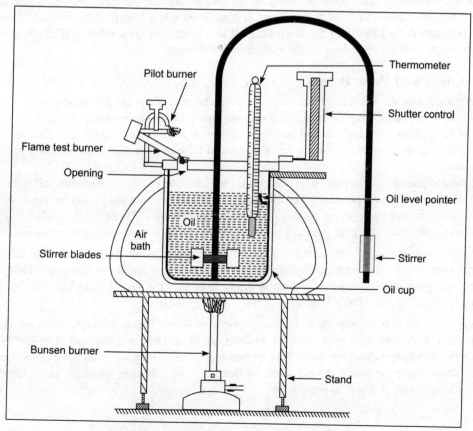

Figure 7.6 Pensky–Marten's flash-point apparatus.

An oil cup (5.5 cm height and 5 cm diameter) fitted with air bath, shutter control, flame exposure device (flame test and pilot burner) and mechanical stirrer placed in an air bath.

Working: The oil cup is filled with oil under test up to the mark. The lid of the oil cup fitted with stirring device, thermometer and flame exposure device fixed on the top. The test flame is lighted and adjusted to 4 mm in diameter. The oil in the oil cup is heated with the help of air bath and the temperature of oil is raised at the rate of 2°C per minute. When the temperature is reached near to the flash point say 30°C, the test flame is dipped into the opening and flash is checked for every two degrees rise of temperature.

When the vapours are ignited for a moment to produce a flash, the temperature is taken as flash point. The oil is further heated to fire point. At fire point the vapours ignited and burn for at least 5 seconds.

7.9.3 Oiliness

Oiliness is very important property of lubricating oil, particularly for boundary line and extreme pressure lubrication as it determines the capacity of oil to stick to surfaces of moving parts of the machine under the conditions of heavy pressure or load. Vegetable oils have good oiliness while mineral oils have very poor oiliness. There is no specific method devised for the experimental determination of oiliness, only relative oiliness is considered while selecting the lubricating oil.

To be a satisfactory lubricant the lubricating oil should possess good oiliness. The oiliness can be improved by adding some long chain hydrocarbon polymers.

7.9.4 Cloud and Pour Points

The petroleum oils do not have fixed freezing point because they are the mixtures of various compounds. When lubricating oil is allowed to cool slowly, it starts appearing cloudy.

"The temperature at which the oil becomes cloudy or hazy in appearance is called its *cloud point*. While, the temperature at which the oil ceases to flow or pour (i.e. stops flowing) is called *pour point*."

Cloud and pour point has great significance for the indication of oils applicability and suitability in cold conditions (low temperature). For example: In the refrigeration plants and air craft engines which may be required to operate at zero or sub-zero temperatures. To be good, a lubricating oil should have low cloud and pour points, i.e. it should not freeze at low operation temperatures.

Experimental determination: It is determined by cloud and pour point apparatus (Figure 7.7). The apparatus consists of a flat bottomed cylindrical jar, (12 cm height and 3 cm diameter). This jar is enclosed in a glass jacket (air jacket) which is firmly fixed in a ice bath (Figure 7.7). The ice bath contains freezing mixture (Ice + $CaCl_2$).

Working: The oil is filled into the test jar to a height $\approx$ 6 cm. The thermometers are introduced into the oil and ice bath. Due to cooling the temperature of oil falls, and a temperature is obtained at which the oil appears cloudy or hazy. This temperature is taken as *cloud point*. On further cooling a temperature reaches at which the oil seizes to flow, i.e. stops flowing when the jar is tilted for 5 s. This temperature is taken as *pour point*.

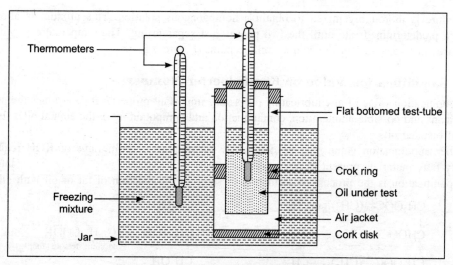

Figure 7.7 Cloud and pour point apparatus.

7.9.5 Aniline Point

"It is defined as the minimum equilibrium solution temperature for equal volumes of aniline and oil sample." It indicates the possibility of deterioration of oil when comes in contact of rubber sealing, packing, etc. Low aromatic contents are desirable in the oil as the aromatic hydrocarbon have a tendency to dissolve natural rubber. Therefore, to be a satisfactory lubricant used in rubber contact or sealing, the oil should have higher aniline point (i.e. low aromatic contents and higher percentage of paraffinic hydrocarbon).

Experimentally, the aniline point is determined with the help of aniline point apparatus (Figure 7.8). The equal volumes of aniline and oil sample are taken in a glass tube and heated.

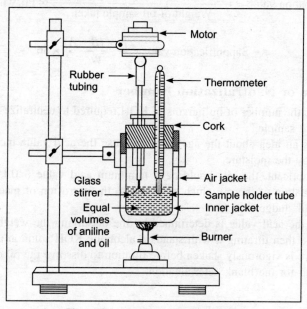

Figure 7.8 Aniline point apparatus.

The mixture is thoroughly mixed to obtain a homogeneous solution. This mixture is allowed to cool at a predetermined rate until the two phases just separate out. The temperature at which the two phases just separate out is taken as aniline point (Figure 7.8).

7.9.6 Saponification Value or Koettsdoerfer Number

The saponification value of any lubricating oil is very important property. It determines the presence of low acids and possible adulteration, contamination and compounding of the animal oils, vegetable oils and blended oils.

"The saponification value may be defined as the number of milligrams of KOH required to saponify fatty materials present in one gram fat or oil."

Saponification is the process of formation of soap by the reaction of fat or oil with alkali.

$$\begin{array}{l} CH_2OOC-(CH_2)_{16}-CH_3 \\ | \\ CHOOC-(CH_2)_{16}-CH_3 + 3KOH \longrightarrow \\ | \\ CH_2OOC-(CH_2)_{16}-CH_3 \\ \text{Glycerol tristearate} \end{array} \quad \begin{array}{l} CH_2OH \\ | \\ CH-OH \\ | \\ CH_2OH \\ \text{Glycerol} \end{array} + C_{17}H_{35}COOK \\ \text{Potassium stearate (soap)}$$

Experimentally, the saponification value is determined by refluxing the mixture of weighed quantity of oil, suitable solvent, e.g. ethyl methyl ketone and alcoholic KOH, and then titrating it with standard acid using phenolphthalein indicator. The refluxing allows the saponification of oil to form soap. The mixture is cooled up to room temperature before titration. The same procedure is followed for the blank determination.

The saponification value is calculated by using the following formula:

$$\text{Saponification value} = \frac{\text{Normality of acid solution} \times (\text{Volume of acid used for blank} - \text{Volume of acid used for sample})}{\text{Weight of oil sample taken}} \times \text{Eq. wt. of KOH}$$

or

$$\text{Saponification value} = \frac{N \times (B - A)}{W} \times 56$$

7.9.7 Acid Value or Neutralization Number

"It may be defined as the number of milligrams of KOH required to neutralize the free acids present in one gram fat or oil sample."

Acid value gives an idea about the age of oil because the acid value increases with time due to hydrolysis of oil by the moisture.

To be a good lubricant, the oil should have minimum acid value (<0.1). High acid value is the indication of oxidation of the oil which may lead to the corrosion of machine parts as well as formation of gum and sludge.

Experimentally, the acid value is determined by the dissolving the weighed quantity of oil in a suitable solvent and then titrating it with standard alcoholic KOH solution using phenolphthalein indicator. The mixture is vigorously shaken before titration to dissolve the oil completely. The same procedure is followed for the blank determination.

The acid value is calculated using the following formula:

$$\text{Acid value} = \frac{\text{Normality} \times (\text{Volume of KOH used for blank} - \text{Volume of KOH used for sample})}{\text{Weight of oil sample taken}} \times \text{Eq. wt. of KOH}$$

or
$$\text{Acid value} = \frac{N \times (B - A)}{W} \times 56$$

7.9.8 Iodine Value or Iodine Number

"It may be defined as the number of grams of iodine equivalent to iodine monochloride added to 100 gm of fat or oil for complete iodination of unsaturation."

$$\begin{array}{l} CH_2OOC(CH_2)_7-CH=CH-(CH_2)_7CH_3 \\ | \\ CHOOC-(CH_2)_7-CH=CH-(CH_2)_7CH_3 + 3I_2 \\ | \\ CH_2OOC-(CH_2)_7-CH=CH-(CH_2)_7CH_3 \end{array} \xrightarrow{HgCl_2} \begin{array}{l} CH_2OOC-(CH_2)_7-CH-CH-(CH_2)_7CH_3 \\ | \qquad\qquad\qquad\qquad\quad | \quad | \\ \qquad\qquad\qquad\qquad\qquad\quad I \quad I \\ CHOOC-(CH_2)_7-CH-CH-(CH_2)_7CH_3 \\ | \qquad\qquad\qquad\qquad\quad | \quad | \\ \qquad\qquad\qquad\qquad\qquad\quad I \quad I \\ CH_2OOC-(CH_2)_7-CH-CH-(CH_2)_7CH_3 \\ \qquad\qquad\qquad\qquad\qquad\quad | \quad | \\ \qquad\qquad\qquad\qquad\qquad\quad I \quad I \end{array}$$

Glycerol trioleate (triolein)   Glycerol trioleate (treble diiodide)

Each oil has its characteristic iodine value and therefore, it is utilized to determine the extent of unsaturation, drying capacity and contamination in any specific animal or vegetable oil. The oil having high iodine value are dried quickly.

To be a good lubricant, the oil should have as low iodine value as possible otherwise during use the oil can be deteriorated to any appreciable extent due to oxidation and polymerization.

Experimentally, the iodine value is determined by Huble's method or Wij's method. In Wij's method, the oil sample is dissolved in CCl_4 and the mixture is titrated with Wij's solution (iodine monochloride, ICl in glacial acetic acid). The ICl equivalent to unsaturation is consumed (one molecule of ICl per double bond). The unreacted ICl oxidizes the KI to iodine (I_2). The I_2 so liberated is titrated with hypo (sodium thiosulphate) solution using starch indicator. The amount of hypo consumed is equivalent to unreacted iodine. The iodine value is calculated using the following formula:

$$\text{Iodine value} = \frac{\text{Normality} \times \text{Volume of } I_2 \text{ (ICl) used}}{\text{Weight of oil sample taken}} \times 100 \times \text{Eq. wt. of } I_2$$

7.9.9 Specific Gravity

"It is the dimensionless quantity which expresses the ratio of density of the oil to the density of water at a specified temperature." Specific gravity helps in identifying unknown oil and determining the nature and type of crude oil from which it is prepared. In USA, the specific gravity is expressed by American Petroleum Institute (API) gravity. The liquids heavier than water are assigned API value less than 10 and liquids lighter than water are assigned API value greater than 10. In this unit, pure water has degrees API of 10 and for the zero a specific gravity of 1.076 has been adopted.

$$\text{Thus, API gravity} = \frac{141.5}{\text{Specific gravity at 6°F}} - 131.5$$

The value 141.5 is the modulus of API scale. It is usually determined at 60°F (15.55°C). Specific gravity of most of the lubricating oil is in the range of 0.85–0.9. The specific gravity is used for the inter-conversion of weight and volume measurement of the lubricating oil.

7.9.10 Carbon Residue

Lubricating oils contain high percentage of carbon in combined from. When the oil is continuously used at high temperatures, on heating it decomposes leaving behind certain amount of carbon which is deposited on machine parts like in IC engine and air compressors.

"The amount of carbon left after the complete combustion of given amount of lubricating oil is called *carbon residue*."

The deposition of carbon in IC engine results in carbonizing of the lubricating oil carried up into the piston ring and combustion chamber. It is also deposited on the spark plug and gives increased compression ratio which leads to the explosion. A good lubricant should deposit least amount of the carbon during its use.

Experimental determination

Carbon residue is determined by Conrodson method. The structure of Conrodson apparatus is shown in Figure 7.9. Weighed quantity (≈ 10 g) of oil under test is taken in the porcelain crucible; the crucible is heated by a Maker burner such that the pre-ignition period is 10 ± 1.5 min. When smoke is observed on chimney the flame is exposed to the side of crucible to ignite the vapours

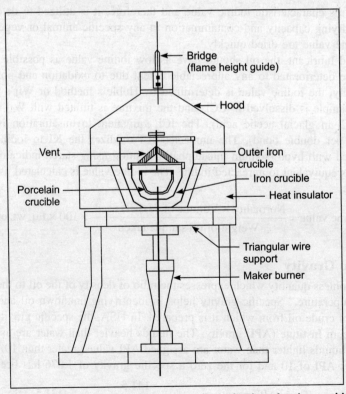

Figure 7.9 Conrodson apparatus for the determination of carbon residue.

with the flame above the chimney. After heating the crucible for total around 30 min, is removed which then allowed for cooling and weighed. The carbon residue is measured as the percentage of the residue of the total weight of oil taken.

7.9.11 Volatility

When the lubricating oil is subjected to high temperature at high load it may volatilize off due to volatile components. The residual lubricating oil may have different properties like higher viscosity and different viscosity index than the original oil. A good lubricating oil should have low volatility.

Experimentally, it is determined by vaporimeter (Figure 7.10). The vaporimeter consists of a furnace, one coiled copper tube and one platinum tray. A known weight of oil is taken in platinum tray and heated by passing the hot air through copper tube at the rate of 2 litre per minute. The volatile components escape out leaving behind the non-volatile components. After one hour, the residue is cooled and weighed. The difference in the weight of original sample and weight after heating is taken for the calculation of carbon residue.

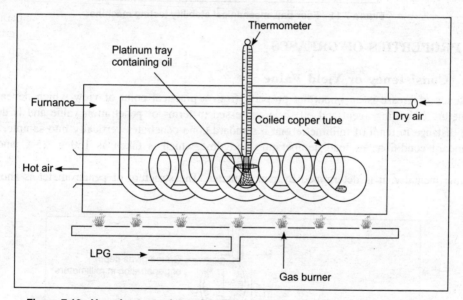

Figure 7.10 Vaporimeter to determine the vaporization stability of a lubricating oil.

7.9.12 Mechanical Stability

The oil used for the lubrication of machine parts should be stable under working conditions of high pressure and load. If it is not so the machine parts can jam and can result in the seizure. So to be good lubricant, the oil should have high mechanical stability.

The mechanical stability of lubricating oil is determined by four-ball extreme pressure test apparatus (Figure 7.11). The apparatus consists of four balls and a base. Three balls are fixed on a base and are stationary (s). The fourth upper ball is rotated (R). The load is gradually increased to examine the scar formation. The maximum load that the balls can bear without squeezing for one minute is considered the maximum mechanical stability of lubricating oil.

This test enables us to find the maximum load that a lubricant can bear safely under working conditions.

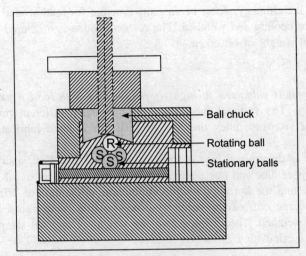

Figure 7.11 Four-ball, mechanical stability testing machine.

7.10 PROPERTIES OF GREASES

7.10.1 Consistency or Yield Value

Consistency of grease is an important property from a practical point of view which depends on the structure of gelling agent and oil. It is expressed in terms of penetration value and is defined as "the distance in tenth of millimetre that a standard cone penetrates vertically into sample, under the standard conditions of load, temperature and time which is taken as 150 g, 25°C and 5 s, respectively".

Experimentally, it is determined by penetrometer. The structure of penetrometer is shown in Figure 7.12.

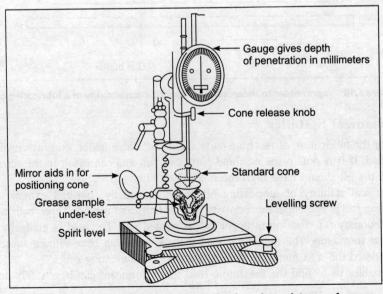

Figure 7.12 Penetrometer for the determination of consistency of grease.

The apparatus is levelled with the help of levelling screws. Standard cone is thoroughly cleaned. The grease under test is filled in the box and placed below the cone on the base. The position of cone is adjusted with the help of mirror in such a way that the tip of the cone just touches the surface of grease sample. The initial reading of the cone is recorded by the position of pointer on the dial and the cone is allowed to penetrate into the grease and final reading is recorded after 5 s. The difference in final and initial readings on dial is taken as the consistency or yield value of grease.

The consistency gives an indication of the applicability of greases at high temperature and high load conditions. To be a good lubricant, the grease should possess adequate consistency.

7.10.2 Drop point

The drop point of grease may be defined as the temperature at which the grease changes from the semisolid to liquid state i.e. it is converted into the form of drop. Drop point gives the upper temperature limit up to which the grease can be used satisfactorily. To be a good lubricant, the grease should have high drop point.

Experimentally, it is determined with the help of drop point apparatus (Figure 7.13). It consists of the following parts: (i) metal cup and lid, (ii) glass jacket, (iii) glass beaker, (iv) stirrer and (v) thermometer.

The grease sample is taken in a metal cup fitted in an air jacket. This assembly then placed in a glass beaker containing water. The temperature of water is raised gradually at the rate of 1°C rise per minute.

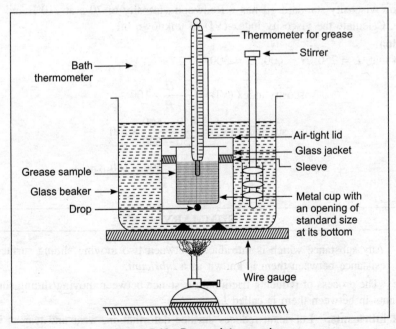

Figure 7.13 Drop point apparatus.

A temperature reached where the grease is converted into the liquid and comes out in the form of drop. The temperature at which the first drop of grease falls through the opening is taken as its drop point.

SOLVED NUMERICAL PROBLEMS

Problems based on viscosity index (VI)

1. Lubricating oil of unknown VI has a viscosity of 80 Redwood (RW) seconds at 210°F and 740 RWs at 100°F. The high viscosity standard oil having a viscosity of 80 RWs at 210°F and 300 RWs at 100°F. The low viscosity standard oil having a viscosity of 80 RWs at 210°F and 820 RWs at 100°F. Calculate the viscosity index (VI) of unkonwn oil.

Solution:
Given that, $L = 820$, $H = 300$, $U = 740$.

$$\text{Viscosity index (VI)} = \frac{L - U}{L - H} \times 100$$

where
U = viscosity of oil under test at 100°F (Gulf oil)
L = viscosity at 100°F of the low viscosity standard oil having a VI of 0 and also having the same viscosity at 210°F

$$\text{Viscosity index (VI)} = \frac{820 - 650}{820 - 300} \times 100$$

$$= \frac{170}{480} \times 100 = \frac{17000}{480} = \mathbf{35.416 \text{ Ans.}}$$

2. An oil of unknown VI has a Redwood viscosity of 70 s at 210°F and 500 s at 100°F. The high viscosity standard oil has a Redwood viscosity of 70 s at 210°F and 600 RWs at 100°F. The low viscosity standard oil has a Redwood viscosity of 70 s at 210°F and 750 RWs at 100°F. Calculate the viscosity index (VI) of unknown oil.

Solution:
Given that, $L = 750$, $H = 600$, $U = 500$.

$$\text{Viscosity index (VI)} = \frac{L - U}{L - H} \times 100$$

$$\text{Viscosity index (VI)} = \frac{750 - 500}{750 - 600} \times 100$$

$$= \frac{250}{150} \times 100 = \mathbf{160 \text{ Ans.}}$$

SUMMARY

Lubricants: Any substance which is introduced between two moving sliding surfaces to reduce the frictional resistance between them is known as a *lubricant*.

Lubrication: The process of reducing frictional resistance between moving/sliding surfaces by the use of lubricants in between them is called *lubrication*.

Functions of lubricants: Lubricant reduces surface deformation, wear and tear as it avoids the direct contact between surfaces, minimizes the loss of energy, avoids the seizure (self-welding) of moving surfaces, and sometimes also acts as a seal.

Mechanism of lubrication: There are three types of lubrication mechanisms, i.e. fluid film or thick film, thin film or boundary line lubrication and extreme pressure lubrication.

Fluid film or thick film: Thickness of lubricant film >1000 Å. Coefficient of friction is very low (0.001–0.03). Hydrocarbon oils and their blends are used in delicate instruments/light machines like watches, clocks, gun, and sewing machines.

Thin film or boundary line lubrication: Thickness of lubricant film <1000 Å. Coefficient of friction is very low (0.05–0.15). The suitable oils are vegetable oils, animal oils and their blends. This mechanism is used in the ball bearings with high load, gears of tractors, rollers, etc.

Extreme-pressure lubrication: This lubrication is used under the conditions very high pressure and speed, excessive heat is generated and a high local temperature is attained. Special additives (extreme pressure additives) are added to mineral oils. This mechanism is used in wire drawing machine, lathe machine, car rear axle, etc.

Classification of lubricants: Liquid lubricants or lubricating oils, semisolid lubricants or greases, solid lubricants and emulsions.

Liquid lubricants or lubricating oils: These lubricants are in the liquid form and reduce friction and wear between two moving/sliding metallic surfaces.

Semisolid lubricants or greases: Lubricating grease may be defined as a semisolid combination of thickening (gelling) agent like soap dispersed throughout liquid lubricating oil.

Uses of greases: They are used where oils cannot remain in place due to high load, low speed intermittent operation, sudden jerk, etc., e.g. rail axle boxes, in bearing and gears working at high temperature and pressure, where oil cannot be maintained in position due to bad seal or intermittent operation.

Solid lubricants: The lubricants which are used either in the dry powder form or mixed with oil or water are called *solid lubricants*. These lubricants reduce friction by separating two moving surfaces under boundary conditions. They are used in air compressors, lathes and other machine shop operations and equipments used for food processing.

Emulsions: An emulsion is a mixture of two or more immiscible liquids which is obtained by thoroughly shaking these liquids together to get a homogeneous mixture. There are two types of emulsions, oil in water type or cutting emulsions and water in oil type or cooling liquid emulsions.

Synthetic lubricants: The lubricants which are prepared/synthesized in the laboratory are known as *synthetic lubricants*. Synthetic lubricants are much better than ordinary lubricants as they can meet the most drastic and severe conditions such as in the aircraft engines at −50°C to 250°C temperature. They are used in hot running bearings and hot rolling mills and die castings.

Biodegradable lubricants: The lubricants which degrade or decompose into simpler molecules by the biological actions like eating, digestion, hydrolysis, enzymatic action after use when comes in contact with micro-organisms and which are non-toxic are called *biodegradable lubricants*. Following are some natural and synthetic lubricants which are decomposed by the bacterial and environmental action—Synthetic biodegradable lubricants and natural biodegradable lubricants

Viscosity and viscosity index (VI): It may be defined as the resistance to flow of liquid. It is the measure of the internal resistance to the flow of liquid which is due to the forces of cohesion between the liquid molecules. The rate at which the viscosity of oil changes with temperature is known as *viscosity index* (VI). Viscosity of lubricating oils is determined with the help of Redwood viscometer.

Flash and fire point: Flash point is the lowest temperature at which the lubricating oil gives off enough vapours that ignite to produce a flash for a moment when a tiny flame is brought near it. While the fire point is that lowest temperature at which the lubricating oil gives off enough vapours which burn (catch fire) continuously for at least five seconds when a tiny flame is brought near it.

Cloud and pour points: The temperature at which the oil becomes cloudy or hazy in appearance is called its *cloud point*. While the temperature at which the oil ceases to flow or pour (i.e. stops flowing) is called *pour point*.

Aniline point: It is defined as the minimum equilibrium solution temperature for equal volumes of aniline and oil sample.

Saponification value or Koettsdoerfer number: The saponification value may be defined as the number of milligrams of KOH required to saponify one gram fat or oil.

Acid value or neutralization number: It may be defined as the number of milligrams of KOH required to neutralize the free acids present in one gram fat or oil sample.

Iodine value or iodine number: It may be defined as the number of grams of iodine equivalent to iodine monochloride added to 100 gm of fat or oil for complete iodination of unsaturation.

Mechanical stability: The maximum load that the balls can bear without squeezing for one minute is considered the maximum mechanical stability of lubricating oil.

Consistency or yield value of grease: "The distance in tenth of millimetre that a standard cone penetrates vertically into sample, under the standard conditions of load, temperature and time which is taken as 150 g, 25°C and 5 s, respectively. Experimentally, it is determined by penetrometer.

Drop point: The drop point of grease may be defined as the temperature at which the grease changes from the semisolid to liquid state, i.e. it is converted into the form of drop. Drop point gives the upper temperature limit up to which the grease can be used satisfactorily.

EXERCISES

1. What do understand by lubricants and lubrication?
2. What are the functions of lubricants?
3. What are the criteria for the selection of lubricants for particular service condition?
4. Describe the various mechanisms of lubrication.
5. Which method is the most suited for the lubrication of delicate instruments? Explain in details.
6. Describe the lubrication mechanism suitable for the lubrication of rail axel and gear of tractor.
7. Explain in details the lubrication mechanism used for the lubrication of lathe and wire drawing machines.
8. Why are the lubricants required?
9. Classify the lubricants on the basis of their physical state.
10. What are liquid lubricants or lubricating oils? How they can meet the requirements of good lubricants?
11. What are the blended oils? How are they used as satisfactory lubricants even in drastic conditions?
12. What are synthetic oils? How they alone can meet all the requirements to be a satisfactory lubricant?
13. What are emulsions? Explain the different types of emulsions and their applications.
14. What is cutting fluid? Give its advantages over normal lubricant.
15. What do you understand by semisolid lubricants? Explain the consistency and drop point of grease.
16. Under what conditions greases are used? Explain the function of soap in grease.
17. What are biodegradable lubricants? Describe the need of these lubricants in present scenario.

18. What are solid lubricants? Explain the types of solid lubricants and give their applications.
19. Explain the following properties of lubricating oils:
    (i) Oiliness
    (ii) Viscosity and VI
    (iii) Flash and fire point
    (iv) Cloud and pour point
    (v) Aniline point
    (vi) Saponification value
    (vii) Iodine value
    (viii) Acid value
    (ix) Vaporisation stability
    (x) Carbon residue
    (xi) Mechanical stability.
20. What is viscosity and viscosity index of lubricating oils? How is the viscosity of lubricating oil determined in the laboratory?
21. What do you understand by flash and fire point? Explain its significance.
22. How is viscosity index calculated for unknown oil?
23. How are the viscosity and viscosity index inter-related?
24. What is ASTM test and how is it carried out?
25. What is specific gravity? Define API gravity.
26. What type of lubrication is applied to delicate instrument? Explain its mechanism.
27. What type of lubricants is used for machineries operating under high pressure and low speed?
28. How do viscosity and viscosity-index inference the selection of lubricants for particular purposes?
29. Write short notes on the following:
    (i) Extreme pressure additives
    (ii) Blended oils
    (iii) Synthetic lubricants
    (iv) Biodegradable lubricants
    (v) Emulsions
    (vi) Solid lubricants
    (vii) Greases
30. Give reason of the following:
    (i) Antioxidants are added to hydrocarbon oils.
    (ii) Lubricant should possess low carbon residue.
    (iii) Synthetic lubricants have added advantage over natural lubricants.
    (iv) The lubricant should possess low carbon residue.
    (v) Graphite acts as good lubricant even on the moon surface.
    (vi) Fatty oils are no longer used as lubricants.
    (vii) Lubricants are used in machine parts.
    (viii) Additives are added in oils.
    (ix) Common lubricants fail to work under extreme pressure conditions.
    (x) Good lubricants should possess low volatility.

CHAPTER 8

POLYMERS AND POLYMERIZATION

8.1 INTRODUCTION

We are living in the era of polymers. These polymers have multipurpose applications, e.g. household articles, clothes, furnitures, automobiles, space aircrafts, biomedical laboratories, etc. So the polymers have became an indispensable part of our lives.

8.2 POLYMERS

"Polymers are macromolecules of very high molecular mass formed by the repeated combination of large number of simple molecules (monomers)."

The process of conversion of monomers into polymers is called *polymerization*.

For example, polyethylene is commonly known as *polymer*, which is formed by the polymerization of ethane molecules.

$$n\text{CH}_2\!=\!\text{CH}_2 \xrightarrow{\text{Polymerization}} (-\text{CH}_2\!-\!\text{CH}_2\!-\!)_n$$

Ethylene (Monomers) → Polyethylene (Polymer)

8.3 DEGREE OF POLYMERIZATION

"The number of repeating units (monomers) involved in the formation of polymer chain is called *degree of polymerization*." In the above example, n is the degree of polymerization. The size of polymers may be decided by the degree of polymerization.

Polymers with high degree of polymerization are called *high polymers* while those with low degree of polymerization are called *oligopolymers*.

8.4 FUNCTIONALITY

The number of reacting sites or functional groups in a monomer molecule is called its *functionality*. For example vinyl chloride monomer is considered as bifunctional as it has two reactive sites on either side of double bond.

$$CH_2=CH \longrightarrow -CH_2-CH- \\ \quad\;\; | \qquad\qquad\quad\;\; | \\ \quad\;\; Cl \qquad\qquad\quad Cl$$

Ethylene (bifunctional)

8.5 TYPES OF POLYMERS

Polymers are classified in different ways as given below:

8.5.1 Classification on the Basis of Origin

On the basis of origin, polymers can be classified as:

(a) *Natural polymers (biopolymers):* These are involved in the life processes and obtained from plants and animals. Examples: starch, cellulose, proteins, natural rubber, etc.

(b) *Synthetic polymers:* These polymers are prepared or synthesized in laboratories. These polymers are also called *man-made polymers*. Examples: polyethylene, polyvinyl chloride (PVC), nylon, synthetic rubber, etc.

8.5.2 Classification on the Basis of Structure

On the basis of structure polymers can be classified as:

(a) *Linear polymers:* In this type of polymers the monomer units are linked together to form linear chains [Figure 8.1(a)]. These polymers are well packed and therefore have high density, high tensile strength and high melting point. Examples: polyethylene, nylon, polyester, etc.

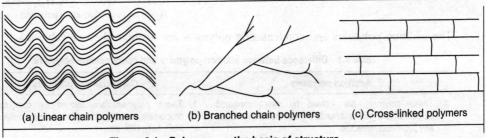

(a) Linear chain polymers   (b) Branched chain polymers   (c) Cross-linked polymers

Figure 8.1 Polymers on the basis of structure.

(b) *Branched chain polymers:* In these polymers, the monomer units are joined to form long chains with side chains or branches of different lengths [Figure 8.1(b)]. These polymers are irregularly packed and, therefore, have low density, low tensile strength and low melting point. Examples: low density polyethylene, starch, etc.

(c) *Cross-linked polymers or three-dimensional polymers:* These are the polymers in which polymeric chains form a network structure [Figure 8.1(c)]. These polymers are hard, rigid and brittle because of network structure. Examples: bakelite, melamine formaldehyde resin, etc.

8.5.3 Classification on the Basis of Nature of Monomer Units

Depending upon the nature of monomer units, polymers can be categorized into the following two categories:

(a) *Homo-polymers:* The polymers obtained by the repeated combination of only one type of monomer molecules are called homo-polymers, e.g. Polyethylene $-(CH_2-CH_2)_n$

(b) *Copolymers:* These are the polymers which are obtained by the repeated combination of two or more types of monomers. Examples: nylon 66 is the copolymer of hexamethylene diamine and adipic acid. Terylene is the copolymer of ethylene glycol and terepthalic acid.

8.5.4 Classification on the Basis of Mode of Synthesis

On the basis of mode of synthesis the polymers are classified into two categories:

(a) *Addition polymers:* The polymers which are formed by direct addition of repeated monomers without the elimination of any by product are called *addition polymers*. The process of formation of these polymers due to repeated combination (addition) of monomer units is called *addition polymerization*. Examples: polyethylene, polypropylene, etc.

$$n\,CH_2{=}CH_2 \xrightarrow{\text{Addition polymerization}} {-}(CH_2{-}CH_2){-}_n$$
$$\text{Ethylene} \qquad\qquad\qquad\qquad \text{Polyethylene}$$

(b) *Condensation polymers:* The polymers which are formed by the combination of monomers with the elimination of simple molecules such as water, alcohol, HCl, NH_3, etc. are called *condensation polymers*. The process of formation of condensation polymers is called *condensation polymerization*. Examples: nylon 66, polyester, polyamide, etc.

$$n\,H_2N{-}(CH_2)_6{-}NH_2 + n\,HO{-}\overset{O}{\overset{\|}{C}}{-}(CH_2)_4{-}\overset{O}{\overset{\|}{C}}{-}OH \xrightarrow{\text{Condensation polymerization}}$$

$$\text{Hexamethylene diamine} \qquad \text{Adipic acid}$$

$${-}(HN{-}(CH_2)_6{-}NH{-}\overset{O}{\overset{\|}{C}}{-}(CH_2)_4{-}\overset{O}{\overset{\|}{C}}){-}_n + H_2O$$
$$\text{Nylon 66 (a condensation polymer)}$$

The addition polymers and condensation polymers are distinguished in Table 8.1.

Table 8.1 Difference between addition polymers and condensation polymers

Addition polymers	Condensation polymers
1. These polymers are formed by direct repeated combination of monomer units without the elimination of any byproduct molecule.	1. These polymers are formed by combination of monomers with the elimination of simple molecules like H_2O, NH_3, HCl, etc.
2. The monomers are generally unsaturated compounds.	2. The monomers generally contain two functional groups.
3. Elimination of simple molecules does not take place.	3. Small molecules like R–OH, H_2O, NH_3, HCl, etc. are eliminated.
4. These are generally formed due to chain growth polymerization.	4. These are formed by step growth polymerization.
5. These are generally thermoplastics. Examples: polyethylene, PVC, polystyrene, etc.	5. These polymers may be either thermoplastic or thermosets. Examples: nylon 66, bakelite, dacron, melamine formaldehyde resin, etc.

8.5.5 Classification on the Basis of Molecular Forces

Applications of polymers in various fields are based upon their mechanical properties such as tensile strength, elasticity, toughness, etc. These properties of polymers depend upon intermolecular forces like van der Waal's forces, hydrogen bonds, dipole–dipole interactions, etc. existing in macromolecule. The polymers have been classified into four types:

(a) *Elastomers:* The polymers which have elastic character like rubber are called *elastomers*. Their elastic character is due to the linking of polymer chains together by weak intermolecular forces. Due to the weak forces these polymers can be easily stretched by applying stress. When the stress is removed the polymers regain their original shape. Examples: natural rubber, SBR, etc.

(b) *Fibres:* These are the polymers which have strong intermolecular forces between the chains. These forces may be either hydrogen bonds or dipole–dipole interactions. Due to close packing of chains these polymers have high tensile strength, low elasticity and sharp melting point. Since these polymers are long, thin and thread-like and hence can be woven in fabrics. Examples: nylon 66, dacron, silk, etc.

(c) *Thermoplastics:* These are the polymers in which intermolecular forces of attraction are neither very strong nor very weak, i.e. intermediate to that of elastomers and thermosets. Thermoplastics are the polymers which are softened on heating and can be moulded into desired shape and become hard on cooling. Examples: polyethylene, polypropylene, polystyrene. The melting point of these polymers can be increased by using plasticizers.

(d) *Thermosets or thermosetting polymers:* The polymers which undergo permanent change on heating, i.e. become hard and infusible on heating are called *thermosetting polymers*. These are prepared in two steps: (i) formation of long chain molecules which are capable of further linking with each other and (ii) heating which causes further reactions and cross-linking between the chains thus producing a cross-linked polymer. Examples: phenol formaldehyde (bakelite) and urea formaldehyde resins.

The thermoplastic polymers and thermosetting polymers are distinguished in Table 8.2.

Table 8.2 Differences between thermoplastic and thermosetting polymers

Thermoplastic polymers	Thermosetting polymers
1. These polymers can be softened repeatedly by heating and hardened on cooling without the change in properties.	1. These polymers cannot be softened or remoulded as they undergo permanent change on melting and set into a new solid which cannot be melted.
2. These can be moulded to any desired shape and can be processed again.	2. These cannot be moulded and cannot be reprocessed easily into desired shape.
3. These are less brittle and soluble in some organic solvents.	3. These are more brittle and insoluble in organic solvents.
4. These are formed by addition polymerization.	4. These are formed by condensation polymerizations.
5. They have usually linear structures. Examples: Polythene, PVC, teflon, etc.	5. They have three-dimensional cross-linked structure Examples: bakelite, UF resin, etc.

8.5.6 Classification on the Basis of Mechanism

On the basis of mechanism the polymers can be classified into following two categories:

(a) *Chain growth polymers:* These are the polymers which are formed by the successive addition of the monomer units to grow the chain through reactive intermediates (free radical, carbonium ion or carbanion). Examples: polyethylene, PVC, Teflon, etc.

(b) *Step growth polymers:* These are the polymers which are formed through a series of independent steps between two monomers initially and then between small polymer and monomer, with the elimination of some simple molecules like H_2O, HCl, etc. Examples: bakelite and terylene, etc.

8.6 TYPES OF POLYMERIZATION

Polymerization is of the following types:

8.6.1 Addition or Chain Polymerization

In this type of polymerizations, the polymer is formed due to the repeated combination of monomer units without any loss of material. The molecular mass of polymer so formed is the whole number multiple of the molecular mass of monomer units. This polymerization takes place by the initial form of some reactive species like free radicals or ions and addition of another molecule with the regeneration of reactive species.

Examples: formation of polythene, polypropylene, teflon, etc.

$$n\,CH_2=CH_2 \xrightarrow{\text{Addition polymerization}} {+CH_2-CH_2+}_n$$

8.6.2 Condensation Polymerization

In this polymerization, the polymers are formed by the combination of simple molecules with the elimination of some small molecules such as H_2O, HCl, CH_3OH, etc. The molecular weight of the polymers so formed is not integral multiple of molecular mass of monomer units.

$$n\,H_2N-(CH_2)_6-NH_2 + n\,HO-\underset{\text{Adipic acid}}{C}(=O)-(CH_2)_4-C(=O)-OH \xrightarrow{\text{Condensation polymerization}}$$

$$\underset{\text{Nylon 66 (a condensation polymer)}}{+HN-(CH_2)_6-NH-C(=O)-(CH_2)_4-C(=O)+_n} + H_2O$$

(Hexamethylene diamine)

This polymerization takes place by intermolecular combination due to the affinity between different functional groups (like OH^-, $-COOH$, NH_2, etc.) present in the monomers.

8.6.3 Coordination Polymerization

"The polymerization which takes place through the addition of monomers to a growing macromolecule through an organometallic active centre is called *coordination polymerization*, and the polymers so obtained are called *coordination polymers.*"

Basically, this polymerization is a form of addition polymerization. This polymerization technique was started in the 1950s with the use of heterogeneous Ziegler–Natta catalyst based on titanium tetrachloride and aluminium co-catalysts such as methyl aluminoxane. Therefore, this polymerization is also known as *Ziegler–Natta polymerization*.

Vinyl polymers have been synthesized successfully by this polymerization technique. It has a great impact on the physical properties of such polymers like polyethylene and polypropylene compared to the same polymers prepared by other techniques such as free radical polymerization. The polymers obtained by this technique tend to be linear and unbranched and have much higher molecular mass and high crystallinity. These polymers are known as *coordination type polymers*. These are also stereo-regular and can be isotactic or syndiotactic instead of just atactic. Due to this tacticity the polymers are crystalline in nature otherwise they will be amorphous. That is why the low density polyethylene (LDPE) is amorphous, and high density polyethylene (HDPE) and even ultra high molecular weight polyethylene (UHMWPE) are crystalline in nature.

Mechanism of coordination polymerization

Although the mechanism of coordination polymerization is not much clear because it takes place on the surface of an insoluble particle which is difficult to investigate experimentally. The following approximate mechanism has been suggested out of several models proposed to explain the action of Ziegler–Natta catalyst taking the example of the synthesis of vinyl polymer.

Ziegler studied the insertion of aluminium alkyls into olefins:

$$\begin{array}{c}R\\ \end{array}\!\!\!\!>\!\!\text{Al—CH}_2\text{—CH}_3 + n\text{CH}_2\!\!=\!\!\text{CH}_2 \xrightarrow{\text{Moderate temperature and pressure}} \begin{array}{c}R\\ \end{array}\!\!\!\!>\!\!\text{Al—CH}_2\!\!-\!\!(\text{CH}_2\text{—CH}_2)_n\!\!-\!\!\text{CH}_3$$

$$\downarrow \text{Oxidation}$$

$$\text{HO—CH}_2\!\!-\!\!(\text{CH}_2\text{—CH}_2)_n\!\!-\!\!\text{CH}_3$$

He obtained the high density polythene (HDP) with high crystallinity by using the insoluble complex of R_3Al and Lewis acid $TiCl_4$ as a catalyst.

Termination is almost done entirely by chain transfer.
By controlling the tacticity he obtained highly crystalline polymers

$$CH_2=CH-CH_3 \xrightarrow{\text{Zeigler-Natta catalyst}} \text{An isotactic chain (highly crystalline)}$$

Overall scheme of coordination polymerization can be shown as:

$$n CH_2=CH(R) \xrightarrow{\text{Metal complex}} n(CH_2-CH(R))_n$$

Monomer → Polymer

Consequences of coordination polymerization

1. This technique is very good for the polymerization of olefinic monomers like ethylene or propylene.
2. It produces linear polymer, with very few branches, e.g. high density polyethylene (HDPE).
3. It is capable of producing homotactic polymers which results in the high crystallinity.
4. Most commercial initiators (catalysts) are insoluble complexes or supported on insoluble carriers.
5. The chains are initiated/started per molecule of the catalyst.
6. The mechanism is very complex and termination is almost entirely done by chain transfer.

8.6.4 Chain Growth Polymerization

It is the process in which the polymers are formed by the successive addition of monomer units to grow the chain through reactive intermediate species like free radical, carbonium ion or carbanion. On the basis of intermediate species the chain growth polymerization can be classified into three categories:

Free-radical polymerization

In this polymerization, the polymers are formed due to the involvement of free radicals (odd electron species). The mechanism of free-radical polymerization involves the following steps:

1. *Initiation:* This step consists of two elementary reactions:
   (a) 1° radical generation from initiator.

   Example: $C_6H_5-C-O-O-C-C_6H_5 \xrightarrow{-CO_2} 2C_6H_5-\overset{\cdot}{C}-\overset{\cdot}{O} \longrightarrow 2\overset{\cdot}{C}_6H_5$

   (b) Combination of the above formed free radical with other single monomer molecule.

   $\overset{\cdot}{C}_6H_5 + CH_2=CH_2 \longrightarrow C_6H_5-CH_2-\overset{\cdot}{C}H_2$

2. *Propagation:* This free radical reacts with the double bond of an unexcited monomer molecule and combines to form a new free radical which is capable of further reactions with a fresh monomer.

$$C_6H_5-CH_2=\overset{\bullet}{C}H_2 + CH_2=CH_2 \longrightarrow C_6H_5-CH_2-CH_2-CH_2-\overset{\bullet}{C}H_2$$
$$\downarrow +n(CH_2=CH_2)$$
$$C_6H_5-(CH_2-CH_2)_n-CH_2-\overset{\bullet}{C}H_2$$

3. *Termination:* In this step, the macro free radical may be converted into an inactive polymer molecule in two different ways, i.e. by coupling or by disproportion.

   Coupling: by the combination of free radical together to give the product:

$$C_6H_5-CH_2-\overset{\bullet}{C}H_2 + \overset{\bullet}{C}H_2-(CH_2-CH_2)_n-C_6H_5$$
$$\downarrow$$
$$C_6H_5-CH_2-CH_2-CH_2-CH_2-(CH_2-CH_2)_n-C_6H_5$$

Disproportionation: A hydrogen radical transferred from β position to another radical centre.

$$C_6H_5-(CH_2-CH_2)_n-CH_2-\overset{\bullet}{C}H_2 + \overset{\bullet}{C}H_2-\overset{\overset{H}{|}}{C}H(CH_2-CH_2)_n-C_6H_5$$
$$\downarrow$$
$$C_6H_5-(CH_2-CH_2)_n-CH_2-CH_3 + CH_2=CH-(CH_2-CH_2)_n-C_6H_5$$

Ionic polymerization

This polymerization is highly specific, i.e. depends upon the nature of the growing polymer centres. This is of two types:

(a) *Cationic polymerization:* In this polymerization, the growth centre is cationic in nature and the monomer molecule adds on the growth centres of the polymer cation. The mechanism of this polymerization is similar to that of free-radical polymerization.

1. *Initiation:* Initiation is done by protonic acid, e.g. HCl, H_2SO_4, etc.; non-protonic acid, e.g. BF_3, $AlCl_3$, $SnCl_4$, etc. or carbonium salt like $Al(C_2H_5)_3$, $Al(C_2H_5)_2Cl$, etc.

$$H_2SO_4 \longrightarrow H^+ + HSO_4^-$$

$$CH_2=CH_2 + H^+ \longrightarrow CH_3-CH_2^+HSO_4^-$$

2. *Propagation:* In this step, more and more monomer units are added up and thus the chain keeps on growing and the counter ion moves all the time along the reactive positive centre in the direction of the growth of the chain.

$$CH_3^--CH_2^+HSO_4^- + CH_2=CH_2 \longrightarrow CH_3-CH_2-CH_2^--CH_2^+HSO_4^- \xrightarrow{nCH_2=CH_2}$$
$$CH_3-CH_2-(CH_2-CH_2)_n-CH_2-CH_2-CH_2^+HSO_4^-$$

3. *Termination:* In this step, the mutual termination can be done by the abstraction of proton from the polymer ion by the counter ion.

$$CH_2-CH_2-(CH_2-CH_2)_n-CH_2-CH-CH_2^+ \; HSO_4^-$$
$$\downarrow \quad H$$
$$CH_2-CH_2-(CH_2-CH_2)_n-CH_2-CH=CH_2 + H_2SO_4$$

(b) *Anionic Polymerization*

In this type of polymerization, the growth centre is anionic in nature and the monomer molecules add on the growth centre of the polymer anion, similar to that of cationic polymerization.

The mechanism of this polymerization involves two steps, i.e. initiation and propagation. The termination reactions hardly occur, if it occurs, the same takes place only in the absence of the impurities.

1. *Initiation:* This polymerization is initiated by alkali metals and alkali metal complexes, e.g. Na, K, Li and their stable complexes with aromatic compounds, liquid NH_3, etc. Lewis base like ammonia, triphenyl methane, aniline, etc. Organometallic compounds like boron alkyls, tetraethyl lead, Grignard reagent, etc. The initiator attacks on monomer molecule and breaks its double bond to give electron directly and form positive counter ion.

$$CH_2=CH + Na^+NH_2 \longrightarrow H_2N-CH_2-CH-Na^+$$
$$\quad\quad | \quad\quad\quad\quad\quad\quad\quad\quad\quad\quad\quad\quad |$$
$$\quad\quad Cl \quad\quad\quad\quad\quad\quad\quad\quad\quad\quad\quad\quad Cl$$

2. *Propagation:* In propagation step, more and more monomeric units are added up to grow up the chain and the counter ion moves along the reactive centre.

$$Na^+NH_2 \longrightarrow H_2N-CH_2-CH^--Na^+ + CH_2=CH \longrightarrow$$
$$\quad\quad\quad\quad\quad\quad\quad\quad\quad\quad | \quad\quad\quad\quad\quad\quad\quad\quad |$$
$$\quad\quad\quad\quad\quad\quad\quad\quad\quad\quad Cl \quad\quad\quad\quad\quad\quad\quad\quad Cl$$

$$H_2N-CH_2-CH-CH_2-CH^--Na^+ \xrightarrow{n(CH_2=CH_2)}$$
$$\quad\quad\quad\quad | \quad\quad\quad\quad |$$
$$\quad\quad\quad\quad Cl \quad\quad\quad\quad Cl$$

$$H_2N-CH_2-CH-(CH_2-CH)_n--CH_2-CH-CH_2CH-Na^+$$
$$\quad\quad\quad\quad | \quad\quad\quad\quad | \quad\quad\quad\quad\quad\quad | \quad\quad\quad\quad |$$
$$\quad\quad\quad\quad Cl \quad\quad\quad\quad Cl \quad\quad\quad\quad\quad\quad Cl \quad\quad\quad\quad Cl$$

3. *Termination:* The termination in anionic polymerization is not generally a spontaneous process and occurs only in the presence of some impurities.

8.7 STEREO-SPECIFIC POLYMERIZATION

The polymerization process in which the polymers of tactic structure (isotactic or syndiotactic) are obtained is called *stereo-specific polymerization*, and the polymers so obtained are called *stereo-specific polymers*. It is also called *coordination polymerization* as it initiated by coordination compounds or metal oxides. This type of polymerization oftenly occurs in dienes and olefins, and catalyzed by organometallic compounds, e.g. Ziegler–Nata Catalyst.

8.8 EFFECT OF STRUCTURE ON PROPERTIES OF POLYMERS

Properties of polymers are strongly correlated with their structure. These properties can be explained as follows:

8.8.1 Strength or Toughness

The strength of the polymers depends on the type and extent of intermolecular forces of attraction. The forces of attraction may be of two types, i.e. (i) primary forces or chemical bonds and (ii) secondary forces or van der Waal's forces of attraction.

Primary forces, i.e. chemical forces are much more stronger than van der Waal's forces. Hence, greater is the number of primary forces greater will the strength of polymers. On the other hand, greater the number of secondary forces less will be the strength of polymers. This order of strength in the various polymers with the magnitude of intermolecular forces of attraction is observed as follows:

$$\underrightarrow{\text{Linear polymers < branched polymers < cross-linked polymers}}$$
$$\text{Magnitude of intermolecular forces}$$

For example: in polyethylene (linear polymer) the secondary forces are greater in between different polymer chains so its strength is low. On the other hand, bakelite (cross-linked polymer) has least or no van der Waal's forces hence it is very strong and tough polymer.

Strength of polymer also depends on degree of polymerization and slippage power. High degree of polymerization leads to high strength of polymers as shown in Figure 8.2.

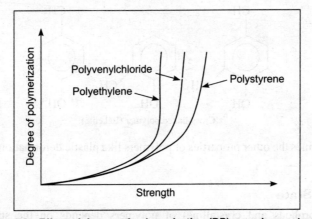

Figure 8.2 Effect of degree of polymerization (DP) on polymer strength.

Figure 8.2 shows that for same degree of polymerization the strength of polystyrene is greater than PVC and polyethylene. It is because the slippage power of one chain of polystyrene over other is very less due to presence of bulky benzene side chain. The order of slippage power is as follows:

$$\underrightarrow{\text{Strength}}$$
$$\text{Polyethylene (PE) < polyvinyl chloride (PVC) < Polystyrene (PS)}$$
$$\underleftarrow{\text{Slippage power}}$$

$$-CH_2-CH_2-CH_2-CH_2-CH_2-CH_2-CH_2-CH_2-$$
$$\text{Polyethylene}$$

$$-CH_2-\underset{Cl}{CH}-CH_2-\underset{Cl}{CH}-CH_2-\underset{Cl}{CH}-CH_2-\underset{Cl}{CH}-$$
$$\text{Polyvinyl chloride (PVC)}$$

Polystyrene (PS)

In cross-linked polymers there are no secondary forces but only chemical bonds, moreover, there is no slippage between polymer chains, hence, their strength and toughness is very high.

Cross-linked polymer (Bakelite)

Strength determines the other properties of polymers like plastic deformation, chemical resistance, workability, etc.

8.8.2 Physical State

Physical state (amorphous and crystalline) of polymers depends on the structure of polymers. The linear polymers, (e.g. polyethylene, polypropylene) and the polymers with low degree of polymerization have amorphous nature. It is because of random arrangement of molecules and polymer chains.

On the other hand, cross-linked polymers and linear polymers with high degree of polymerization have crystalline nature because of regular alignment of polymer molecules and chains. Cross-linked polymers like bakelite, UF resin, etc. are typically crystalline solids (Figure 8.3).

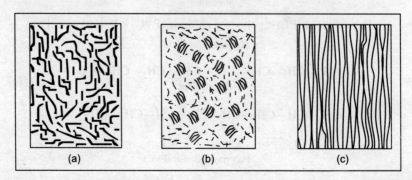

Figure 8.3 (a) Amorphous polymer, (b) crystallites embedded in amorphous matrix and (c) crystalline polymer.

In most of the polymers the crystallites are embedded in amorphous matrix. The crystalline nature of polymers imparts sharp melting point, high rigidity, high strength and high heat resistance, but completely crystalline polymer acquires brittleness.

The polymers of amorphous nature have low softening point, low strength but good moulding capacity.

8.8.3 Crystallinity

Crystallinity is the measure of degree of orderly arrangement of polymer molecule with each other. The portion of polymer molecule where the polymer molecules and chains are arranged in some regular way is called *crystallite*. Crystallinity is higher in the fibres and cross-linked polymers.

8.8.4 Plastic Deformation

It is the property of deformation of polymer material on applying heat, pressure or both. The linear polymers without branched structure and cross-linking have greater plastic deformation because in linear polymers the polymer chains are bound with each other by weak van der Waal's forces of attraction. These weak forces of attraction can easily be overcome on applying temperature and pressure and results in permanent deformation of the polymer molecules at weaker points. These materials (usually thermoplastics) can easily be moulded into desired shape on injecting into the mould in molten state at high pressures.

The cross-linked polymers like bakelite, UF resin, etc. do not undergo deformation on heating because they possess only primary forces (covalent bonds). If they are subjected to very high temperature, results into complete damage.

8.8.5 Elasticity

It is the characteristic that allows the polymers to retrieve its original shape after releasing the deforming stress. The cross-linked polymers (e.g. UF resin) have least or no elasticity and linear polymers or copolymers have low elastic property. While the polymers having coiled structure (elastomers) possess high elasticity as they can be stretched on applying stress due to their coiled structure. When the deforming stress is removed, they regain their original shape. The coiled structure of polymer is amorphous while stretched form is crystalline in nature. For example, natural rubber, buna-S, etc. show greater elasticity due to their coiled structure.

8.8.6 Chemical Reactivity

Chemical reactivity of a polymer depends on the nature of functional group present on it. It is said that like attracts like while unlike repels. Hence, aliphatic polymers (e.g. PE, PVC, etc.) readily affected/attacked by aliphatic solvents (petrol, CCl_4, etc.) and aromatic polymers (polystyrene) dissolve in aromatic solvents (benzene, toluene, etc.). The polymers having polar groups (–OH, –COOH) usually affected by polar solvents (alcohol, ketones, etc.) but show greater resistance towards non-polar solvents. On the other hand, non-polar polymers are readily attacked by non-polar solvents.

The polymers having double bonds in their chains are readily oxidized by oxygen or ozone. The polymer containing amide and ester linkages are vulnerable to hydrolysis.

In general, the polymers of amorphous nature have low chemical resistance while crystalline polymers show higher chemical resistance. Cross-linked polymers are least affected even by strong chemicals.

8.8.7 Electrical Properties

Most of the polymers are usually insulator of electricity because they are able to bear up a potential difference between different points of a given piece with the passage of very small current and low energy.

The dielectric constant of non-polar polymers is due to only electronic polarization while the dielectric constant of polar polymers is due to both electronic as well as dipole polarization. The extent of dipole polarization decreases with increase in frequency.

8.9 SOME COMMERCIALLY IMPORTANT POLYMERS: THERMOPLASTICS

These are the linear, long chain polymers having moderate intermolecular forces of attraction. They can be reshaped or remoulded again and again as they melt on heating.

8.9.1 Polyvinyl Chloride (PVC)

PVC is the addition polymer of vinyl chloride.

Preparation

It is prepared by heating vinyl chloride emulsion in presence of small quantity of benzoyl peroxide or hydrogen peroxide. The mixture of water emulsion of vinyl chloride and peroxide is heated in autoclave under pressure, a hard mass of PVC is obtained.

$$n\underset{\substack{|\\ Cl \\ \text{Vinyl chloride}}}{CH_2=CH} \xrightarrow{\text{Polymerization}} \underset{\substack{|\\ Cl \\ \text{Polyvinyl chloride (PVC)}}}{(CH_2-CH)_n}$$

The vinyl chloride (monomer) so used is prepared by the reaction of acetylene with HCl gas in presence of $HgCl_2$ (as catalyst) at 65°C.

$$n\underset{\text{Acetylene}}{CH\equiv CH} + HCl \xrightarrow[650°C]{HgCl_2} \underset{\substack{|\\ Cl \\ \text{Vinyl chloride}}}{CH_2=CH}$$

Properties

1. Pure PVC is a tough flexible plastic material.
2. It is colourless, odourless, non-inflammable and chemically inert material.
3. Its softening point is 148°C.
4. It has good resistance to light, atmospheric oxygen, inorganic acids and alkalis.
5. It has good electrical resistance.
6. It has excellent resistance to oil and weathering agents.
7. It is soluble in methyl chloride and tetrahydro-furan.
8. Plasticized PVC is soft, flexible and more economical. It possesses greater flux resistance, greater electrical resistance and abrasion resistance.

Applications

Hard (unplasticized PVC): It is most widely used synthetic plastic.

1. It is used in making sheets which are employed for the preparation of cycle and motor cycle mudguards, safety helmets, rain coats, radio and TV cabinets, electric insulators, etc.

2. It is used in making bottle and packing materials for consumable liquids like fruits, squash, detergents, etc.
3. It is used in acid recovery plants and hydrocarbon plants due to its good chemical resistance.
4. It is used in making pipes for irrigation, drains and guttering.
5. It is used in making window frames.

Soft (plasticized) PVC
1. It is used in making thin sheets which are employed in making plastic rain coats, baby rain coat wears, etc.
2. It used for sheathings in cables.
3. It is used in making ladies hand bags, bathroom curtains, table clothes.
4. It is also used in making garden hose and kitchenwares.

8.9.2 Polyvinyl Acetate (PVA)
Polyvinyl acetale (PVA) is the polymer of vinyl acetate.

Preparation
It is prepared by heating the suspension of vinyl acetate in the presence of small amount of benzoyl peroxide as a catalyst.

$$n\text{CH}_2=\underset{\underset{\text{Vinyl acetate}}{\text{OCOCH}_3}}{\text{CH}} \xrightarrow{\text{Benzoyl peroxide}} \underset{\underset{\text{Polyvinyl acetate (PVA)}}{\text{OCOCH}_3}}{-(\text{CH}_2-\text{CH})_n-}$$

Vinyl acetate is prepared by the reaction of acetylene with acetic acid in the presence of mercuric acetate as a catalyst.

$$\text{CH}\equiv\text{CH} + \text{CH}_3\text{COOH} \xrightarrow{(\text{CH}_3\text{COO})_2\text{Hg}} \text{CH}_2=\underset{\text{OCOCH}_3}{\text{CH}}$$

Properties
1. It is colourless and transparent material.
2. It is amorphous in nature and has low softening point ($\approx 28\ °C$).
3. It becomes light yellow above 120°C.
4. The articles of PVA are distorted even at room temperature, high pressure and on applying tensile force.
5. It is fairly soluble in organic solvents, oils and fats.
6. It has good resistance to atmospheric oxygen, water, heat and chemicals.

Applications
1. It is harmless if taken orally so it is used in making chewing gums.
2. It is used as adhesive (gum), water-based emulsion, paints, etc.
3. It is used in bonding papers, lather, textiles, etc.
4. It is used in making gramophone records, card boards, wrapping papers, etc.
5. It is used in the lining of acid containers and in finishing textiles and fabrics.

8.9.3 Polytetrafluoroethylene (PTFE) or Teflon or Fluon

Teflon is the polymer of tetrafluoroethylene.

Preparation

It is prepared by heating the water emulsion of tetrafluoroethylene under pressure in presence of benzoyl peroxide.

$$nC(F)(F)=C(F)(F) \xrightarrow{\text{Polymerization}} \left[-\underset{\underset{F}{|}}{\overset{\overset{F}{|}}{C}} - \underset{\underset{F}{|}}{\overset{\overset{F}{|}}{C}} - \right]_n$$

Tetrafluoroethylene → Polytetrafluoroethylene

Properties

1. The teflon is very hard, rigid and soapy in touch.
2. It has good electrical resistance, heat resistance and abrasion resistance.
3. Its softening point is high ($\approx 360°C$) even then it has good moulding capacity at high pressures.
4. It is not affected by chemicals, even strong acids but affected by hot alkali metals and hot fluorine.

Applications

1. It is used in making gaskets, tanks, tank lining, pump parts, packing materials, etc.
2. It is used in electric wire and cable wire insulation.
3. It is used as insulating material for transformers, electric motors, etc.
4. It is used in impregnating the glass fibre, textiles and paper.
5. It is used in making non-lubricating bearings, magnetic rotor and stop-cocks of burettes as it is non-adherents and non-sticky.
6. It is used as protective coatings and glaze of automobile body.

8.10 THERMOSETTING POLYMERS (THERMOSETS)

These are the cross-linked polymers which once set (solidified) during moulding cannot be softened and reshaped even by strong heating due to their have cross-linked structure.

8.10.1 Phenol Formaldehyde (PF) Resin or Bakelite

Bakelite is phenolic resin. It is the cross-linked, condensation, copolymer of phenol and formaldehyde.

Preparation

Bakelite is prepared by the condensation reaction of phenol with formaldehyde in presence of small amount of acid or alkali (as catalyst). Reactions complete in three steps. Initially, phenol and formaldehyde react to from *ortho* and *para*-hydroxy methylphenol compounds.

[Phenol + HCHO reaction scheme producing o-Hydroxy methyl phenol and p-Hydroxy methyl phenol]

The *ortho* and *para*-hydroxyl methyl phenol polymerizes to give linear polymer novolac.

[Structural scheme showing condensation to form Novolac (a linear polymer)]

The linear polymer novolac on heating with hexamethylenetetramine gives three-dimensional cross-linked polymer, bakelite.

[Structure of Cross-linked polymer (bakelite)]

Properties

1. Bakelite is brittle, hard, rigid, infusible material. It does not melt even at high temperatures but burns if heated strongly.
2. It is transparent and light pink in colour.

3. It has high scretch and abrasion resistance.
4. It is insoluble in water and many organic solvents due to its cross-linked structure.
5. It has very high electrical resistance.
6. It has good resistance to acids and oxidizing agents, but readily affected by strong alkalis due to presence of phenolic (acid) group.

Application

1. Due to its high heat resistance and good electrical insulation properties, it is used in making electric switches, plugs, switch boards, etc.
2. It is used as adhesive for grinding wheels and brake linings.
3. It is used in impregnating papers, fabrics, wood, etc.
4. It is used for the preparation of decorative laminates and electrical printed circuits.
5. It is used for the production of ion exchange resins.
6. It is used in making the handles of cookers, heaters, irons, sauce pans, etc.
7. It is used in making radio, TV cabinets, computer cabinets, safety helmets, etc.
8. It is used as paints and varnishes, and protective coatings.

8.10.2 Urea Formaldehyde (UF) Resin

Urea formaldehyde (UF) resin is the amino resin. It is the cross-linked polymer of urea and formaldehyde.

Preparation

It is prepared by the copolymerization of urea and formaldehyde. Mixture of two parts urea and one part formalin (40% formaldehyde solution) is heated at $\approx 500°C$ in a steel vessel in acidic/basic medium. Initially, it gives mono and dimethylol urea which on condensation give linear polymer. This linear polymer on heating produces cross-linked polymer, urea formaldehyde (UF) resin.

$$O=C\begin{matrix}NH_2\\NH_2\end{matrix} \xrightarrow{HCHO} O=C\begin{matrix}NHCH_2OH\\NH_2\end{matrix} \text{ Monomethylol urea}$$

$$\downarrow HCHO$$

$$O=C\begin{matrix}NH_2\\NH_2\end{matrix} \xrightarrow{2HCHO} O=C\begin{matrix}NHCH_2OH\\NHCH_2OH\end{matrix} \text{ Dimethylol urea} \xrightarrow{\text{Condensation polymerization}}$$

$$\begin{matrix}-N-CH_2-N-CH_2-N-CH_2-\\|\quad\quad\quad|\quad\quad\quad|\\CO\quad\quad CO\quad\quad CO\\|\quad\quad\quad|\quad\quad\quad|\\-N-CH_2-N-CH-N-CH_2-\end{matrix}$$

Properties

1. Urea formaldehyde (UF) resin is colourless, transparent material.
2. It is hard, infusible, having better tensile strength than bakelite.
3. It has high resistance to the chemicals, solvents, moisture, grease, etc.
4. It has good resistance to atmospheric oxygen and fire.
5. It has good abrasion and erosion resistance.
6. It has excellent electrical insulation properties.

Applications

1. It is used as an electric insulator and in making switches, plugs, switchboards, etc.
2. It is used in the manufacture of containers, household's appliances, bottle caps, etc.
3. It is used as adhesives in plywood and furniture industries.
4. It is used as anion exchanger in water purifier.
5. It is used as binder in the grinding wheels.
6. It is used in paper industries to improve the wet strength of paper.
7. It is used in finishing of cotton textile and to improve the quality of textiles.

8.11 ELASTOMERS

The polymers which have elastic character like rubber are called *elastomers*. Their elastic character is due to the linking of polymer chain together by weakest intermolecular forces and spring-like (coiled) structure. Due to the weak forces, these polymers can be easily stretched (4–10 times) by applying stress. When the stress is removed the polymer regains its original shape.

Natural rubber consists of basic material latex (obtained from rubber tree *Havea Brasiliensis* and *guayule*) which is a dispersion of isoprene.

$$CH_2=C-CH=CH_2 \xrightarrow{\text{Polymerization}} \text{Natural rubber (poly-isoprene)}$$
$$\underset{CH_3}{|}$$
Isoprene

8.12 VULCANIZATION OF RUBBER

Vulcanization is the process of hardening of rubber materials by introducing the cross-linking between chains and thus preventing the intermolecular movement of rubber coiling (springs). This technique was discovered by Charles Good Year in 1839. The vulcanization of rubber is done by treating it with sulphur, H_2S, S_2Cl_2, etc. Vulcanization improves the property like tensile strength, abrasion resistance, chemical resistance, tackiness, retraction, etc. of the rubber.

8.13 SYNTHETIC RUBBER

Elastomer is any man-made rubber like vulcanizable polymer which can be stretched to at least twice of its length. On removing stress it acquires its original shape. Synthetic rubbers possess better properties than natural rubber, like high tensile strength, high chemical resistance, high heat resistance, high ageing resistance and high elasticity.

8.13.1 Styrene Butadiene Rubber (SBR) or Buna-S

Styrene butadiene rubber (SBR) is the most important type of synthetic rubbers. It is the copolymer of styrene and butadiene.

Preparation

It is prepared by copolymerization of butadiene (75%) and styrene (25%). The emulsion of butadiene and styrene is heated at 50°C in presence of cumin hydroperoxide as a catalyst.

$$x\text{H}_2\text{C}=\text{CH}-\text{CH}=\text{CH}_2 + y\text{H}_2\text{C}=\text{CH}-\text{C}_6\text{H}_5 \xrightarrow{\text{Cumin hydroperoxide}} [-(\text{H}_2\text{C}-\text{CH}=\text{CH}-\text{CH}_2)_x-(\text{CH}_2-\text{CH}(\text{C}_6\text{H}_5))_y-]$$

Butadiene + Styrene → Polybutadiene co-styrene (SBR)

It can be vulcanized due to double bond like natural rubber by S or S_2Cl_2.

Properties

1. Styrene butadiene rubber (SBR) possesses high tensile strength.
2. It has good abrasion resistance and high load bearing capacity.
3. It has low oxidation resistance due to the presence of double bond, therefore, it gets oxidized in presence of trace of O_3 in atmosphere.
4. It swells in oils and solvents.
5. It can be vulcanized by sulphur or S_2Cl_2.

Applications

1. SBR is used in making motor tyres.
2. It is used in making gaskets, floor tiles, shoe soles.
3. It is used in tank linings and carpet backing.
4. It is used in cables wire insulation.
5. It is used as adhesive.

8.13.2 Nitrile Rubber (NBR or GR-N or Buna-N)

Nitrile Rubber (NBR) is the copolymer of butadiene and acrylonitrile.

Preparation

It is prepared by the reaction of butadiene (75%) and acrylonitrile (25%) in the form of emulsion.

$$x\text{CH}_2=\text{CH}-\text{CH}=\text{CH}_2 + y\text{CH}_2=\text{CH}(\text{CN}) \xrightarrow[\text{copolymerization}]{\text{Benzoyl peroxide}} -(\text{CH}_2-\text{CH}=\text{CH}-\text{CH}_2)_x-(\text{CH}_2-\text{CH}(\text{CN}))_y$$

Butadiene + Acrylonitrile (vinyl cyanide) → (NBR) polybutadiene co-acrylonitrile

Properties

1. Because of the presence of CN group and its hydrocarbon nature, NBR possesses extraordinary resistance to oils, acids and salts, but are affected by alkalis.
2. It has good tensile strength.
3. It swells in non-polar organic solvents up to some extent.
4. It has good resistance to abrasion, heat and sunlight.
5. It can also be vulcanized like natural rubber.

Applications

1. It is used in making conveyer belts.
2. It is used in the manufacture of automobile parts and high altitude aircraft components.

3. It is extensively used in lining of fuel tanks and creamery equipments.
4. It is used in the manufacture of gasoline hoses, gaskets, printing rollers, oil resistance foam, etc.
5. It is used (as latex) for impregnating textiles, paper and leather.
6. It is used as adhesive.

8.14 BIOPOLYMERS

"The polymers which are found in or obtained from living organisms are called *biopolymers* and the process of production of biopolymers by the combination of biomolecules (monomers) is called *biopolymerization*". Examples of biopolymers are given in Table 8.3

Table 8.3 Biopolymers and their monomers

Biopolymer	Monomer
Starch, Cellulose	Saccharide (sugar)
Peptides, Proteins	Amino acids
DNA, RNA	Nucleotides

8.14.1 Characteristics of Biopolymers

Following are the general characteristics of biopolymers:
1. Biopolymers are found in or obtained from living organisms.
2. The biopolymers control the life processes in living organisms.
3. Biopolymers are also known as renewable polymers as some of them are obtained from plant materials which can be grown continuously on year to year basis.
4. Biopolymers have potential to reduce carbon emission and minimize the quantity of CO_2 in the atmosphere.
5. Most of the biopolymers are biodegradable, i.e. degraded or decomposed by enzymatic hydrolysis or oxidation.
6. Some of the biopolymers are compostable.
7. Some biopolymers have plastic properties and hence can be used as the substitute of plastic, e.g. polylactic acid, poly 3-hydroxy butyrate, polyglycolate, etc.
8. Biopolymers (bioplastics) are environmentally friendly.
9. Some of the biopolymers show surface chemical properties, e.g. biopolymer solids like fibrinogen, BSA, etc.

8.14.2 Types of Biopolymers

On the basis of monomers from which they are formed, the biopolymers can be classified into the following types:

(a) *Sugar-based biopolymers (carbohydrates):* The polymers which are obtained by the sugar (saccharide as a basic material) known as *sugar-based biopolymers*. These are also called *polysaccharides*. They can further be classified into two sub-categories:

1. *Starch biopolymers:* Starch is a natural polymers of simple saccharide units (α-glucose) joined by glycosidic linkage, having linear (amylase) or branched (amylopectin) structure. It occurs in the grains like wheat, maize, etc. potato and in plant tissues, from which it can easily be extracted in large quantities. Starch can be modified into the forms which can be melted and deformed thermoplastically. The material so obtained is thus suitable for the formation of articles by injection moulding and extruding similar to that of conventional plastic forming processes.

    Properties and applications of starch: Since the thermoplastic starch materials are soluble in water, hence they are not of appropriate use for packaging liquids. Although they can sustain only brief contact with water. They have good oxygen barrier properties.

    Starch-based biopolymers are harmless for the human body and, therefore, are used as foodstuff, basic nutrients, medicine and paper and textile sizing. They are also used in the fermentation process to produce alcohol.

2. *Cellulose biopolymers:* Cellulose is also the natural polymer having thousands of monosaccharide units joined together by glycoside linkage (at C_1 and C_4), hence it is also a polysaccharide giant biopolymer.

    Properties and applications: Cellulose is water insoluble and is readily affected by temperature. It can be converted into hard plastic and cellophane film for different uses. It can be converted into long threads which are used in textile industries. It is used in paper industries to increase the strength, wet resistance and glow of the papers. Cellophane is used in packaging for CDS, confectionary and cigarettes.

    Other examples of carbohydrates are polyhydroxybutyrate and polylactides formed by the bacterial fermentation of sucrose or starch which are used for medical applications.

(b) *Amino acids-based biopolymers (proteins):* Proteins are the biopolymers of amino acids (monomers) which are joined together by peptide linkage between the carboxyl and amino groups of adjacent amino acid residues. Proteins are also known as *polypeptides*. Amino acids in proteins may be arranged in two ways, i.e. in a linear chain (fibrous proteins) and folded into spheroid shape (globular proteins).

Properties and applications: Proteins are water insoluble and denatured by strong acids or bases and at high temperatures. Proteins are the life materials as the maximum parts of the body made up of proteins. For example: keratin protein (skin, nails hairs, etc.), myosin protein (muscles), collagen protein (nerves), haemoglobin protein (blood), enzymes and hormones protein, etc.

1. *Nucleic acids:* The basic material for the nucleic acid is nucleotides (polyester chain) as the backbone. The polynucleotide is derived from phosphoric acid and sugars. The nucleic acids are responsible for the transfer of heredity characters from one generation to other. Moreover, they play a vital role in the biosynthesis of proteins. There are two types of nucleic acids: deoxyribonucleic acid (DNA) and ribonucleic acid (RNA).

    Deoxyribonucleic acid (DNA): Chemically, DNA have deoxyribose sugar and consists of two long polymers of simple units called *nucleotides*, with backbones made of sugars and phosphate groups joined by ester linkage. These two strands run in opposite directions to each other and are therefore antiparallel. DNA has double stranded helical structure. Four types of molecules called *bases* are attached to each sugar. The sequence of bases in DNA is: adenine $\rightarrow$ guanine $\rightarrow$ cytosine $\rightarrow$ thymine.

This nucleic acid contains the genetic instructions used in the development and functioning of all the known living organisms and some viruses. The main role of DNA molecules is the long-term storage of information. DNA is often compared to a set of blueprints or a recipe or a code, since it contains the instructions needed to construct other components of cells, such as proteins and RNA molecules. The DNA segments that carry this genetic information are called *genes*.

Ribonucleic acid (RNA): Chemically, RNA has the ribose sugar and consists of a long chain of nucleotide units. It has linear and single stranded structure. Each nucleotide consists of a nitrogeneous base, a ribose sugar, and a phosphate. The sequence of bases in RNA is: adenine, guanine, cytosine and uracil.

The main function of RNA is the biosynthesis of proteins. Other functions are transfer of information (messenger RNA), translation of the information (tRNA).

(c) *Synthetic-based biopolymers:* Synthetic compounds derived from petroleum can also be a starting point for biodegradable polymers, e.g. aliphatic aromatic copolyesters. These polymers have technical properties resembling those of low density polyethylene (LDPE). Although these polymers are produced from synthetic starting materials. They are fully biodegradable and compostable.

Applications: The relatively high price of biodegradable polymers of synthetic substances, e.g. aliphatic and aromatic copolyesters, has prevented them from reaching a large scale market. The best known application is for making substrate mats.

8.14.3 Benefits of Biopolymers

Besides the above applications the biopolymers have several economic and environmental advantages. Biopolymers could also prove an asset to waste processing. For example, replacing and elimination of plastic scraps, etc. Biopolymers can provide an image advantage, e.g. biodegradable packaging.

The major advantage of these polymers is that they can be composted. But the biodegradability of raw materials does not necessarily mean that the product or package made from them, (e.g. coated paper) is itself compostable. Biopolymers can also be used in waste processing. The basic differences between normal polymers and biopolymers are described in Table 8.4.

Table 8.4 Differences between normal polymers and biopolymers

Normal Polymers	Biopolymers
1. Polymers are obtained from non-living materials.	1. Biopolymers are found in or obtained from living organisms.
2. Usually they are not involved in life processes.	2. The biopolymers control the life processes in living organisms.
3. These polymers have much simpler and more random structures.	3. These polymers have well defined complex structure like proteins have primary, secondary and tertiary structures.
4. They have definite molecular mass.	4. Molecular mass distribution may be missing.
5. Most of the polymers are non-degradable.	5. Most of the biopolymers are degradable.
6. They are usually environmental unfriendly.	6. Most of the biopolymer are environmental friendly.

8.15 BIODEGRADABLE POLYMERS

The polymers which degrade or decompose by the biological actions like eating, digestion, hydrolysis, enzymatic action of macro and microorganisms are called *biodegradable polymers*, while the process of production of biodegradable polymers is called the *biodegradable polymerization*.

Many natural and synthetic polymers are biodegradable. Examples: polyglycols and synthetic esters are decomposed by the bacterial action. Other bio and environmentally degradable polymers include poly(hydroxyalkanoate) of the PHB–PHV (polyhydroxybutyrate–polyhydroxyvalerate) class, polyesters and natural polymers particularly, modified polysaccharides, e.g. starch, cellulose, and chitosan.

8.15.1 Classification of Biodegradable Polymers

These are divided into two categories:

(a) *Natural biodegradable polymers:* These polymers are obtained by living organisms, i.e. plants and animals. Examples: proteins, carbohydrates, natural rubber, lignin, etc.

(b) *Synthetic biodegradable polymers:* These polymers are synthesized in the laboratories. Examples: polyesters based on polylactides (PLA), polyanhydride, polyglycolide (PGL), polycaprolactone (PCL), and polyhydroxy alkanoate.

8.15.2 Conditions of Biodegradation

Following are the essential requirements of polymer biodegradation:

(a) *Nature of biodegradable polymer:* The biodegradable polymer must have the following characteristics:

1. *Presence of heteroatom or functional group:* A polymer based on a heteroatom backbone tends to confer biodegradability. It should have suitable functional group and chemical linkage such as anhydride, ester, ether, etc. which are readily attacked by micro-organisms (enzymes) and cleaved at these positions.

   For example, polyesters based on polylactides (PLA), have been converted into polyanhydride, polycaprolactone (PCL), etc. as ester linkage is susceptible to hydrolysis (biodegradation).

2. *Hydrophilic nature:* The polymer should be hydrophilic in nature. Greater the hydrophilic nature of polymer greater will be the rate of biodegradation.

3. *Molecular mass:* The polymers should have low molecular mass as small chains are readily attacked by the micro-organism, i.e. degraded easily.

(b) *Presence of microorganisms:* The presence of micro-organisms as well as their contact with polymer is must for biodegradation process. The enzymes present in micro-organisms decompose the polymer backbone and convert into simpler molecules.

(c) *Favourable environmental conditions:* The adequate amount of moisture, oxygen, concentration of salts, light and optimum temperature and pressure favour the biodegradation.

8.15.3 Mechanism of Biodegradation of Polymers

Sometimes the macro-organisms eat the polymers. These polymers are digested in their intestine by enzymatic actions. Biodegradation can take place in the following steps:

1. Micro-organisms present in the water, soil, solid waste or sewage attack on the heteroatom of polymer backbone, when come in contact with them.

2. The depolymerases enzymes present in the living cells decompose the polymers into simpler molecules to give a mixture of oligomers, dimers and monomers.

3. This mixture may further degrade or decomposed in two ways (Figure 8.4):

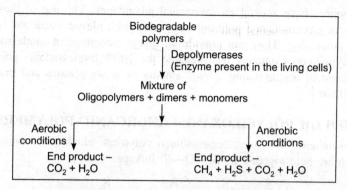

Figure 8.4 Reaction pathway of a degradation polymer.

(a) *Under aerobic condition (presence of O_2):* The end product on complete degradation is CO_2 and H_2O.
(b) *Under anaerobic conditions (absence of O_2):* The end product on complete degradation is the mixture of CH_4, H_2S, CO_2 and H_2O.
These end products may contain some non-degradable material like metal compounds. The compounds CH_4, H_2S, CO_2 and H_2O are released in the atmosphere leaving behind the non-degradable material.

The polyester is cleaved at C—OR bond by Micrococcus roseus bacteria and polypropylene glycol and PVA are attacked at C—OH bond by pseudomonas bacteria in presence of salt at pH ≈ 7.0.

$$\underset{\text{Triester}}{CH_3CH_2-\overset{\overset{\displaystyle COOR}{|}}{\underset{\underset{\displaystyle COOR''}{|}}{C}}OOR'} + 3H_2O \xrightarrow{\text{Micrococcus roseus}} \underset{\text{Triacid}}{CH_3CH_2-\overset{\overset{\displaystyle COOH}{|}}{\underset{\underset{\displaystyle COOH}{|}}{C}}OOH} + ROH + R'OH + R''OH \text{ Alcohols}$$

8.15.4 Applications and Limitations of Biodegradable Polymers

1. Biodegradable polymers are used commercially in both the tissue engineering and drug delivery field of biomedicine.
2. They are used in dental devices, orthopaedic fixation devices and tissue engineering scaffold.
3. They are used in biodegradable vascular stents.
4. Polyhydroxy butyrate is used in the production of shampoo containers (bottles).
5. Polylactic acid (PLA) is used in drug delivery system, wound clips and some agricultural applications.

The biodegradable polymers also have some limitations like they are rarely available and expensive having low malleability and pliability.

8.15.5 Benefits of Biodegradable Polymers

Biodegradable polymers have several environmental advantages. The use of these polymers can solve the problems of environmental pollution associated with plastic waste and have also proved an asset to waste processing. They can provide an image advantage of biodegradable packaging and composting. Symphony Company has prepared the 100% biodegradable polyethylene which can solve the problem of accumulating a huge amount of waste plastics and therefore the space available in landfill sites.

8.16 SILICONES OR POLYSILOXANES (INORGANIC POLYMERS)

Silicones are the synthetic, inorganic, organosilicon polymers which are constituted of repeated R_2SiO units, which are held together by Si—O—Si linkage.

$$\begin{array}{c} R \\ | \\ -O-Si-(O-Si-)_n-O-Si- \\ | \\ R \end{array} \quad \begin{array}{c} R \\ | \\ \\ | \\ R \end{array} \quad \begin{array}{c} R \\ | \\ \\ | \\ R \end{array}$$

These are derived from alkyl chlorosilanes and also by phenyl chlorosilanes. The name to these polymers is given as silicones (R_2SiO) as they resemble to ketones in their empirical formula.

On the basis of structure, silicones may be classified as linear, branched and cross-linked silicones. These polymers may be liquid, viscous liquids, rubber like, semisolids or solids.

8.16.1 Preparations of Silicones

Silicones are prepared by hydrolysis of alkyl chlorosilanes. These alkyl chlorosilanes are prepared by the reactions of methyl chloride with silicon.

$$CH_3Cl + Si \longrightarrow CH_3-\underset{\underset{CH_3}{|}}{\overset{\overset{CH_3}{|}}{Si}}-Cl + CH_3-\underset{\underset{Cl}{|}}{\overset{\overset{CH_3}{|}}{Si}}-Cl + Cl-\underset{\underset{Cl}{|}}{\overset{\overset{CH_3}{|}}{Si}}-Cl$$

Methyl chloride Silicon   Monochlorosilane   Dichlorosilane   Trichlorosilane

and also by the reaction of silicon tetrachloride with Grignard reagent:

$$SiCl_4 + CH_3MgCl \longrightarrow CH_3-\underset{\underset{CH_3}{|}}{\overset{\overset{CH_3}{|}}{Si}}-Cl + CH_3-\underset{\underset{Cl}{|}}{\overset{\overset{CH_3}{|}}{Si}}-Cl + Cl-\underset{\underset{Cl}{|}}{\overset{\overset{CH_3}{|}}{Si}}-Cl$$

Silicon   Grignard   Monochlorosilane   Dichlorosilane   Trichlorosilane
tetrachloride reagent

This mixture of different organosilicon chloride is fractionally distilled to get particular organosilicon chlorides. These chlorides are then polymerized by careful, controlled hydrolysis as follows:

1. Hydrolysis of monochlorosilane gives linear silicone dimer:

$$2CH_3-\underset{\underset{CH_3}{|}}{\overset{\overset{CH_3}{|}}{Si}}-Cl \xrightarrow[-2HCl]{+2H_2O} 2CH_3-\underset{\underset{CH_3}{|}}{\overset{\overset{CH_3}{|}}{Si}}-OH \xrightarrow{-H_2O} CH_3-\underset{\underset{CH_3}{|}}{\overset{\overset{CH_3}{|}}{Si}}-O-\underset{\underset{CH_3}{|}}{\overset{\overset{CH_3}{|}}{Si}}-CH_3$$

    Monochlorosilane   Monohydroxylsilane   Siliconedimer
                              (unstable)

2. Dichlorosilane yields very long chain polymers:

$$nCH_3-\underset{\underset{Cl}{|}}{\overset{\overset{CH_3}{|}}{Si}}-Cl \xrightarrow[-2HCl]{+2H_2O} nCH_3-\underset{\underset{OH}{|}}{\overset{\overset{CH_3}{|}}{Si}}-OH \xrightarrow{-H_2O} \left[\underset{\underset{CH_3}{|}}{\overset{\overset{CH_3}{|}}{Si}}-O\right]_n$$

    Dichloride silane   Dihydroxy silane   Linear silicone polymer
                      (unstable)

3. Trichlorosilane gives cross-linked silicones polymers

$$nCl-\underset{\underset{Cl}{|}}{\overset{\overset{CH_3}{|}}{Si}}-Cl \xrightarrow[-3HCl]{+3H_2O} OH-\underset{\underset{OH}{|}}{\overset{\overset{CH_3}{|}}{Si}}-OH \xrightarrow[-H_2O]{Polymerization}$$

Trichlorosilane   Trihydroxy silane
                 (unstable)

Cross-linked polymer
(Solid silicone)

8.16.2 Properties of Silicones

1. The physical state of silicones depends upon the chain length, cross-linking and size of alkyl and aryl groups. Their state vary from oily liquid to hard resins.
2. They have high heat resistance and water resistant.
3. Viscosity of silicone oils does not change with temperature, i.e. they have high VI.
4. They have excellent resistance to the most of the chemical reagent but can be cleared by Grignard's reagent, alkyl lithium and $LiAlH_4$.
5. They have good electrical insulation property.

8.16.3 Uses of Silicones

1. Silicone fluids are used as lubricants over a wide temperature range (–40°C to 200°C) as they are not affected much by temperature variations.

2. Silicon oils are also used in high temperature oil bath.
3. Silicon oils are used for finishing of leather and textiles.
4. These are also used as cosmetics and antifoaming agent.
5. Silicon greases are used as lubricants under very high as well as very low temperature conditions.
6. Silicone greases are also used as adhesive, sealing materials, protective coatings and insulating material.
7. Silicone rubbers are used in making the tyres of aircraft, gaskets, seal, adhesive, paints etc. to make water resistant glass, fabric, wool, etc.
8. Solid silicones are used as high voltage/temperature insulators, glass wood etc.

8.16.4 Types of Silicones

On the basis of physical state, silicones may be classified into following types:
(a) Liquid silicones or silicone fluids
(b) High polymer silicones or silicone greases
(c) Silicone rubber (easterners)
(d) Solid silicones or silicone resins

(a) *Liquid silicones or silicone fluid:*   These are low molecular mass, short range, straight chain polymers, prepared by hydrolysis of dichlorosilane.

Properties
1. Silicone fluids possess low surface tension.
2. These are soluble in organic solvents.
3. They possess high thermal stability, antifoam properties, excellent dielectric constant, good lubrication property even at high temperature.

Uses
1. Silicone fluids are used as lubricant within a wide range of temperature (–40°C to 200°C).
2. They are used in water proofing.
3. They are used for finishing of fabric, paper, leather, etc.
4. They are used in cosmetics.
5. They are used as polish additives.
6. They are used as insulator in electric appliances.

(b) *High polymer silicones or silicone gums/greases:*   These are high molecular mass linear polymers. These are prepared by equilibration of cyclic siloxanes in the absence of functional groups and by adding some fillers like silica, carbon black, soap, etc. to silicone fluids.

Properties
1. These are the modified form of the silicone fluids.
2. They are stable at wide temperature range (–100 to 450ºC).
3. They possess good water repellent property.
4. They have high heat resistance.

Uses
1. They are used as antifoaming agent.
2. They are used as lubricants under very high and very low temperature conditions.
3. They are used as water repellent in the textile industry.
4. They are used as adhesives.

(c) *Silicone rubbers or elastomers:* Silicone rubber is obtained by compounding of high molecular mass polydimethyl silane with fillers (finely divided silica), cross-linking agent (benzoyl perxide) and other additives as per requirements.

Properties
1. The silicone rubbers have high thermal stability.
2. They possess excellent resistance to dilute acids, alkalis, common oils, boiling water, atmospheric oxygen and sunlight.
3. They can be used within high temperature range (–90°C to 250°C) as they remain flexible with this temperature range but at very high temperature they get decomposed to give non-conducting silica.
4. They are good water repellents.
5. They are good electric insulators.

Uses of silicone rubbers
1. Silicone rubbers are used in making tyres of fighter aircrafts.
2. They are used as sealing and insulating material in search light of ships, washing machines, electric blanket, etc.
3. They are used as adhesive in electronics industry.
4. They are used in making gaskets and seals.
5. They are used in making artificial heart valves, transfusion tubing and surgical padding.
6. They are used as paints and varnishes.

(d) *Solid silicones or silicone resins:* These are highly cross-linked polysilanes having high molecular masses. These are prepared by bi and trifunctional silanes (dichloro and trichlorosilanes).

Properties
1. They are hard, rigid, thermosetting materials.
2. They possess excellent resistance to heat and chemicals.
3. They are very good electrical insulators.

Uses of Silicone resins:
Silicone resins are used:
1. They are used for coating of metals on cooking pans, glass, wood, etc.
2. They are used in making high voltage insulating foam.
3. They are used for fabric finishing and lamination.

8.17 DETERMINATION OF MOLECULAR MASS OF POLYMER

The polymers are the mixture of molecules of different molecular masses. Three main types of molecular mass of polymers have been recognized:

8.17.1 Number Average Molecular Mass ($\overline{M}_n$)

It is determined by the measurement of colligative properties such as depression in freezing point (cryoscopic), elevation in boiling point (ebullimetry), osmotic pressure and lowering of vapour pressure.

"Number average molecular mass $\overline{M}_n$ is defined as the total mass (W) of all the molecules in a polymer sample divided by the total number of molecules present."

Thus, the number average molecular mass is given by

$$\overline{M}_n = \frac{W}{\sum N} = \frac{\sum N_i M_i}{\sum N_i}$$

where N_i = number of molecules and M_i = molecular mass.

8.17.2 Weight Average Molecular Mass $\overline{M}_w$

It measures the molecular size. It may be defined as:

$$\overline{M}_w = \frac{\sum w_i M_i}{\sum w_i}$$

where w_i is the weight fraction of molecules of mass M_i and M_w. Weight average molecular mass is always greater than the number average molecular mass (Figure 8.5).

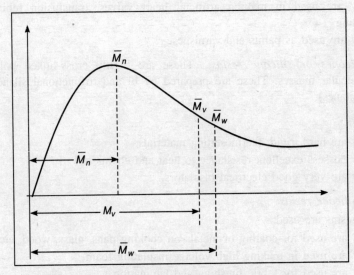

Figure 8.5 Distribution of molecular mass in polymer.

8.17.3 Viscosity Average Molecular Mass $\overline{M}_v$

It is obtained by viscosity measurement. It is equal to weight average molecular masses.

$$\overline{M}_v = \frac{\sum w_i M_i}{\sum w_i}$$

where w_i is the weight fraction of molecules of mass M_i and M_w.

8.18 POLYMER COMPOSITES

A polymer composite may be defined as any multiphase material consists of at least one polymer and other polymeric or non-polymeric materials.

Polymer composites consist of body constituent (matrix phase in bulk) and structural constituents (dispersed phase).

Matrix phase: It may be metal, ceramic or polymer. Matrix phase plays the following roles:
1. The matrix phase binds the fibres together.
2. It keeps the reinforcing fibres in the proper position and orientation.
3. It acts as medium to transfer the applied load to the dispersed phase.
4. It protects the individual fibres from chemical action.

Dispersed phase: Dispersed phase determines the structure of composite materials. It may be fibre (e.g. glass, carbon or aramid fibre) flakes (e.g. mica flakes), whiskers (e.g. graphite, Al_2O_3, SiC, etc.) or particulate (metallic or non-matallic).

Some important polymer composites are described as follows.

8.18.1 Fibre Reinforced Plastics (FRP)

Fibre reinforced plastics (FRP) consists of fibres like glass, graphite, Al_2O_3, polyamides, cotton, etc.) as reinforcing materials in the from of dispersed phase. The type of fibre used depends upon the desired properties of the composite material.

FRP are prepared by reinforcing the fibre material with a resin matrix (PVC, polyamide, melamine resin, bakelite, silicon resin, etc.) under high temperature and pressure conditions. Some fibre reinforced plastics are:

(a) *Glass–fibre reinforced plastic (GRP):* GRP is prepared by reinforcing the spun fibres of borosilicate glass with the matrix of polyester nylon, etc. GRP have low density, high tensile strength, excellent chemical and corrosion resistance, and high impact resistance but limited working temperature and low stiffness.

   GRP is generally used in storage tanks, plastic pipes, automobile parts and transportation industries.

(b) *Carbon fibre reinforced plastics (CRP):* These are prepared by reinforcing high performance carbon fibres with epoxy resin or polyesters. CRP possesses high thermal stability and excellent resistance to corrosion along with lighter weights. These are used in the body, wings and stabilizer of helicopter, wings of military and commercial aircrafts, sports equipments, finishing roads, etc.

(c) *Aramid fibre reinforced composite (ARC):* These are prepared by reinforcing short fibres or long fibres of aromatic polyamides (aramides) like kevlar and nomex.

1. *Short fibres reinforced composites:* They possess inherent toughness, strength, and thermal stability and also possess high wear and tear resistant. They are used in making automobile clutches and breaks.
2. *Long fibres reinforced composites:* They possess ductility like metals. They can absorb energy like compressive collapse. They are used in aircraft industry for primary structure and components. They are also used in manufacturing thermal cuts and ballistics.

8.18.2 Polymer Blends

"The mixed materials of two or more incompatible polymers in which the polymeric chains of the constituent polymers are held together by van der Waal's forces of attraction, hydrogen bonding, etc. are called polymer blends".

Blending improves the resistance to abrasion and erosion, and prevents polymers from degradation. Important examples of polymer blends are:

(a) *Acrylonitrile butadiene styrene (ABS) plastic:* These are made by blending acrylonitrile–styrene polymer with butadiene–styrene rubber.

Properties: It has attractive appearance. It has high impact strength, high softening point, high moulding capacity and bright reflecting surface (electroplated ABS).

Uses: Acrylonitrile butadiene styrene (ABS) plastics are used in panels, door bands, door covers, etc. Electroplated ABS is used for making reflectors, name plates, etc.

(b) *Polydimethylphenylene (PDP):* These polymer blends possess dielectric properties over a wide range of temperatures, resistance to hydrolysis and low water absorption capacity. These are used in electrical industries, radio and television parts, and automobile parts.

(c) *Polycarbonate (PC):* Acrylonitrile–butadiene–styrene (ABS) blends possesses high temperature resistance and workability. These are used in electrical housings and machine parts.

(d) *Nylon-6-polycarbonate blends:* These are used in making sports equipments and transport containers.

8.18.3 Polymer Alloys

"These are the compatible mixtures of two or more polymers which interact chemically in the presence of compatibles, (e.g. chlorosulphonated polyethylene) under specific set of conditions."

Alloying is generally done for obtaining a specific combination of properties like workability, impact resistance, abrasion resistance, etc.

Examples: ABS–PC alloy—used in making electronic printers, food mixtures, typewriters, helmets, grills, etc.

ABC–PVC alloy—used in making TV components, electronic switch devices, dash boards, etc.

SUMMARY

Polymers: The process of conversion of monomers into polymers is called *polymerization*.

Degree of polymerization: The number of repeating units (monomers) in the polymer chain so formed is called degree of polymerization.

Types of Polymers

On the basis of origin: Natural polymers and synthetic polymers.

On the basis of structure: Linear chain polymers, branched chain polymers, cross-linked polymers.

On the basis of nature of monomer units: Homopolymers and copolymers.
On the basis of mode of synthesis: Addition polymers and condensation polymers.
On the basis of molecular forces: Elastomers, fibres, thermoplastics and thermosets.
On the basis of mechanism: Chain growth polymers and step growth polymers.

Types of polymerization: (1) Addition or chain polymerization, (2) condensation polymerization and (3) coordination polymerization.

Thermoplastics: These are the linear, long chain polymers having moderate intermolecular forces of attraction, they can be reshaped as melt on heating and therefore can be remoulded again and again. Examples: polyvinyl chloride (PVC).

Thermosetting polymers (thermosets): These are the cross-linked polymers which once set (solidified) during moulding cannot be softened again and reshaped even by strong heating. They have cross-linked structure.

Elastomers: The polymers which have elastic character like rubber are called *elastomers*.

Natural rubber: Natural rubber is consists of basic material latex (obtained from rubber tree (Havea brasiliensis and Guayule) which is a dispersion of isoprene.

Vulcanization of rubber: Vulcanization is the process of hardening of rubber materials by introducing the cross-linking between chains and thus preventing the intermolecular movement of rubber coiling (springs).

Synthetic rubber: Elastomer is any man-made rubber like vulcanisable polymer which can be stretched to at least twice of its length. On removing stress (face), it acquires its original shape.

Biopolymers: The polymers which are obtained by living organisms are called *biopolymers,* and the process of production of biopolymers by the combination of biomolecules (monomers) is called *biopolymerization.*

Biodegradable polymers: The polymers which degrade or decompose by the biological actions like eating, digestion, hydrolysis, enzymatic action of macro and micro-organisms are called *biodegradable polymers,* while the process of production of biodegradable polymers is called the *biodegradable polymerization.*

Silicones or polysiloxanes (inorganic polymers): Silicones are the synthetic, inorganic, organosilicone polymers constitute of repeated R_2SiO units held together by Si—O—Si linkage.

High polymer silicones or silicon gums: These are high molecular mass linear polymers. These are prepared by equilibration of cyclic siloxanes in the absence of non-functional groups.

Silicone rubber or elastomers: Silicone rubber is obtained by compounding of high molecular mass polydimethyl silane with fillers (like finely divided SiO_2), cross-linking agent (benzoyl peroxide) and other additives as per requirements.

Solid silicones or silicone resins: These are highly cross-linked polysilanes having high molecular mass. These are prepared by bi and trifunctional silanes (dichlaro and trichlorosilanes).

Polymer composites: A polymer composite may be defined as any multiphase material consists of at least one polymer and other polymeric or non-polymeric material.

Polymer blends: "The mixed materials of two or more incompatible polymers in which the polymeric chains of the constituent polymers are held together by van der Waal's forces of attraction, hydrogen bonding, etc. are called polymer blends."

Polymer alloy: "It is a compatible mixture of two or more polymers which interact chemically in the presence of compatibles under specific set of conditions."

EXERCISES

1. What is polymer and polymerization? Give examples.
2. Explain the types of polymers on different basis giving examples of each.
3. What is meant by functionality in a monomer?
4. Describe the various types of polymerization giving suitable examples.
5. What are the different types of mechanisms of the polymerization? Explain free radical polymerization mechanism in detail.
6. Classify the polymers on the basis of structure, i.e. linear, branched and cross-linked polymers. Give suitable examples.
7. Distinguish between addition and condensation polymerization.
8. What do you understand by sterio-specific polymerization? Explain with examples.
9. Explain the effect of structure on properties of polymers.
10. Explain the preparation, properties and uses of the following thermoplastic polymers:
    (i) polyvinyl chloride (PVC),
    (ii) polyvinyl acetate (PVA),
    (iii) teflon (PTFE).
11. Distinguish between the following:
    (a) Addition (chain growth) and condensation (step growth) polymerization
    (b) Thermoplastic and thermosetting
    (c) Free radical and ionic polymerization
    (d) Homopolymers and copolymers
    (e) Simple polymers and biopolymers
    (f) Simple polymers and biodegradable polymers.
12. What are resins? Give the details of various types of resins with examples.
13. Give reasons of the following:
    (i) Ionic polymerizations are carried out at low temperatures.
    (ii) Chain transfer to polymer in free radical polymerization leads to branching and cross-linking.
    (iii) Free radical polymerizations are carried out in nitrogen atmosphere.
    (iv) Oxygen can act as both initiator or inhibitor in free radical polymerization of vinyl monomers.
    (v) Molecular weight increases with time in free radical polymerization.
    (vi) Only very small amount (0.1%) of quinine is added as inhibitor during preparation and storage of vinyl monomers.
    (vii) There is continuous elimination of byproducts during condensation polymerization.
    (viii) In suspension polymerization method, the polymer separates out in the form of pearls or beads.
    (ix) In bulk polymerization, thermal control is difficult.
    (x) All simple molecules are not monomers.
    (xi) The functionality of phenol is three.
    (xii) Collision between two growing chains results in a dead polymer.
    (xiii) Is it impossible to assign an exact molecular weight to a polymer.
    (xiv) Fillers are used in compounding of plastics.

(xv) Addition polymerization leads to the formation of linear polymers.
(xvi) Elastomers undergo reversible elongations.
(xvii) PVC is tougher and stronger than polythene?
14. Discuss the preparation, properties and uses of phenolic resins.
15. Differentiate between thermoplastics and thermosetting resins. Give two examples for each type.
16. Justify the following statements:
    (i) Thermoplastics become soft on heating.
    (ii) Thermoset articles cannot be reshaped by application of heat and pressure.
17. What are polymer composites. How are they prepared? Explain their applications.
18. What are the general characteristics of composites? Discuss the advantages and limitations of composite materials.
19. Write general characteristics of polymers in comparison with common organic compounds?
20. Classify the following into plastics, elastomers and fibres:
    (i) PVC,
    (ii) polystyrene
    (iii) natural rubber,
    (iv) nylon,
    (v) silicon rubber
    (vi) tereline.
21. Write the preparation, properties and uses of the following polymers:
    (i) PVC,
    (ii) PVA,
    (iii) teflon,
    (iv) bakelite,
    (v) UF resin,
    (vi) SBR,
    (vii) NBR,
    (viii) silicone rubber,
    (ix) protein
    (x) DNA and RNA,
    (xi) carbohydrates,
    (xii) cellulose, (xiii) carbohydrates.
22. What are elastomers? Give the preparation and uses of GR-N and silicone rubber.
23. What do you understand by biopolymers and biopolymerization? Explain the sources and applications of various biopolymers.
24. What is a rubber? Distinguish between a natural rubber and an elastomer.
25. Describe the preparation, properties and uses of any two commercially important elastomers.
26. Describe the composition and uses of polymer composites.
27. What are elastomers? Differentiate between a natural rubber and an elastomer.
28. Describe the preparation, properties and uses of any two commercially important elastomers.

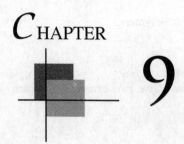

CHAPTER 9

ANALYTICAL METHODS

9.1 INTRODUCTION

Analytical chemistry is an effective and accurate tool for determining qualitative and quantitative compositions of substances and chemical species.

Analytical chemistry is integrated with other scientific and industrial disciplines.

"Analytical chemistry may be defined as the branch of science which deals with the study of determination of composition of materials in terms of the elements or compounds contained".

Sometimes, it is also considered as the art of determining the quality and purity of the materials.

9.2 TYPES OF CHEMICAL ANALYSIS

Chemical analysis is of two types:
  (a) *Qualitative Analysis:* It deals with the study to find out what a substance is composed of.
  (b) *Quantitative Analysis:* It deals with the study to find exactly how much the components are present in a substance.

In analytical chemistry, we get the information about the qualitative and quantitative composition of substances and chemical species. Analytical methods can be broadly classified into following four categories:

1. Gravimetric methods
2. Volumetric methods
3. Optical methods
4. Electrical methods
5. Separation methods.

9.2.1 Gravimetric Methods

It is the technique which is based on the weight changes in the sample during heating. It involves the separation of a substance from the solid or the solution of the weighed sample into a pure weighable form of a stable compound of known *percentage composition*.

This analysis can be carried out by: precipitation, electro-deposition and volatilization.

9.2.2 Volumetric Methods

In this technique, the volume of titrant (known solution) used for given volume of unknown solution is measured. This technique is also known as *titration*. Thus,

"The titration is the volumetric technique which is used to determine the strength of unknown solution by measuring the volume of known (standard) solution used with the measured volume of unknown solution using a suitable indicator or instruments for detecting the end point."

Titrations are of the following types:

(a) *Acid–base titrations:* Example—titration of hydrochloric acid against sodium hydroxide solution.

$$H^+ + Cl^- + Na^+ + OH^- \longrightarrow + Na^+ + Cl^- + H_2O$$

(b) *Redox titrations:* Example—titration of an iron ore against potassium dichromate solution.

$$6Fe^{2+} + Cr_2O_7^{2-} + 14H^+ \longrightarrow 6Fe^{3+} + 2Cr^{3+} + 7H_2O$$

(c) *Complexometric titrations:* Example—titration of hard water against EDTA solution.

$$H_2EDTA^{2-} + Ca^{2+}/Mg^{2+} \longrightarrow Ca/Mg - EDTA^{2-} + 2H^+$$

(d) *Precipitation titrations:* Titration of silver nitrate solution against hydrochloric acid.

$$Ag^+ + NO_3^- + H^+ + Cl^- \longrightarrow AgCl \downarrow + H^+ + NO_3^-$$

9.2.3 Optical Methods

These methods are based on the interaction behaviour of matter (sample) with electro-magnetic radiations (EMR). Most important optical properties which can be correlated with concentration are:

1. The absorption or emission of radiant energy
2. The bending of radiant energy
3. The scattering of radiant energy
4. The delayed emission of radiant energy.

9.2.4 Electrical Methods

These methods involve electronic instruments which are used to measure or produce electrical phenomena. Flow of current as a function of time, potential developed or required, availability to pass current, resistance, etc. are the important properties which are related to the reaction taking place.

9.2.5 Separation Methods

These methods are based on separation of useful components from useless, and purification of the compounds, e.g. chromatographic techniques, solvent extraction, etc.

9.3 THERMOGRAVIMETRIC ANALYSIS (TGA)

"Thermogravimetric analysis (TGA) is an analytical technique used to determine a material's thermal stability and its fraction of volatile components by monitoring the weight change that occurs as a specimen is heated."

9.3.1 Types of TGA

It is of two types:
  (a) *Dynamic TGA:* It involves the measurement of weight loss of the sample with continuously increasing the temperature as a function of time.
  (b) *Static or isothermal TGA:* In this analysis, the sample is maintained at constant temperature over a certain period of time and during which the weight loss of the sample is measured.

The measurement is normally carried out in air or in an inert atmosphere, such as helium or argon and the weight loss is recorded as a function of increasing temperature. In the presence of air the oxidation may be very fast, therefore sometimes, the measurement is performed in a lean oxygen environment (1–5% O_2 in N_2 or He) to slow down oxidation.

Apparatus

The key features of a thermogravimetric analysis kit are as follows (Figure 9.1):

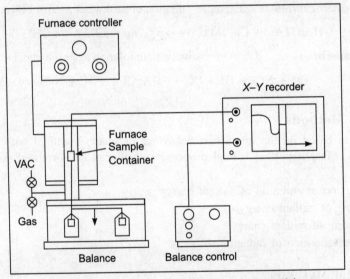

Figure 9.1 A modern thermobalance (schematic).

1. *The sample holder assembly:* It consists of a sample holder surrounded by a metal block. The sample is contained in a small crucible made up of materials such as pyrex, silica, nickel or platinum, depending on the temperature and nature of the test involved.
2. *Furnace:* It consists of a metal block fitted with two heating coils in order to supply the heat constantly. The furnace should have sufficiently hot zone and able to respond rapidly to be commanded from temperature programmer.
3. *Temperature programmer:* It is the electronic control device which controls the temperature of the furnace automatically in order to obtain constant heating rate.
4. *Recording system:* It is the device which has a low inertia to authentically reproduce variations in the experimental set up.

Principle and Method

The principle of the technique can be illustrated by the weight loss curve of a compound, calcium oxalate hemihydrates ($CaC_2O_4 \cdot H_2O$) as shown in Figure 9.2.

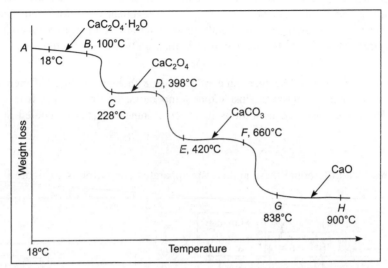

Figure 9.2 Weight loss curve (thermogramme) for the calcium oxalate.

It is clear from Figure 9.2 that when the calcium oxalate hemihydrate, $CaC_2O_4 \cdot H_2O$ is heated in a thermobalance at a predetermined rate, the water is evolved starting at point A (18 °C), this point is called *minimum weight loss temperature*. The compound remains stable up to point B (100°C). The break in the curve and weight loss from point B to C is observed due to the elimination of water molecule to form CaC_2O_4. This CaC_2O_4 remains stable up to point D (398°C). On further heating the CaC_2O_4 decomposes at point D (398 °C) to give $CaCO_3$ and results in the weight loss up to point E (420°C) due to elimination of CO_2. This $CaCO_3$ is stable up to point F (660 °C) and decomposes again at point F to give CaO. Weight loss is observed from point F to point G (838°C) due to the elimination of CO_2. It is clear from the curve that CaO is stable after point G (838 °C) and exist as simple oxide.

Applications of TGA

1. TGA is used in automatic gravimetric analysis.
2. It is used in the study of new weighing compositions in gravimetric analysis and determination of their thermal stability range.
3. It is used in the weighing of substances which are unstable at ambient temperature and absorb CO_2 and O_2 at room temperature.
4. It is used in the determination of the composition of the complex mixture.
5. It is used in the determination of the purity and thermal stability of analytical reagents including primary and secondary standards.
6. It is used in the correction of errors in classical gravimetric analysis.
7. It is used in the study of the sublimation behaviour of various substances.
8. It is used in the study of suitability of preparation of materials in relation to the methods of their preparation.

9.4 DIFFERENTIAL THERMAL ANALYSIS (DTA)

In this technique the heat effect is associated with chemical and physical changes of a substance which are recorded at a linear rate.

Differential thermal analysis (DTA) is defined as the technique which involves the heating or cooling of a test sample and an inert reference under identical conditions and measurement of temperature difference of a sample and thermally inert material (alumina) as a function of sample's temperature or time. This differential temperature is then plotted against temperature or time.

Or

It may also be defined as the technique of recording the temperature difference between a substance and a reference material against either temperature or time as the two specimens are subjected to identical temperature conditions in an environment heated or cooled at a controlled rate.

Apparatus

The key features of a differential thermal analysis apparatus are as follows (Figure 9.3):

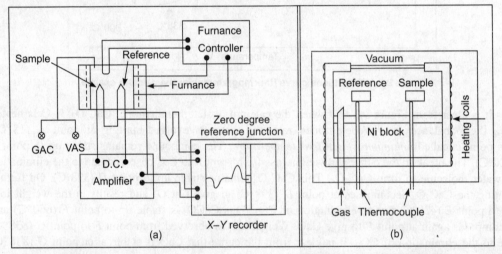

Figure 9.3 (a) Schematic illustration of a DTA cell, (b) Structure of a furnance.

- (a) *Sample holder:* The sample holder assembly consists of a thermocouple each for the sample and reference, surrounded by a block to ensure an uniform heat distribution. The sample is contained in a small crucible designed with an indentation on the base to ensure a snug over the thermocouple bead. The crucible may be made up of materials such as pyrex, silica, nickel or platinum, depending on the temperature and nature of the tests involved.
- (b) *Furnace:* It consists of a metal block fitted with two heating coils in order to supply the heat constantly. The furnace should be stable and sufficiently hot zone and must be able to respond rapidly to be commanded from temperature programmer.
- (c) *Temperature programmer:* Electronic control device which controls the temperature of the furnace automatically in order to obtain constant heating rate.
- (d) *Recording system:* It is the device which has a low inertia to authentically reproduce variations in the experimental set up.

Principle

When a sample is subjected to the heat change the changes in the sample which lead to the absorption or evolution of heat can be detected relative to the inert reference. The thermograme of DTA shows peaks or bands. The *endothermic changes* like vaporization, absorption, desorption, fusion, sublimation, etc. appear below zero whereas the *exothermic changes* like oxidation, decomposition, etc. appear above zero of differential temperature. Differential temperatures can also arise between two inert samples when their response to the applied heat treatment is not identical. A DTA curve can be used as a finger print for identification purposes. For example, in the study of clay where the structural similarity of different forms renders diffraction experiments which is difficult to interpret. The area under a DTA peak can be due to the enthalpy change and is not affected by the heat capacity of the sample. The DTA curve for clay mineral halloysite, $Al_2Si_2H_5(OH)_4$ is shown in Figure 9.4.

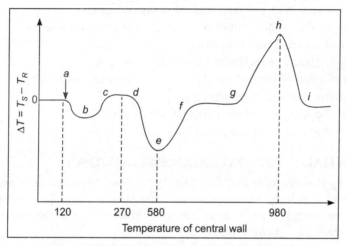

Figure 9.4 DTA curve for clay mineral halloysite.

In this figure, 'a' represents the point where water begins to evaporate off and the heat absorbed results in the thermocouple junction in the sample to delay in temperature behind the junction in the inert material. The trough 'abc' in the curve shows the completion of the evaporation of water. Deeper loop 'def' in the curve shows the decomposition of mineral into water, Al_2O_3 and SiO_2. Sharp peak 'ghi' in the curve is due to the liberation of big amount of heat during the crystallization of Al_2O_3. It is obvious that the upward loop results from the exothermic change while downward loop results from the endothermic change.

For the sample block constructed from a metal such as nickel, containing cylindrical sample holder. S.L. Boersma in 1955 derived the following equation for peak area.

$$\text{Peak area } (A) = \int_{T_1}^{T_2} \Delta T \cdot dt = \frac{m \cdot \Delta H}{g \cdot \lambda} \qquad (9.1)$$

where T_1 = time at the start of the peak, T_2 = time at the end of the peak, ΔT = the differential temperature, ΔH = heat of transition per unit volume, λ = thermal conductivity of the sample and dt = differential operation, g = geometrical factor of the apparatus, m = mass of the sample.

Boersma suggested the use of metal sample and reference cups in which the temperature difference was measured from outside the sample and reference material. Under these conditions the peak area will depend upon the heat of transformation and calibration factor of the instrument.

Hence, heat of reaction, ΔH can be given as

$$\Delta H = \frac{\psi}{m} \int_{T_1}^{T_2} \Delta T \cdot dt \qquad (9.2)$$

where ψ = experimentally determined constant and m = mass of the sample.

Applications of DTA

1. In the study of phase transition.
2. In the determination of melting point and boiling point of organic compounds.
3. In the characterization of polymers and other organic compounds.
4. In industries for the quality control of materials like cement, glass, etc.
5. In the study of ceramics and minerals.
6. In the study of effects of radiations on polymeric materials.
7. In the determination of thermal stability of compounds or complexes.
8. In the qualitative analysis of polymeric mixtures.
9. In analytical chemistry to test the purity of substances.
10. In the study of specific heat and heat of reaction.

9.5 DIFFERENTIAL SCANNING CALORIMETRY (DSC)

This technique is very closely related to DTA. "The DSC may be defined as the instrumental approach in which the sample as well as reference material subjected to a closely controlled programmed temperature. The heat energy is added to or subtracted from the sample or reference container is recorded as a function of temperature or time.

Apparatus

The apparatus of the DSC is identical to that of the DTA. The main difference is that the DSC apparatus consists of individual heaters for sample and reference materials (Figure 9.5).

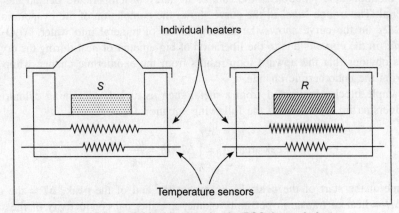

Figure 9.5 Individual heaters for the DSC thermobalance.

The main parts of this apparatus are:
1. *Sample holder:* The sample holder assembly consists of a thermocouple each for the sample and reference.
2. *Furnace:* It consists of individual heaters for the sample and reference materials fitted in a metal block.
3. *Temperature programmer:* It is the electronic control device which controls the temperature of the furnace automatically in order to obtain constant heating rate.
4. *Recording system:* It is the device which has a low inertia to authentically reproduce variations in the experimental set up.

Principle and method

This technique is similar to that of the DTA. In this method, the sample and reference material also subjected to a closely controlled programmed temperature. The heat energy is added to or subtracted from the sample or reference containers in order to maintain both reference and sample at the same temperature. The heat energy is recorded as a function of temperature or time.

Whenever, the temperature difference between the sample and reference material is noticed by thermocouple due to exothermic or endothermic reaction, heat is added to the cooler system so that the temperature equality is maintained and the required heat is continuously recorded as a function of temperature or time. As this energy, input is precisely equivalent in magnitude to the energy absorbed or evolved in the particular transition; a recording of this balancing energy yields a direct calorimetric measurement of the transition energy.

A DSC thermogram is almost similar to that of DTA thermogram. A +ve sign is given in case of heat supplied to the sample and –ve sign in case of heat supplied to the reference material. The peak area of DSC thernogram can be calculated with the help of following equation:

$$\text{Peak area } (A) = k' \cdot m \, \Delta H \tag{9.3}$$

where k' = proportionality constant which is independent of temperature, m = mass of sample and ΔH = heat of reaction.

or
$$A \propto m \, \Delta H \tag{9.4}$$

Thus, the heat of reaction ΔH can be determined by using the following equation:

$$\Delta H = \frac{A}{k'm} \tag{9.5}$$

Applications of DSC

1. In the determination of melting and boiling point of the organic compounds.
2. In checking the purity (within 1% error) of the compounds as each pure compound has its own characteristic thermogram.
3. In the characterization of organic compounds.
4. In the determination of enthalpy of transitions.
5. In the determination of purity and composition of drugs (Figure 9.6).

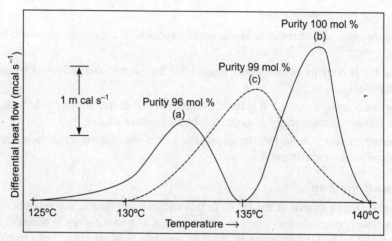

Figure 9.6 DSC curve of phenacetine drug.

DTA and DSC are differentiated in Table 9.1.

Table 9.1 Difference between DTA and DSC.

DTA	DSC
1. DTA deals with the measurement of temperature difference between the sample and reference as a function of sample's temperature or time.	1. DSC deals with the measurement of heat added or subtracted from the sample or reference as a function of temperature or time.
2. Size of sample may be 5–20 mg.	2. Size of sample may be 2–10 mg.
3. Secondary power source is not required.	3. Secondary power source is required as to keep the sample or reference at equal temperature.
4. Area of DTA peaks are not exactly equal to the heat loss or heat gain.	4. Area of DSC peaks are directly related to the heat loss or heat gain (enthalpy change).
5. Graph is plotted between ΔT and T.	5. Graph is plotted between $\Delta H/dt$ and T.
6. Optimum temperature range is −190 to 1000°C.	6. Optimum temperature range is −170 to 750°C.

9.6 SPECTROSCOPY

"Spectroscopy is the branch of science which deals with the study of interaction of electromagnetic radiations with matter and its properties by investigating light, sound or particles that are emitted, absorbed or scattered by the matter under investigation."

Historically, the spectroscopy is referred to a branch of science in which *visible light* was used for theoretical studies on the structure of matter and for qualitative and quantitative analysis.

Recently, however, the definition has been broadened as new techniques have been developed that utilize not only visible light, but many other forms of electromagnetic and non-electromagnetic radiations like microwaves, radiowaves, X-rays, electrons, phonons, sound waves and others.

It is the most powerful tool available for the study of atomic and molecular structures.

9.6.1 Properties of Electromagnetic Radiations (EMR)

The electromagnetic radiations are the radiations in which the electric field and the magnetic field oscillate perpendicular to each other and both are perpendicular to the direction of propagation of wave.

There are some general properties shared by all the forms of electromagnetic radiations:
1. Electromagnetic radiations (EMR) are propagated by oscillating electric and magnetic fields at right angles to each other.
2. They travel with a constant velocity of 3×10^8 m s^{-1} in vacuum.
3. They are not deflected by electric or magnetic field.
4. They can show interference or diffraction.
5. They can travel through empty space.
6. The speed of light is constant in space. All forms of light have the *same* speed $c = 299,800$ km s^{-1} in space.
7. White light comprises of different colours (wavelengths). When white light is passed through a prism or diffraction grating, it is spread out into all of its different colours.

Wavelength (λ): Wavelength of light is defined as the distance between two successive crests or troughs. Wavelength of visible light, determines its colours, is inversely related to its energy. The radiations with short wavelength have high energy and vice versa [nanometre (nm) or centimetre (cm)] (Figure 9.7).

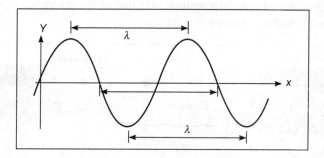

Figure 9.7 Wavelength of an EM wave.

1 Å = 0.1 nm = 10^{-8} cm = 10^{-10} m. Radio wavelengths are often measured in centimetres: 1 cm = 10^{-2} m. Visible light has wavelength range of 4000–8000 Angstroms (Å).

Frequency (v): It is the number of crests of the wave that pass by a point every second, i.e. the number of waves per second.

$$\text{Frequency } (v) = \frac{\text{Speed of light (cm s}^{-1})}{\text{Wevlength (cm)}} = \frac{c}{\lambda} \quad (9.6)$$

Frequency is measured in units of *cycles per second, i.e. Hertz* (Hz): 1 hertz =1 wave crest s^{-1}.

Velocity (c): The distance travelled by the wave in one second is called velocity of the wave.

Velocity, $c = v\lambda$ (velocity of light $c = 3.0 \times 10^{10}$ cm s^{-1} = 3.0×10^8 m s^{-1})

Wave Number ($\overline{v}$): It is the number of waves spread in a one centimetre length.

$$\text{Wave number, } \overline{v} = \frac{1}{\lambda} \quad (9.7)$$

Radiations with large number of waves have high energy and vice versa.

9.6.2 Origin of Spectra

Spectrum can be originated from atoms or molecules due to the energy changes.

(a) *Atomic spectra:* According to quantum mechanics the energy levels of all the systems are quantized and are designated by appropriate quantum numbers. These energy levels are obtained by solving the time independent Schrödinger wave equation. Suppose a photon of frequency v, falls on a molecule in the ground state and its energy is exactly equal to the energy difference between two energy levels, then the molecule undergoes a transition from the lower energy level to higher energy level with the absorption of photon of energy hv. The spectrum so obtained due to the absorption of radiations by the system is called *absorption spectrum*. If the molecule falls from a higher energy level, i.e. excited state to the ground state with the emission of a photon of energy hv, the spectrum so obtained is called *emission spectrum*.

According to Bohr's atomic model, an electron passing from a lower orbit to a higher orbit must absorb light with energy exactly equal to the difference in energies between the two orbits. Similarly, for an electron to move from a higher orbit to a lower one, it must emit light corresponding to energy exactly equal to the difference in energies between these orbitals. This model allows a simple understanding of the absorption or emission of light by atoms that explained line spectra (Figure 9.8).

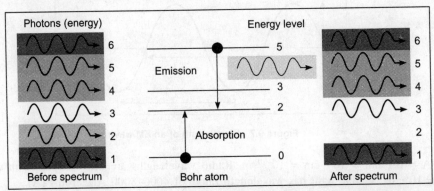

Figure 9.8 Electronic transition and absorption and emission spectra according to Bohr's model.

(b) *Molecular spectra:* The molecular spectra arises from the three types of transitions, i.e. rotational, vibrational and electronic transitions. According to the Born–Oppenheimer, approximation, the total energy of the molecule is given by Eq. (9.8):

$$E = E_t + E_r + E_v + E_e \tag{9.8}$$

Thus, the total energy of the molecule is the sum of translational energy (E_t), rotational energy (E_r), vibrational energy (E_v) and electronic energy (E_e) (Figure 9.9). Except translational energy, all energies are quantized. Hence,

$$E_t \gg E_r \gg E_v \gg E_e$$

The total energy of molecule, $E = E_r + E_v + E_e$.

As the translational energy is very small so it is neglected. Molecular spectra in various regions can be given by various energy levels. The rotational spectra of a molecule is observed in microwave region, the vibrational spectra in infrared (IR) region, the electronic spectra in ultraviolet (UV) and visible region of electromagnetic spectrum.

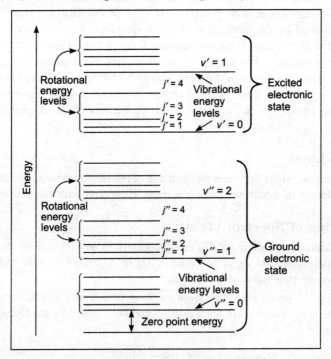

Figure 9.9 Energy level diagram showing various molecular energy levels.

9.6.3 Selection Rule

The molecular spectra are governed by the selection rule which specify the changes in the quantum number accompanying the particular electronic transition. This rule is the backbone of spectroscopy and obtained by the quantum theory of interaction of radiations. A *selection rule* or *transition rule* is a constraint on the possible transitions of a system from one state to another. Selection rules have been derived for electronic, vibrational and rotational transitions. For a diatomic molecule such as NO_2, CO_2, NO_2, etc. the selection rule can be written as:

Pure rotational transition:
$$\Delta J = \pm 1 \tag{9.9}$$

where J is the rotational quantum number.

Pure vibrational transition:
$$\Delta v = \pm 1 \tag{9.10}$$

where v is the vibrational quantum number.

This rule, however, is not obeyed by all transitions.

There are two types of electronic transitions, i.e. allowed transitions and forbidden transitions. In general, the allowed transitions are stronger than forbidden transitions.

(a) *Allowed transition:* The spectral transitions which obey the *selection rule* are *called allowed transitions*. Transition between two electronic states is allowed to take place according to selection rules, associated with group theory. The probability of a transition between states m and n produced by the interaction of electromagnetic radiation with an atomic system is proportional to the square of the magnitude of the matrix elements of the electric dipole.

(b) *Forbidden transitions:* The transitions which violate the selection rule are called *forbidden transitions*. Transition between different energy levels of some atoms or ions in a quantum mechanical system that is not allowed to take place because of selection rules. In practice, forbidden transitions can occur, but they do so with much lower probability than allowed transitions.

9.6.4 Spectral Lines

Spectral lines are dark or bright lines in a uniform and continuous spectrum, resulting from an excess or deficiency of photons in a narrow frequency range compared with the nearby frequencies.

9.6.5 Broadening of Spectral Lines

In the study of transitions in atomic spectra, and definitely in any type of spectroscopy, it has been observed that transitions are not exactly "sharp". There is always a finite width to the observed spectral lines. There are two types of broadening:

(a) *Natural linewidth or lifetime broadening:* This broadening arises from the uncertainty in energy of the states involved in the transition and determined by the Heisenberg's uncertainty principle, i.e.

$$\Delta E \cdot \Delta t \geq h/4\pi \quad (9.11)$$

where ΔE is the uncertainty in energy and Δt is the uncertainty in life-time of the energy level, h is the plancks constant.

For a photon the energy is $E = h\nu$, therefore,

$$\Delta E = h \cdot \Delta \nu \quad (9.12)$$

Hence, the natural linewidth $\Delta \nu$ is given by

$$\Delta \nu \geq (4\pi \cdot \Delta t)^{-1} \quad (9.13)$$

This source of broadening is important in nuclear spectra, such as Mossbauer spectra, but it is rarely significant in atomic spectroscopy. A typical lifetime for an atomic energy state is about 10^{-8} s, corresponding to a natural linewidth of about 6.6×10^{-8} eV.

(b) *Doppler broadening:* This broadening arises when the molecule under investigation has a velocity relative to the observer or observing instrument. For atomic spectra in the visible and UV region, the limit on resolution is often set by Doppler broadening. With the thermal motion of the atoms, those atoms travelling towards the detector with a velocity v, will have transition frequencies, which differ from those of atoms at rest, by the Doppler shift.

Since the thermal velocities are non-relativistic, the Doppler shift in the angular frequency is given by the following equation:

$$\omega = \omega_0 \left(1 \pm \frac{v}{c}\right) \quad (9.14)$$

where ω = frequency for an atom at rest.

From the Boltzmann distribution, the number of atoms with velocity v in the direction of the observed light is given by

$$n(v)\, dv = N \sqrt{\frac{m_0}{2\pi kT}}\, e^{-m_0 v^2 / 2kT}\, dv \tag{9.15}$$

where N = total number of atoms, m_0 = atomic mass.

9.6.6 Chromophores

"The covalently unsaturated group responsible for electronic absorption is called *chromophore*." The colour usually appears in an organic compound if it contains certain unsaturated groups.

When a molecule absorbs certain wavelengths of visible light and transmits or reflects others, the molecule will have a colour.

"A chromophore is a region in a molecule where the energy difference between the two different molecular orbitals falls within the range of the visible spectrum." Visible light that hits the chromophore can thus be absorbed by an electron exciting from its ground state to an excited state.

According to the modern definition "the chromophore is any isolated functional group that exhibit characteristic absorption in the UV–visible region. These are the organic functional groups which undergoes $n \rightarrow \pi^*$ and $\pi \rightarrow \pi^*$ electronic transitions, and are conveniently classified as chromophoric groups or chromophores". A compound containing a chromophore is called a *chromogen*.

9.6.7 Types of Chromopores

There are two types of chromophores:

(a) *Independent chromophore:* When a single chromophore is sufficient to impart colour to the compound is called *independent chromophore*. For example, azo groups. —N=N—, nitrosogroup, —NO, and *o*- and *p*-quinonoid group, etc. are independent chromophores.

(b) *Dependent chromophore:* When more than one chromophores are required to produce colour in the chromogen is called *dependent chromophore*.

For example, >C=O group, >C=C< group, etc. Acetone having one ketonic group is colourless whereas diacetyl having two ketonic groups is yellow and triketopantane having three ketonic groups is orange. It is due to the presence of number of unpaired electrons. In acetone, $n = 0$, in diacetyl, $n = 3$ and in triketopantane, $n = 5$.

9.6.8 Auxochrome

It is a group of atoms with non-bonded electrons attached to a chromosphere, which when attached to a chromophore, modifies both the wavelength and intensity of absorption, i.e. the ability of that chromophore to absorb light. Example: —OH, —NH$_2$, aldehydes.

If these groups are in direct conjugation with the pi-system of the chromophore, they may increase the wavelength at which the light is absorbed and as a result intensify the absorption.

It increases the colour of any organic compound. For example, benzene does not display colour as it does not have any chromophore but nitrobenzene is pale yellow in colour because of the presence of nitro group. *para*-Hydroxynitrobenzene exhibits a deep yellow colour. Here, an auxochrome (—OH) is conjugated with the chromophore —NO$_2$.

There are mainly two types of auxochromes:
1. Acidic —COOH, —OH, —SO$_3$H
2. Basic —NHR, —NR$_2$, —NH$_2$, —CONH$_2$

The presence of an auxochrome in the chromogen (a substance capable of converting itself into a pigment or dye) molecule is essential to make a dye. Auxochromes generally deepen the colour of a chromogen, but cannot by themselves impart the colour to a compound. Auxochrome groups are of two types:

(a) *Bathochromic groups:* Those groups which deepen the colour (displacement to longer wavelengths) of the chromogen are called *bathochromic groups*, i.e. yellow → orange → red → purple → violet → blue → green.

Examples: primary, secondary or tertiary amino groups.

Bathochromic shift: The increase in absorption intensity due to the shifting of wavelength from lower to higher region (lower energy) is called *bathochromic shift*. This is also called red shift and caused due to the substitution or solvent effect (polarity).

(b) *Hypsochromic groups:* Those groups which diminish or weaken the colour (displacement to shorter wavelengths) of the chromogen are called *hypsochromic groups*. Examples: acetylation of —OH or NH$_2$ group, i.e. OCOCH$_3$ and —NHCOCH$_3$ causes hypsochromic shift.

Hypsochromic shift: The decrease in absorption intensity due to the shifting of wavelength from higher to lower region (higher energy) is called *hypsochromic shift*. This is also called *blue shift* and causes due to the substitution or solvent effect (polarity).

9.6.9 Types of Molecular Spectra

The various types of molecular spectra given by various species in different regions are given below:

(a) *Rotational spectra (microwave region):* Rotational spectra results from the transitions between the rotational energy levels of a gaseous molecule when radiations are absorbed in microwave region. The molecules like HCl, CO, NO, H$_2$O vapours, etc. which possess dipole moment, show this type of spectrum. While homonuclear diatomic molecules like H$_2$, N$_2$, etc. do not show these spectra. This spectrum occurs in the spectral region 1–100 cm^{-1}.

(b) *Vibrational and vibrational-rotational spectra (infrared region):* These spectra are obtained due to the transitions between the vibrational energy levels of a molecule on the absorption of radiations in the infrared region. Infrared spectra are shown by the molecules when vibrational motion is accompanied by a change in dipole moment of the molecule. The wavelength range for these spectra is 500–4000 cm^{-1}.

(c) *Raman spectra:* This spectrum is related to vibrational and/or rotational transitions in the molecules in a different manner. In this spectra, scattering of radiations is measured in place of absorption of radiations.

A beam of monochromatic radiations is allowed to fall on the sample in visible region and intensity of scattered radiation is observed at right angles to the incident beam. When the intensity of scattered radiations is same as that of incident radiations, it is called *Rayleigh scattering* and if there is small difference in the wavelength of incident radiations and scattered radiations, it is called *Raman scattering*. Raman spectra are observed in the wavelength region of 12,500–25,000 cm^{-1}.

(d) *Electronic spectra:* These specta originated from the electronic transitions in a molecule by the absorption of radiations in the visible and ultraviolet regions. Electronic spectra in the visible region range from 12,500–25,000 cm^{-1} and in the ultraviolet region from 25,000 to 70,000 cm^{-1}. The electronic spectrum of molecule is highly complex as it is associated with vibrational and rotational transitions.

(e) *Nuclear magnetic resonance (NMR) spectra:* These spectra originated due to the transitions induced between the nuclear spin energy levels of a molecule in an applied magnetic field. NMR is occurred in the radio frequency regions, i.e. 5–100 MHz.

(f) *Electron spin resonance (ESR) and electron paramagnetic resonance (EPR) spectra:* These spectra originated from the *transitions* induced between the electron spin energy levels of a molecule in an applied magnetic field. This spectrum is shown by the molecules which possess unpaired electrons, e.g. free radicals and transition metal ions. This spectrum is occurred in the microwave region, i.e. 2–9.6 GHz.

(g) *Mossbauer spectra or nuclear gamma resonance fluorescence (NRF) spectra:* These spectra resemble with NMR spectra. These are originated from the absorption of high energy γ-photons having frequency around 10^{13} MHz by the nuclei.

9.6.10 Types of Spectroscopy on the Basis of Nature of Sample

Generally, the spectroscopy is divided into two categories:

(a) *Atomic Spectroscopy:* This spectroscopy deals with the study of interaction of electromagnetic radiations with atoms which are usually in their lower energy state, i.e. ground state.

(b) *Molecular Spectroscopy:* This spectroscopy deals with the study of interaction of electromagnetic radiations with molecules. This interaction results into the transition between rotational and vibrational energy levels in addition to the electronic transitions which are usually in their lower energy state, i.e. ground state.

9.6.11 Types of Spectroscopy on the Basis of Absorption and Emission of Radiations

There are three main types of spectroscopy:

(a) *Absorption spectroscopy:* This spectroscopy uses the range of electromagnetic spectra in which a substance absorbs the EMR.

"The technique which deals with the study of absorption of electromagnetic radiation by matter is called *absorption spectroscopy*. In this spectroscopy, the intensity of a beam of light measured before and after interaction with a sample, i.e. intensity of incident and transmitted radiations is compared."

In atomic absorption spectroscopy, the sample is atomized and then light of a particular frequency is passed through the sample's vapour. After calibration, the amount of absorption can be related to the concentrations of various metal ions through the Beer–Lambert law. The method can be automated and is widely used to measure concentrations of ions such as sodium, potassium and calcium in blood and urine.

On the basis of the wavelength range of incident beam: The spectroscopy can be classified into:

1. Infrared spectroscopy
2. Near infrared spectroscopy
3. Microwave spectroscopy
4. UV–visible spectroscopy.

The absorption of radiations by any sample is governed by the Beer–Lambert law. The plot of amount of radiation absorbed versus wavelength for a particular compound is referred to as the absorption spectrum.

(b) *Emission spectroscopy:* This spectroscopy uses the range of electromagnetic spectra in which a substance emits the radiations. The substance first absorbs energy and then radiates this energy as light. This energy can be from a variety of sources, including collisions (either due to high temperatures or otherwise), and chemical reactions.

9.6.12 Some Common Terms Used in Laws of Absorption

(a) *Radiant power (P):* The rate at which the energy is transferred in a beam of radiant energy, i.e. radiant flux, is called *radiant power*.

(b) *Transmittance (T):* It is defined as the fraction of incident light at a specified wavelength that passes through a sample. Mathematically, it is the ratio of the intensity of the radient power passed out through the sample (P) to the intensity of incident radient power (P_0) i.e. Transmittence, $T = P/P_0$.

(c) *Absorbance (A):* It is the logarithm of the base 10 of the reciprocal of the transmittance, i.e. $A = \log 10\ (1/T) = \log 10$ or $A = \log P_0/P$. Alternate name of this term is *extinction* or *optical density*.

(d) *Absorptivity (a):* It is the ratio of the absorbance to the product of concentration and optical path length. It is a constant characteristic of the substance and wavelength. Alternate name of this term is *extinction coefficient* or *absorbance index*.

(e) *Molar absorptivity or molar extinction coefficient or molar absorbance index (ε):* It is the measurement of extent of light absorption by a chemical species at a given wavelength. If the absorptivity is measured in unit of $mol^{-1}\ cm^{-1}$ is called *molar absorptivity*. It is denoted by symbol, epsilon (ε). Thus, molar absorptivity, $\varepsilon = A/\ x \cdot C$. Alternate name of this term is *molar extinction coefficient* or *molar absorbance index*.

The ratio of the absorbance to the product of concentration and length of optical path is a constant characteristic of the substance and wavelength.

Significance of Molar Absorptivity (ε)

Molar absorptivity is a constant for a particular substance, so if the concentration of the solution is halved so is the absorbance, which is exactly expected.

Absorbance, $\qquad\qquad A = \varepsilon\ x \cdot C \quad$ or $\quad \varepsilon = A/x \cdot c \qquad\qquad$ (9.16)

This can be stated as "ε is a measure of the amount of light absorbed per unit concentration".

Let us take a compound with a very high value of molar absorptivity, say 100,000 $l\ mol^{-1}\ cm^{-1}$, which is in a solution in a 1 cm path length cuvette and gives an absorbance of 1.

$$\varepsilon = 1/x\ C \text{ therefore, } C = 1/100{,}000 = 1\ \varepsilon_1^{-5}\ mol\ l^{-1}$$

9.6.13 Laws of Absorption: Beer–Lambert Law

Absorption of monochromatic radiation (light) by homogeneous, transparent medium is governed by the following laws:
(a) Lambert–Beer's law and
(b) Beer's law

(a) *Lambert's law:* It is also known as Lambert–Bouguer's law. This law states that when a beam of monochromatic radiation is allowed to pass through a transparent medium, the rate of decrease of intensity of radiation with thickness of the homogeneous absorbing medium is directly proportional to the intensity of incident radiations.

Mathematically,

$$-\frac{dI}{dx} \propto I$$

or
$$\frac{dI}{dx} = -K \cdot I \tag{9.17}$$

where dI is a small change in the intensity of light when passed through a medium of small thickness dx and I is the intensity of incident light before absorption. K is the absorption coefficient; its value depends upon the nature of absorbing medium (Figure 9.10).

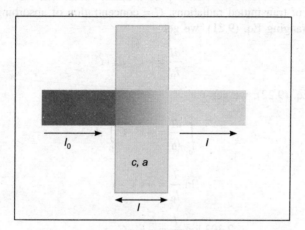

Figure 9.10 Absorption of light through a cuvette.

Equation (9.17) can be rearranged as

$$\frac{dI}{I} = -K \cdot dx \tag{9.18}$$

Equation (9.18) can be integrated within the intensity limit $I_0 \rightarrow I$ and thickness limit $0 \rightarrow x$

$$\int_{I_0}^{I} \frac{dI}{I} = -K \int_{0}^{x} dx$$

or
$$\ln \frac{I}{I_0} = -K \cdot x \tag{9.19}$$

On taking antilog of Eq. (9.19), we get

$$\frac{I}{I_0} = e^{-K \cdot x}$$

or
$$I = I_0 e^{-K \cdot x} \tag{9.20}$$

where K is the absorption coefficient which is equal to $k/2.3026$ and ratio I/I_0 is called *transmittance*, the ratio I_0/I is called *opacity*, and $\log I_0/I = A$ is called *absorbance*. Equation (9.20) is the mathematical form of Lambert's law.

(b) *Beer's law:* This law gives the relationship between the intensities of incident radiations of light and absorption capacity and concentration of solution.

"This law states that the rate of decrease of intensity of incident radiations with the thickness of absorbing medium (solution) is directly proportional to the intensity of incident radiations as well as the concentration of the solution."

Mathematically,

$$-\frac{dI}{dx} \propto I_0 \cdot C \quad \text{or} \quad -\frac{dI}{dx} = k' I_0 \cdot C \tag{9.21}$$

where dI is the decrease in intensity of incident radiations, dx = thickness of medium, I = intensity of transmitted radiations, C = concentration of absorbing solution.

On rearranging Eq. (9.21), we get

$$\frac{dI}{I_0} = -k' \cdot dx \cdot C \tag{9.22}$$

Integrating Eq. (9.22), we get

$$\int_{I_0}^{I} \frac{dI}{I_0} = -k' \cdot C \int_{0}^{x} dx \tag{9.23}$$

or

$$\ln \frac{I}{I_0} = -k' C \cdot x \tag{9.24}$$

or

$$2.303 \log \frac{I}{I_0} = -k' \cdot C \cdot x$$

or

$$I = I_0 e^{-k' C \cdot x} \tag{9.25}$$

Equations (9.24) and (9.25) can be written as

and

$$I = I_0 \, 10^{-a \cdot C \cdot x} \tag{9.26}$$

where

$$a = k'/2.303 \tag{9.27}$$

The value of a depends upon the units of C and x. If C = mol^{-3} dm^{-3} and x = cm then $a = \varepsilon$ (molar absorption coefficient or molar absorptivity or molar extinction coefficient)

$$I = I_0 \, 10^{-\varepsilon \cdot C \cdot x} \tag{9.28}$$

From Eq. (9.16), if ε is the constant and x is also constant (as in case of matched cell used in colorimeter or spectrophotometer), then

$$A \propto C \tag{9.29}$$

We know that $\dfrac{I}{I_0} = T$ (transmittance). Therefore, from Eq. (9.16)

$$A = \varepsilon \cdot C \cdot x = -\log T$$

or
$$A = \log \frac{1}{T}$$
or
$$A = -\log T \quad (9.30)$$

and molar absorptivity,
$$\varepsilon = \frac{A}{C \cdot x} \quad (9.31)$$

If $C = 1$ mol dm^3 and $x = 1$ cm, then $E = A$ dm^3 mol^{-1} cm^{-1}.

Thus, the molar absorptipity may be defined as the specific absorption coefficient for the 1 mol dm^{-3} concentration and 1 cm path length.

For a known value of absorbance (A) for a sample we can determine the concentration (C) of the species of known absorptivity with the help of Beer's law.

Validity of Beer's law

Beer's law is valid under the following conditions:
1. The absorbers must act independently of each other.
2. The absorbing medium must be homogeneously distributed in the interaction volume and must not scatter the radiation.
3. The incident radiation must consist of parallel rays, each passing through the same length in the absorbing medium.
4. The incident radiation should preferably be monochromatic or have at least a width that is more narrow than the absorbing transition.
5. The incident flux must not influence the atoms or molecules; it should only act as a non-invasive probe of the species under study. In particular, this implies that the light should not cause optical saturation or optical pumping, since such effects will deplete the lower level and possibly give rise to stimulated emissions.

If any of these conditions are not fulfilled, there will be deviations from Beer's law.

We prefer to express the Beer–Lambert law using absorbance as a measure of the absorption rather than %T (Figure 9.11). This can be explained as follows. We know that the

Absorbance, $\quad A = \varepsilon \cdot C \cdot x$ and $\quad (9.32)$
Transmittance, $\quad \%T = 100 \; I_l/I_0 = e^{-\varepsilon \cdot C \cdot x}$
$\quad (9.33)$

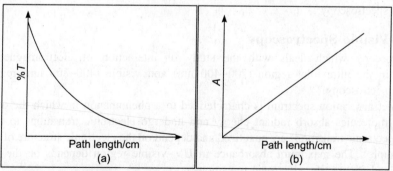

Figure 9.11 (a) Plot of path length and % transmittance, and (b) Plot of path length and absorbance.

$A = \varepsilon \cdot C \cdot x$, tells us that absorbance depends on the total quantity of the absorbing compound in the light path through the cuvette. If we plot absorbance against concentration, we get a straight line passing through the origin (0, 0) (Figure 9.12). Note that the law is not obeyed at high concentrations. This deviation from the law is not dealt with here.

The linear relationship between concentration and absorbance is both simple and straightforward, which is why we prefer to express the Beer–Lambert law using absorbance as a measure of the absorption rather than %T.

Limitations of Beer–Lambert law

1. The law tends to break down at very high concentrations, especially if the material is highly scattering, i.e. it governs the absorption behaviour of dilute solutions only.
2. It is applicable to only non-monochromatic radiations. If the light is especially intense, nonlinear optical processes can also cause variances.

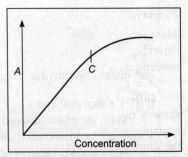

Figure 9.12 Linear relationship between concentration and absorbance.

9.7 SPECTROSCOPY AS AN ANALYTICAL TOOL

Spectroscopy is excellent means to know the chemical composition of a given sample as well as the relative concentrations of the several complex compounds for which a scale or calibration curve, must be created using several known concentrations for each compound of interest. The plot of concentration versus absorbance so obtained is fit by hand or using curve-fitting software. This gives a mathematical formula for the determination of the concentration in the sample. This process is repeated for each compound in a sample to obtain model of several absorption spectra added together to reproduce the observed absorption. This way we can measure the *chemical composition of comets* without actually bringing samples onto the earth.

9.8 TYPES OF ABSORPTION SPECTROSCOPY

On the basis of wavelength range of the radiation being used in the incident beam, the absorption spectroscopy can be classified into following main categories:

1. Infrared (IR) spectroscopy
2. Near infrared spectroscopy
3. Microwave spectroscopy
4. Ultraviolet (UV) spectroscopy
5. Nuclear magnetic resonance (NMR) spectroscopy
6. Electron spin resonance (EPR) spectroscopy
7. Mass spectroscopy.

9.8.1 UV–Visible Spectroscopy

"The spectroscopy which deals with the study of interaction of electromagnetic radiations with matter in the ultraviolet region (200–400 nm) and visible (400–800 nm) region is called UV–visible spectroscopy."

Ultraviolet absorption spectrum is characterized to a phenomenon in which the outer electrons of atoms or molecules absorb radiant energy and undergo electronic transitions to higher energy levels. The electronic transitions are quantized and depend on the electronic structure of the absorbing medium (sample). The maximum absorbance in UV–visible region depends on the magnitude of energy involved for specific electronic transitions.

The ultraviolet–visible spectroscopy is a very suitable tool for the analysis of several inorganic and organic species. The absorption spectrum of a sample containing atoms or simple molecules composed of series of sharp and well defined lines corresponding to the limited number of permitted electronic transitions.

Principle

Many molecules absorb ultraviolet or visible light. The absorbance of a solution increases as shrinking of the beam increases. Absorbance is directly proportional to the path length b, and the concentration c, of the absorbing species. According to Beer's law.

$A = \varepsilon bc$, where ε is a constant of proportionality, called the *absorptivity*.

Different molecules absorb radiation of different wavelengths. An absorption spectrum will show a number of absorption bands corresponding to structural groups within the molecule. For example, the absorption that is observed in the UV region for the carbonyl group in acetone is of the same wavelength as the absorption for the carbonyl group in diethyl ketone.

More precisely, the absorption spectroscopy is based on the absorption of photons by one or more substances present in a sample, which can be a solid, liquid or gas, and successive promotion of electron(s) from one energy level to another in that particular substance.

Electronic Transitions

The absorption of UV or visible radiation corresponds to the excitation of outer electrons. There are three types of electronic transition which can be considered:

(a) Transitions involving π, σ and n electrons.
(b) Transitions involving charge-transfer electrons.
(c) Transitions involving d and f electrons (not mentioned in this chapter).

When an atom or molecule absorbs energy, electrons are promoted from their ground state to an excited state. In a molecule, the atoms can rotate and vibrate with respect to each other. These vibrations and rotations also have discrete energy levels, which can be considered as being packed on top of each electronic level.

(a) *Absorbing species containing π, σ and n electrons:* Absorption of ultraviolet and visible radiation in organic molecules is restricted to certain functional groups (*chromophores*) that contain valence electrons of low excitation energy. The spectrum of a molecule containing these chromophores is complex. This is because the superposition of rotational and vibrational transitions on the electronic transitions gives a combination of overlapping lines. This appears as a continuous absorption band.

Possible *electronic* transitions of π, σ and n electrons are: (Figure 9.13).

1. $\sigma \to \sigma^*$ *Transitions:* An electron in a bonding σ orbital is excited to the corresponding anti-bonding orbital. The energy required is large. For example, methane (which has only C—H bonds, and can only undergo $\sigma \to \sigma^*$ transitions) shows an absorbance maximum at 125 nm. Absorption maxima due to $\sigma \to \sigma^*$ transitions are not seen in typical UV–vis spectra (200–700 nm).

2. $n \to \sigma^*$ *transitions:* Saturated compounds containing atoms with lone pairs (non-bonding electrons) are capable of $n \to \sigma^*$ transitions. These transitions usually need less energy than $\sigma \to \sigma^*$ transitions. They can be initiated by light whose wavelength is in the range 150–250 nm. The number of organic functional groups with $n \to \sigma^*$ peaks in the UV region is small.

3. $n \to \pi^*$ *and* $\pi \to \pi^*$ *transitions:* Most absorption spectroscopy of organic compounds is based on transitions of n or π electrons to the π^* excited state. This is because the absorption peaks for these transitions fall in an experimentally convenient region of the spectrum (200–700 nm). These transitions need an unsaturated group in the molecule to provide the ☐ electrons.

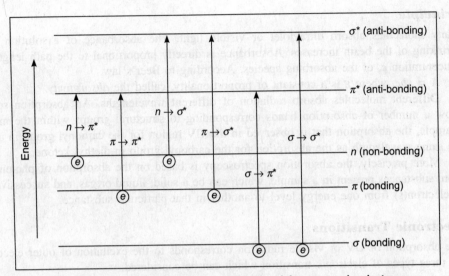

Figure 9.13 Different electronic transitions involving π, σ and n electrons.

4. *Molar absorptivities from $n \rightarrow \pi^*$ transitions:* These are relatively low and range from 10 to 100 l mol^{-1} cm^{-1}. $\pi \rightarrow \pi^*$ transitions normally give molar absorptivities between 1000 and 10,000 l mol^{-1} cm^{-1}. $n \rightarrow \pi^*$ transitions normally give molar absorptivities between 1000 and 10,000 l mol^{-1} cm^{-1}.

The solvent in which the absorbing species is dissolved also has an effect on the spectrum of the species. Peaks resulting from $n \rightarrow \pi^*$ transitions are shifted to shorter wavelengths (*blue shift*) with increasing solvent polarity. This arises from increased solvation of the lone pair, which lowers the energy of the n orbital. Often (but *not* always), the reverse (i.e. *red shift*) is seen for $\pi \rightarrow \pi^*$ transitions.

(b) *Charge-Transfer Absorption:* Many inorganic species show charge–transfer absorption and are called *charge-transfer complexes*. For a complex to demonstrate charge-transfer behaviour, one of its components must have electron donating properties and another component must be able to accept electrons. Absorption of radiation that involves the transfer of an electron from the donor to an orbital associated with the acceptor. Molar absorptivities from charge-transfer absorption are large (> 10,000 l mol^{-1} cm^{-1}).

Spectrometer

The instrument which is used for the production observation and recording of spectra is called spectrometer. Depending on various modifications as per requirements it is also know as spectrograph, photometer and spectrometer. The spectrometer consists of following components:

1. Sample   2. Sample holder   3. Analyzer   4. Detector   and   5. Recorder.

A diagram of the components of a typical spectrometer is shown in Figure 9.14. The functioning of this instrument is relatively simple. A light beam from light source (visible and/or UV light) is passed through a prism or diffraction grating which is separated into its component wavelengths. Each monochromatic (single wavelength) beam in turn is split into two equal intensity beams by a half-mirrored device. Magenta coloured beam, the sample beam passes through a small transparent container (cuvette) containing a solution of the compound being studied in a transparent solvent.

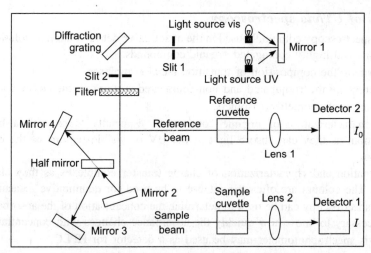

Figure 9.14 Schematic diagram of a spectrometer.

The other beam (blue coloured), the reference beam passes through an identical cuvette containing only the solvent. The intensities of these light beams are then measured by electronic detectors and compared. The intensity of the reference beam, which should have suffered little or no light absorption, is defined as I_0. The intensity of the sample beam is defined as I. Different compounds may have very different absorption maxima and absorbance. The absorbance of a sample is proportional to its molar concentration in the sample cuvette which is known as the *molar absorptivity*. Hence, *Molar absorptivity,* $\varepsilon = A/c \cdot l$.

This value of I is used to compare the spectra of different compounds. From the graph (Figure 9.15), it is clear that compound 2-propanal, 3-(4-dimethyl amino) phenyl absorbs at $\lambda_{max} = 255$ nm and 395 nm in the visible region and appears as orange in colour. Molar absorptivities may be very large for strongly absorbing compounds ($\varepsilon > 10{,}000$) and very small if absorption is weak ($\varepsilon = 10\text{--}100$).

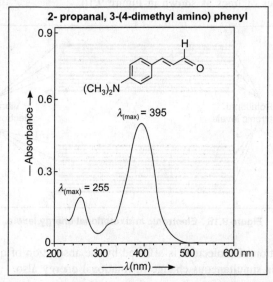

Figure 9.15 Absorption spectrum for unsaturated aldehyde.

Applications of UV/vis spectroscopy

1. UV/vis spectroscopy commonly used in the quantitative determination of solution of transition metal ions and highly conjugated organic compounds.
2. To determine the composition of transition metal complexes.
3. To differentiate the conjugated and non-conjugated systems in the compounds.
4. Identification of geometrical isomers.
5. In the characterization of organic compounds, especially those with a high degree of conjugation, as they can absorb light in the UV or visible regions of the electromagnetic spectrum.
6. Determination and characterization of charge transfer complexes as they also give rise to colours. The colours are often too intense to be used for quantitative measurement.
7. UV/vis spectroscopy can be used to determine the concentration of the absorber in a solution. It is necessary to know how quickly the absorbance changes with concentration.
8. A UV/vis spectrophotometer may be used as a detector for HPLC.

9.8.2 Vibrational–Rotational Spectra–Infrared Spectroscopy

Vibrational spectra are observed in the infrared (IR) region. Homonuclear diatomic molecules such as H_2, O_2, N_2, etc., do not possess a permanent dipole moment and also the stretching of the bond between the two atoms change the dipole moment from zero. Hence, they do not show IR spectra. While heteronuclear diatomic molecules such as CO, NO, CN, HCl, etc. possess a dipole moment which changes when the bond length changes. Hence, they show IR spectra. Thus, important requirement for a molecule to show an infrared spectrum is that the dipole moment of the molecule must change during the vibration.

"This spectroscopy deals with the study of interaction of EMR with the material in the infrared (IR) region. This spectrum is given by the heteronuclear diatomic molecules possessing permanent dipole moment and polyatomic molecules with or without permanent dipole moment."

Such types of spectra arise when a molecule absorbs in infrared region and changes takes place in both the rotatory moment and the atomic vibration within the molecule at the same time. Such spectra consist of a series of lines as shown in Figure 9.16.

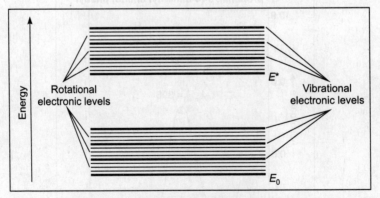

Figure 9.16 Electronic and rotational energy levels.

The vibrational motion of molecules is affected by the absorption of quanta q in IR region and in the gas phase there is simultaneous change in rotational energy also.

The molecule is not rigid. It is supposed to be the flexible, comprising a system of balls having different masses arranged according to actual space geometry of molecules and tied with springs of varying strengths. The balls and springs correspond to atoms and chemical bonds, respectively. Molecular vibration takes place due the absorption of IR radiation and displacement of the atoms. Such change in the bond length produces a change in the dipole moment of a heteronuclear molecule. This results in the exchange of energy when the oscillating dipole couples with the electrical field of radiations of suitable frequency.

There may be two types of vibration in heteronuclear diatomic molecules: (i) stretching vibration and (ii) bending vibration. Simple diatomic molecules have only one bond which may stretch (Figure 9.18).

More complex molecules have many bonds, and vibration can be conjugated leading to infrared absorption at characteristic frequencies that may be related to chemical groups. For example, the atom in a CH_2 group (organic compounds) can vibrate in six different ways: symmetrical and anti-symmetrical stretching, scissoring, rocking, wagging and twisting as shown in Figure 9.17.

Molecular vibrations					
Stretching vibration		Bending vibration			
		In-plane		Out of plane	
Symmetrical stretching	Asymmetrical stretching	Scissoring	Rocking	Wagging	Twisting

Figure 9.17 Different types of molecular vibrations.

Harmonic oscillator model

To explain the theory of vibrational–rotational spectra, consider a diatomic molecule with atomic masses m_1 and m_2 joined by a chemical bond (Figure 9.18). This bond vibrates as a one-dimensional simple harmonic oscillator (SHO). According to Hook's law, the oscillation in which the force rendering to restore an atom to its original position is directly proportional to the displacement of atom from that position.

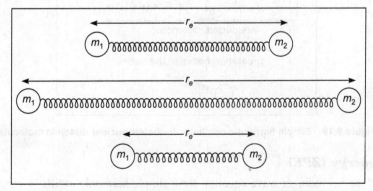

Figure 9.18 Stretching vibration of a diatomic molecule.

Classically, the vibrational frequency (v) of a mass point (m) connected by a spring of force constant (k) is given by

$$v = \frac{1}{2\pi}\sqrt{\frac{k}{m}} \qquad (9.34)$$

If the masses m_1 and m_2 of diatomic molecule vibrate back and forward relative to their centre of masses in opposite directions, the m_1 and m_2 reach the extreme of their respective motions at the same time, the vibrational frequency of the molecule is given by a relation analogous to that of Eq. (9.34) with m replaced by the reduced mass μ.

$$v = \frac{1}{2\pi}\sqrt{\frac{k}{\mu}} \; s^{-1} \qquad (9.35)$$

$$\left(\text{reduced mass}, \mu = \frac{m_1 \times m_2}{m_1 + m_2}\right)$$

or wavenumber,

$$\bar{v} = \frac{1}{2\pi c}\sqrt{\frac{k}{\mu}} \qquad (9.36)$$

where c is the velocity of light.

The potential energy of the simple harmonic oscillator as a function of displacement from the equilibrium relationship is given by the parabolic Hooke's law equation, i.e.

$$V(x) = \frac{1}{2}k(r - r_e)^2 = \frac{1}{2}k \cdot x^2 \qquad (9.37)$$

where $x = (r - r_e)$ is the displacement, r_e is the equilibrium bond length. The Hooke's law energy diagram is shown in Figure 9.19

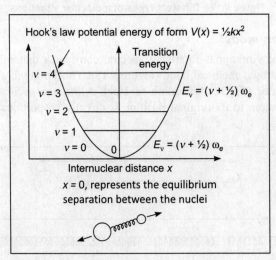

Figure 9.19 Simple harmonic oscillator for heteronuclear diatomic molecule.

Zero-point energy (ZPE)

The solution of the Schrödinger wave equation for a simple harmonic oscillator gives the quantized vibrational energy levels.

$$E_v = (V + \tfrac{1}{2})\, h\nu, \qquad v = 0, 1, 2, 3, \ldots, \qquad (9.38)$$

where v is the vibrational quantum number and ν is the vibrational frequency given by Eq. (9.38).

It is observed that the energy of the lowest vibrational level of the oscillator is not zero but is equal to $\tfrac{1}{2} h\nu$. This is called zero-point energy. The energy levels of the simple harmonic oscillator are equally spaced and equal to $h\nu$. The energy can be converted from joules to cm^{-1}, it is divided by hc:

$$G(v) = E_v/hc = (v + \tfrac{1}{2})\, \nu/c = (v + \tfrac{1}{2})\, \omega_e \qquad (9.39)$$

where ν is the vibrational frequency. $G(v)$ is called the vibrational term and c is the velocity of light.

The selection rule from a vibrational transition in the simple harmonic oscillator is

$$\Delta v = \pm 1 \qquad (9.40)$$

By means of the selection rule, the frequency of the vibrational transition is given by

$$\bar{\nu} = G(v \rightarrow v + 1) \qquad (9.41)$$

$$\bar{\nu} = (v + 1 + \tfrac{1}{2})\, \omega_e - (v + \tfrac{1}{2})\, \omega_e$$

$$\bar{\nu} = \omega_e \qquad (9.42)$$

Most of the molecules are in the ground vibrational state ($v = 0$) at the room temperature, so that the important transition is only that which takes place from $v = 0$ to $v = 1$. The vibrational frequency corresponding to this transition is called *fundamental vibrational* frequency. Diatomic molecules have only one-vibrational frequency, i.e. stretching vibrational frequency.

Applications

1. IR spectroscopy is widely used in both the research and industry as a simple and reliable technique for the dynamic measurement and quality control.
2. It is specially used in forensic analysis both criminal and civil cases.
3. In the identification of unknown compounds.
4. In the identification of presence of conjugations or hydrogen bonding in the compounds.
5. In the study of reaction mechanism.
6. In the study of synthesis as well as degradation of polymers.
7. In the study of nature of bond and its strength.
8. In the study of semiconductor and microelectronics.

9.8.3 Electronic Spectroscopy

"The electronic spectroscopy is the analytical technique which deals with the measurement of changes in the intensity of radiations due to the electronic transitions in atoms of the sample."

The electronic band spectra of molecules are generally very complex and are investigated with considerable difficulty both in emission and in absorption. They are observed in the ultraviolet and visible regions of the electromagnetic spectrum. Electronic spectra are complex as they arise due to transition between two electronic states which are almost invariably accompanied by simultaneous transitions between the vibrational and rotational energy levels as well. Hence, electronic spectra have vibrational fine structure as well as rotational fine structure (Figure 9.16).

Principle

According to the Born–Oppenheimer approximation, the total energy of a molecule in the lower (ground) state is given by:

$$E = E_e + E_v + E_r \qquad (9.43)$$

Translational energy is neglected as it is not quantized. Here, E_e, E_v and E_r are the electronic, vibrational and rotational energies, respectively. Assuming that the Born–Oppenheimer approximation is valid in the upper (excited) state as well, the excited state energy E' is given by

$$E' = E'_e + E'_v + E'_r \qquad (9.44)$$

The energy change for an electronic transition is given by

$$\Delta E = E' - E = \Delta E_e + \Delta E_v + \Delta E_r \qquad (9.45)$$

Considerable simplification of spectra results by recognizing that

$$\Delta E_e \gg \Delta E_v \gg \Delta E_r \qquad (9.46)$$

The frequency for the electronic transition is given by Bohr–Planck relation, viz.,

$$\bar{v} = \frac{\Delta E}{hc} = \frac{\Delta E_e + \Delta E_v + \Delta E_r}{hc} \text{ cm}^{-1} \qquad (9.47)$$

Equation (9.45) shows how an electronic transition possess the vibrational and rotational fine structure.

Frank–Condon principle

It is a very useful and guiding principle for the investigation of vibrational structure of electronic spectra. "Frank–Condon principle states that an electronic transition takes place so rapidly that a vibrating molecule does not change its internuclear distance appreciably during the transition."

This principle also states that "an electronic transition would occur only when the atoms in the lower electronic state are in their extreme position."

This principle is, to a first approximation, true since the electrons move so much faster than the nuclei and do not change their position, hence an electronic transition may be represented by a vertical line on a plot of potential energy versus internuclear distance (Figure 9.20).

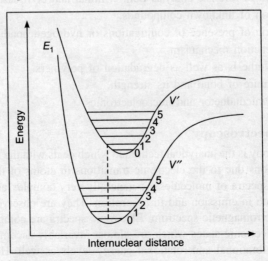

Figure 9.20 Electronic transition in a diatomic molecule.

The electronic transition for a diatomic molecule can be explained with the help of Franck–Condon principle as shown in Figure 9.20. In this figure two potential energies plots for the molecule in the ground electronic state (e_0) and in the first electronic excited state are shown. As the bonding in the excited state is weaker than in the ground state, the minimum for the excited state occur at a slightly greater nuclear distance than the corresponding minimum in the ground electronic state.

According to quantum mechanics, it is known that the molecule is in the centre of the ground vibration level of the ground electronic state. Hence, according to the Franck–Condon principle the most probable electronic transition for a photon falling on the molecule takes place from $v = 0$ to $v' = 2$. The electronic transition to other excited states occur with lower probabilities and therefore their relative intensities are smaller than the intensities of the $0 \rightarrow 2$ transitions (Figure 9.20). The electronic spectra of polyatomic molecules show greater degree of complexity. The electronic spectra for carbonyl group involve promotion of the electrons in a sigma and pi orbitals in the ground state to the sigma anti-bonding (σ^*) and pi anti-bonding (π^*) molecular orbitals in the excited states as shown in Figure 9.13.

Applications

1. The electronic spectroscopy is often used in physical and analytical chemistry for the identification of substances through the spectrum emitted from them or absorbed in them.
2. It is used to study the electronic structure of compounds and complexes.
3. It is used to study the metal–ligand ratio in complexes.
4. This spectroscopy has great scope in astronomy and remote sensing. Most of the large telescopes have spectrographs, which are used either to measure the chemical composition and physical properties of astronomical objects or to measure their velocities from the Doppler shift of spectral lines.
5. It is used in the study of conjugation in organic compounds.

9.8.4 Spectrophotometry

"The spectrophotometry is the technique which involves the measurement of the extent to which the radiant energy is absorbed by a chemical system as a function of the wavelength (or frequency) of the radiation as well as absoption measurements at a pre-determined wavelength".

Spectrophotometry is the quantitative study of electromagnetic spectra. It involves the use of a spectrophotometer.

A spectrophotometer is an optical device used for measuring light intensity as a function of the colour, or more specifically, the wavelength of light. There are many kinds of spectrophotometers.

Following are the criteria to classify the spectrophotometer:
1. The wavelengths they work with.
2. The measurement techniques they use.
3. How they acquire a spectrum.
4. The sources of intensity variation they are designed to measure.
5. The spectral bandwidth and linear range.

There are two major classes of spectrophotometers: (1) Single beam spectrophotometers and (2) Double beam spectrophotometers.

A single beam spectrophotometer measures the absolute light intensity and a double beam spectrophotometer measures the ratio of the light intensity on two different light paths, i.e. sample and reference. Although ratio measurement is easier, and generally stable. Single beam instruments have advantages as they can have a larger dynamic range.

Salient features of single beam spectrophotometer

Single beam spectrophotometer has the following component parts (Figure 9.21):

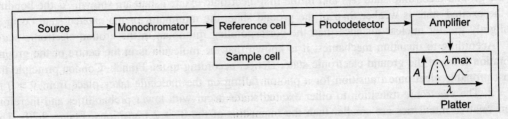

Figure 9.21 Schematic diagram of single beam spectrophotometer.

1. *Source of EMR:* Following EMR sources are used for different wavelength regions:
   (a) H_2 discharge tube for UV region.
   (b) Tungsten filament for visible region.
   (c) Electrically heated rod of rare earth oxide for IR region.
2. *Wavelength controller:* It consists of colour filter, light filter, prism and monochromator.
3. *Sampleholder:* It is a test tube or quartz cell.
4. *Photodetector and receptor unit:* It consists of a phototube which detects and coverts radiant energy into measurable electrical signals.
5. *Amplification and recording unit:* The electrical signals are amplified and recorded on computer monitor or graphical output.

Salient features of double beam spectrophotometer

The salient features of double beam spectrophotometer are almost same except that a single beam spectrophotometer measures the absolute light intensity and a double beam spectrophotometer measures the ratio of the light intensity on two different light paths, i.e. sample and reference. Spectrophotometer has the following component parts (Figure 9.22):

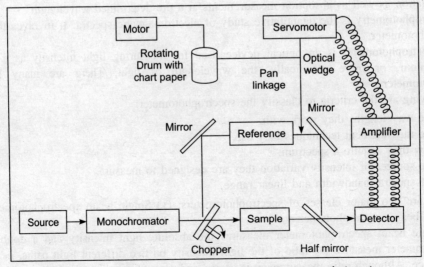

Figure 9.22 Schematic diagram of double beam spectrophotometer.

Analytical Methods ♦ 283

In addition to the components of single beam spectrophotometer like, source of EMR, wavelength controller, sample holder, photodetector and receptor unit, amplification and recording unit and reference holder the double beam spectrophotometer has the following components:
1. *Motor:* The motor rotates the servomotor through rotating drum.
2. *Rotating drum:* This drum is connected to servomotor and rotates it.
3. *Servomotor:* Servomotor drives the optical wedge into the beam of reference.
4. *Optical wedge:* The wedge reduces the radiant power in a smoothly progressive manner while moving.

Working: The working of spectrophotometer can be explained for multi-component analysis. The concentration of two or more coloured solutions can be measured by using proper setting where each coloured species exhibit maximum absorptivity at a wavelength where the other components are optically transparent. Following procedure can be used to analyze the system having two components:
1. The absorbance–wavelength spectrophotometric curve for standard solution of each component can be determined using the same reference solution.
2. Two wavelengths at which it gives the maximum differences in absorbances are decided.
3. A plot of absorbance against concentration for each component is obtained in order to confirm the validity of Beer's law.
4. Absorptivities for each component at the selected wavelength are calculated.
5. Known mixture of two components is prepared and the absorbances at two wavelengths are measured and a graph is plotted between the observed absorbance at two wavelengths against the calculated absorbance. In case of straight line the absorbance is additive.
6. Additive absorbance is calculated and the equations are solved simultaneously for different concentrations.

Applications of spectrophotometry
1. Identification of chemical substances.
2. Multi-component analysis.
3. Photometric titrations.
4. Determination of metal–ligand ratio in a complex.
5. Determination of *pK* value of an indicator.
6. Confirming the constitution of compounds.
7. Measuring the concentration of solutions.
8. Determination of structure of inorganic complexes.
9. Study of H^+ ion concentration, i.e. pH.
10. Study of *ortho* and *para* forms of H_2 molecules.

9.8.5 Flame Photometry (Flame Emission Spectroscopy)
"Flame photometry is the technique or instrumental approach which deals with the examination of radiations (energy) emitted from a substance when suitably excited by introducing it into the flame with the help of atomizer. The intensity of emitted radiations depends on the concentration of the solution of the substance."

Principle and method

In this technique, the excitation is brought about by spraying the sample into the hot flame with the help of atomizer. The metal atoms into the flame get excited to emit the radiations that are the characteristics of individual element.

The flame photometer consists of following component parts (Figure 9.23):

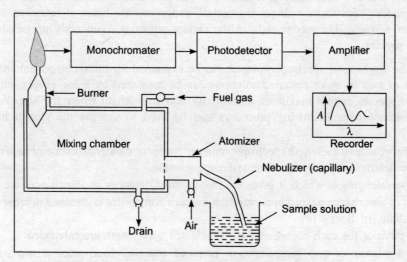

Figure 9.23   Schematic block diagram of flame photometer.

1. *Mixing chamber:* This chamber is fitted with one nebulizer and fan which allows the mixing of sample solution with LPG and air, and after atomizing the sample, air mixture is sent to the flame.
2. *Flame:* Flame is generated by burning the LPG. The colour (blue) and height (up to the slit window) of flame is maintained by controlling the LPG and air ratio in the mixture.
3. *Wavelength controller:* It consists of lens, slit, prism and monochromator (grating).
4. *Photo detector and receptor unit:* It consists of a phototube which detects and converts radiant energy into measurable electrical signals.
5. *Amplification and recording unit:* The electrical signals are amplified and recorded on computer monitor or graphical output.

The process of flame photometry involves the following steps:

1. The sample solution is prepared by dissolving it in the water or organic solvent.
2. The sample is introduced into the flame through mixing chamber (atomizer) under controlled conditions. In the mixing chamber the sample, fuel, gas and air are mixed and then sent to the flame.
3. The solvent of the sample solution get vaporized leaving behind the salt particles.
4. The salt is subsequently vaporized and dissociate into atoms.
5. Metal atoms get excited by absorbing the heat from flame.
6. The radiation from the flame enters into dispersing device (lens, slit and grating) in order to isolate the desired reason of spectrum.

7. The intensity of isolated radiations is measured by a photodetector (phototube) and convert it into electrical signals. The intensity of electrical signals from phototube is poor so these are sent to the amplifier to make them readable. These signals then displayed on readout device or plotter.

Working: After carefully calibrating the photometer with the solution of known concentration and composition, it is possible to correlate the intensity of a given spectral line of the unknown with the amount of an element present that emits the particular radiations. Atomization of solution permits the distribution of the solution throughout the body of the flame and entire sample or a portion is introduced into the flame.

Applications

1. Flame photometry is used to analyze wide variety of materials like soils, cement, glass, etc.
2. The most important application of flame photometery is in the analysis of Na^+, K^+ and Ca^{2+} particularly in biological fluids (e.g. blood serum, urine, oils extracts, etc.) natural waters, plant materials and tissues.
3. Li, Na, K, Mg, Ca, Ba, B, Al, Cu, Cr, Ce, Fe, Pb, Mn, Ru and Sr can be determined commonly.
4. Some metallurgical products can also be analyzed by flame photometry. For example, alkali and alkaline earth metals in catalyst, alloys, etc. can be determined.
5. Determination of TEL and Mn in gasoline stock, accurately.
6. Determination of metal additives in lubricating oils.
7. Analysis of agriculture products.

9.9 CONDUCTOMETERIC ANALYSIS

"The analysis based on the measurement of conductivity of the solution is called conductometric analysis." Two factors, i.e. speed of ions and their concentration are responsible for the conductivity of solutions.

9.9.1 Some Common Terms Used in Conductometric Analysis

(a) *Ohm's law:* "It states that the resistance is the property of a substance impeding the flow of current through it and the current strength I flowing through a conductor at uniform temperature is directly proportional to the potential difference V across it and inversely proportional to the resistance R." Mathematically,

$$I = \frac{V}{R} \tag{9.48}$$

(b) *Resistance:* It is the obstruction to the flow of current. It is directly proportional to the length l and inversely proportional to the cross-section area (a) of the conductor (solution), i.e.

$$R \propto \frac{l}{a}$$

or

$$R = \rho \frac{l}{a} \tag{9.49}$$

Unit of resistance is ohm (Ω).

(c) *Specific resistance:* The proportionality constant (ρ) in Eq. (9.49) is called specific resistance or resistivity.

If $\qquad l = 1$ cm, $a = 1$ cm^2, then $R = \rho \qquad$ (9.50)

Thus, the specific resistance may be defined as the resistance of a conductor (solution) of 1 cm length and having area of cross-section equal to 1 cm^2.

Unit of specific resistance is ohm cm (Ω cm).

(d) *Conductance or conductivity (C):* It is the measure of the ease with which the current flows through the conductor (solution). It is the reciprocal of resistance, i.e.

$$C = \frac{1}{R} \qquad (9.51)$$

Unit of conductance is ohm^{-1} or mho.

(e) *Specific conductance (κ):* It may be defined as the conductance of a solution of 1 cm length and having cross-section area, 1 cm^2. Or it is the conductance of 1 cm^3 of a solution of electrolyte.

Thus $\qquad \kappa = \dfrac{1}{\rho} \qquad$ (9.52)

Unit of specific conductance is ohm^{-1} cm^{-1}.

(f) *Equivalent conductivity (Λ_{eq}):* It is defined as the conductance or conducting power of all the ions produced by dissolving one gram equivalent of an electrolyte in one litre of the solution.

Thus, equivalent conductivity, $\qquad \Lambda_{eq} = \dfrac{\kappa \times 1000}{C} \qquad$ (9.53)

where C is the concentration of the solution in gram equivalents per litre. Unit ohm^{-1} cm^2 eq^{-1}.

(g) *Molar conductivity (Λ_m):* It is defined as the conductance or conducting power of all the ions produced by dissolving one gram mole of an electrolyte in one litre of the solution.

Thus, equivalent conductivity, $\qquad \Lambda_m = \dfrac{\kappa \times 1000}{M} \qquad$ (9.54)

where M is the concentration of the solution in moles per litre. Unit: ohm^{-1} cm^2 mol^{-1}

(h) *Cell constant (x):* As the value of distance between two electrodes, (l) and cross-section area (a) of electrodes may not exactly 1 cm and 1 cm^2, respectively. It is therefore necessary to calculate a factor to have a standard value for the conductivity cell. This factor is called *cell constant.*

Thus $\qquad R = \rho \cdot \dfrac{l}{a} \qquad$ (9.55)

or $\qquad R = \rho \cdot x$

where $\qquad x = \dfrac{l}{a} =$ cell constant, $\rho =$ specific conductivity

Hence, cell constant, $\qquad x = \dfrac{\text{Specific conductivity}}{\text{Observed conductivity}} \qquad$ (9.56)

9.9.2 Conductometric Titrations

"The titration in which the end point is detected by conductivity measurement is called *conductometeric titration.*"

Conductometric titrations are used to find the end point of a titration in volumetric analysis. This method is based on the principle that electrical conductance of a solution depends upon the number (concentration) and mobility of the ions. In this titration, conductance values recorded during the titration and plotted against volume of titrant added. The point of intersection of straight lines so obtained gives the end point.

A conductivity cell dipped in a measuring solution is placed in the inverting input path of an operational amplifier. When AC voltage of constant amplitude and suitable frequency is applied to the system, then for a given feed back resistance R_f, the output e_o is linearly proportional to the conductance of the solution g_i.

Principle and working

The titration is carried out in a conductivity cell. The titrant is added from the burette. Conductivity readings corresponding to various increment of titrant are plotted against volume of titrant. This can be explained by taking the example of acid–base titration. At the beginning of the titration when the NaOH is not added in the HCl solution, it shows maximum conductivity because *H^+ ions possesses highest mobility of any ion*. As NaOH added the concentration of H^+ ions decreases so that the conductivity of the solution decreases rapidly (Figure 9.24).

The solution at neutralization, i.e. at the end point contains only Na^+ and Cl^- ions and will have a minimum conductance. Now if a little NaOH is added at the neutralization the conductivity again increases owing to the presence of the OH^- ions, since OH^- ions have second highest mobility.

Initially, a high value of conductance is obtained as HCl is highly ionized and addition of NaOH to it neutralizes the H^+ ions resulting into formation of water and decrease in conductance.

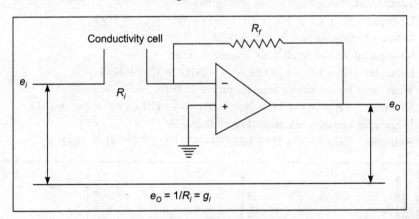

Figure 9.24 Circuit diagram of conductivity meter.

$$H^+ + Cl^- + Na^+ + OH^- \longrightarrow Na^+ + Cl^- + H_2O$$

After the complete neutralization, further addition of NaOH will increase the conductance due to highly mobile OH^- ions. Thus, the conductance will be minimum at equivalence point. The graph obtained in this case is V-shaped as shown in Figure 9.25 and point of interaction corresponds to the end point. The strength of unknown solution is calculated using normality equation.

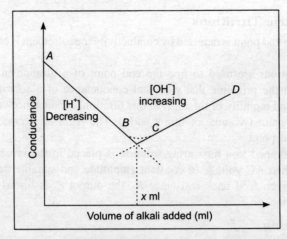

Figure 9.25 Conductometric titration of a strong acid (HCl) against a strong base (NaOH).

$$\underset{\text{HCl}}{N_1 V_1} = \underset{\text{NaOH}}{N_2 V_2}$$

or

$$N_1 = \frac{N_2 V_2 (x \text{ ml})}{V_1}$$

$$\text{Strength} = \text{Normality} \times \text{Equivalent weight} = a \text{ gl}^{-1}. \tag{9.57}$$

9.9.3 Types of Conductometric Titration

1. Acid-based titrations
   (a) Titration of strong acid versus strong base:
       Example: $H^+ + Cl^- + Na^+ + OH^- \longrightarrow + Na^+ + Cl^- + H_2O$
       Nature of curve shown in Figure 9.25.
   (b) Strong acid versus weak base [Figure 9.26(a)]
       Example: $H^+ + Cl^- + NH_4OH \longrightarrow + NH_4^+ + Cl^- + H_2O$
   (c) Weak acid versus strong base [Figure 9.26(b)]
       Example: $CH_3COO^- + H^+ + Na^+ + OH^- \longrightarrow CH_3COO^- + Na^+ + H_2O$
   (d) Weak acid versus weak base [Figure 9.26(c)]
       Example: $CH_3COO^- + H^+ + NH_4OH \longrightarrow CH_3COO\ NH_4 + H_2O$

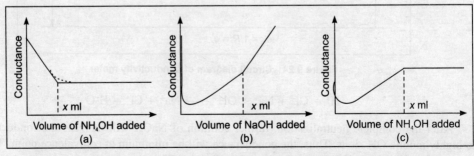

Figure 9.26 (a) Strong acid (HCl) versus weak base (b) Weak acid (CH_3COOH) versus strong base and (c) Weak acid (CH_3COOH) versus weak base.

2. Redox titration (Figure 9.27)
   Example: $6Fe^{2+} + Cr_2O_7^{2-} + 14H^+ \longrightarrow 6Fe^{3+} + 2Cr^{3+} + 7H_2O$
3. Complexometric titration (Figure 9.28)
   Example: $H_2EDTA^{2-} + Ca^{2+}/Mg^{2+} \longrightarrow EDTA-Ca/Mg + 2H^+$
4. Precipitation Titration (Figure 9.29)
   Example: $Ag^+ + NO_3^- + K^+ + Cl^- \longrightarrow AgCl \downarrow + K^+ + NO_3^-$

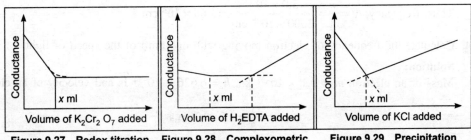

Figure 9.27 Redox titration (Fe^{2+} versus $K_2Cr_2O_7$).
Figure 9.28 Complexometric titration (Ca^{2+}/Mg^{2+} versus H_2EDTA^{2-}).
Figure 9.29 Precipitation titration ($AgNO_3$ versus KCl).

9.9.4 Advantages of Conductometric Titration

1. Conductometric titrations are more accurate than usual titrations.
2. These can be used even for the titration of very dilute solutions.
3. These can be used for titration of coloured solutions.
4. These can be used for titration of even weak acid or weak bases.
5. In this type of titrations no any indicator is required as the end point is determined graphically.
6. Conductometric titrations can be used to measure the salinity of sea water in oceanographic work.

SOLVED NUMERICAL PROBLEMS

1. Calculate the frequency of radiation whose wavelength is 450 nm. Also express the wavelength in terms of wavenumber.
   Solution:
   Given that wavelength, $\lambda = 450$ nm $= 450 \times 10$ Å $= 4 \times 10^{-5}$ cm

   Now frequency, $\nu = \dfrac{c}{\lambda} = \dfrac{3 \times 10^{10} \text{ cm s}^{-1}}{4.5 \times 10^{-5}} = 6.66 \times 10^{-14}$ cycles s^{-1}.

   Wavenumber $\bar{\nu} = \dfrac{1}{\lambda} = \dfrac{1}{4.5 \times 10^{-5} \text{ cm}} = 2.2 \times 10^4$ cm^{-1}

2. Calculate the wavenumber of radiation whose wavelength is 6 μm.
   Solution:
   Given that wavelength $\lambda = 6$ nm $= 6 \times 10^{-7}$ cm

Wavenumber, $\bar{v} = \dfrac{1}{\lambda} = \dfrac{1}{6 \times 10^{-7} \text{ cm}} = \mathbf{1.66 \times 10^5 \text{ cm}^{-1}}$

3. Calculate the frequency of radiation in Hertz whose wavelength is 3000 Å.

   Solution:
   Given that wavelength, $\lambda = 3000$ Å $= 3000 \times 10^{-8}$ cm

   Thus, frequency, $v = \dfrac{c}{\lambda} = \dfrac{3.0 \times 10^{10}}{3000 \times 10^{-8} \text{ cm}} = \mathbf{1.66 \times 10^5 \text{ cm}^{-1}}$

4. Calculate the frequency of electron moving with one third of the speed of light.

   Solution:
   Mass of an electron $m = 9.1 \times 10^{-31}$ kg, $h = 6.626 \times 10^{-34}$ Js and velocity of an electron,

   $$v = \dfrac{1}{3} = 3.0 \times 10^8 \text{ m s}^{-1}$$

   Thus, frequency $\lambda = \dfrac{h}{mv} = \dfrac{6.626 \times 10^{-34}}{9.1 \times 10^{-31} \times \dfrac{1}{3} \times 3 \times 10^8} = \mathbf{7.28 \times 10^{-12} \text{ m}}$

5. The frequency of calculate the wavelength in nm of the strong yellow line in the spectrum of sodium whose frequency is 5.09×10^4 s^{-1} ($c = 3.0 \times 108$ m s^{-1}).

   Solution:
   Frequency $v = \dfrac{c}{\lambda}$ or $\lambda = \dfrac{c}{v} = \dfrac{3.0 \times 10^8}{5.09 \times 10^{14} \text{ cm}} = 5.894 \times 10^{-6}$ m $= \mathbf{589.4 \text{ nm}}$

6. The value of Planck's constant is 6.62×10^{-27} ergs per atom. Calculate its value in kcal mol^{-1}.

   Solution:
   1 mol $= 6.023 \times 10^{23}$ and 1 kcal $= 4.184 \times 107 \times 103$ ergs

   Thus, $h = \dfrac{6.62 \times 10^{-27} \times 6.023 \times 10^{23}}{4.184 \times 10^7 \times 10^3} = \mathbf{9.527 \times 10^{-14} \text{ kcal s mol}^{-1}}$

7. Wavelength of an electromagnetic radiation is 700 nm which is visible in red light. Calculate: (a) the frequency and wavenumber of this radiation, (b) the wavelength in Å.

   Solution:
   (a) wavelength,

   $$\lambda = 700 \text{ nm} = 700 \times 10^{-19} \text{ m}$$

   Now, $v = \dfrac{c}{\lambda} = \dfrac{2.998 \times 10^8 \text{ ms}^{-1}}{700 \times 10^{-9} \text{ m}} = \mathbf{4.28 \times 10^{14} \text{ s}^{-1}}$

   $\bar{v} = \dfrac{1}{\lambda} = \dfrac{1}{700 \times 10^{-9} \text{ m}} = \mathbf{1.43 \times 10^6 \text{ m}^{-1}}$

   (b) $\because$ 1 nm $= 10$ Å $\therefore$ 700 nm $= 700 \times 10 = \mathbf{7000 \text{ Å}}$.

8. Convert 2000 Å to: (i) cm, (ii) cm^{-1}, (iii) eV, (iv) erg, (v) cal and (vi) Hz.

Solution:

(i) Å to cm:

$$2000 \text{ Å} = 2000 \times 10^{-8} = \mathbf{2.00 \times 10^{-5} \text{ cm}} \qquad (\because 1 \text{ Å} = 10^{-8} \text{ cm})$$

(ii) Å to cm^{-1}:

$$\frac{1}{\text{cm}} = \text{cm}^{-1} \quad \therefore \quad \frac{1}{2.00 \times 10^{-5}} \times = \mathbf{5.00 \times 10^4 \text{ cm}^{-1}}$$

(iii) Å to eV

$$2000 \text{ Å} = \frac{5.00 \times 10^4}{8068.3} = 6.197 \text{ eV} \qquad (\because 2 \text{ eV} = 8068.3 \text{ cm}^{-1})$$

(iv) Å to erg:

$$2000 \text{ Å} = 2.00 \times 10^{-5} \text{ cm}$$

We know that the energy of radiation,

$$E = \frac{hc}{\lambda} = \frac{6.626 \times 10^{-27} \times 3.0 \times 10^{10}}{2.00 \times 10^{-5} \text{ cm}} = \mathbf{9.94 \times 10^{-12} \text{ erg}}$$

(v) Å to cal:

$$2000 \text{ Å} = 6.20 \text{ eV} = 6.20 \times 23063 = 1.43 \times 10^5 \text{ cal mol}^{-1} \quad (\because 1 \text{ eV} = 23063 \text{ cal mol}^{-1})$$

(vi) Å to Hertz:

Frequency, $$\nu = \frac{c}{\lambda} = \frac{3.00 \times 10^{10} \text{ cm s}^{-1}}{2.00 \times 10^{-5} \text{ cm}} = \mathbf{1.5 \times 10^{15} \text{ Hz}}$$

9. The percentage transmittance of an aqueous solution of unknown compound is 25% at 25 °C and 400 nm wavelength for a 5.0×10^{-5} M solution in a cell of 2.00 cm thickness. Calculate: (i) Absorbance (A), (ii) Molar extinction coefficient (ε).

Solution:

Given that, transmittance, $T = 25\%$, i.e. $T = \frac{I}{I_0} = \frac{25}{100}$, thickness of cell, $l = 2$ cm and concentration of solution, $c = 5.0 \times 10^{-5}$.

(i) Absorbance $\qquad A = \log \frac{I_0}{I} = \log \frac{100}{25} = \log 4 = \mathbf{0.6020}$

(ii) Molar extinction coefficient (ε)

Absorbance $A = \varepsilon \cdot l \cdot c \quad \therefore \quad \varepsilon = A/l \cdot c$

$\therefore$ Molar extinction coefficient, $\varepsilon = \dfrac{0.6020}{2 \times 5.0 \times 10^{-5}} = \dfrac{6.020 \times 10^{-1}}{1.0 \times 10^{-4}} = \mathbf{6.020 \times 10^3 \text{ cm}^{-1}}$.

10. An aqueous solution of a coloured compound has a molar absorptivity (ε) of 3000 at 500 nm. Calculate the absorbance (A) and % transmittance (T) of a 3.40×10^{-4} M solution in a cell of 1.00 cm thickness is used.

Solution:

We know that, absorbance, $A = \varepsilon \cdot l \cdot c = 3200 \text{ l mol}^{-1} \text{ cm}^{-1} \times 1.00 \text{ cm} \times 3.40 \times 10^{-4} = 1.09$

Now we know that Absorbance, $A = -\log[T]$.

$\% T = 1/A = \text{antilog } 1/1.09 = \text{antilog } 0.917 = \mathbf{0.826\,\%}$.

11. A solution contained in a cell of thickness 2 cm transmits 20% of incident light. Calculate the concentration of the solution. The molar extinction coefficient, $\varepsilon = 3500 \text{ dm}^3 \text{ mol}^{-1} \text{ cm}^{-1}$.

Solution:

Given that transmittance, $T = 20\%$, i.e. $T = \dfrac{I}{I_0} = \dfrac{20}{100}$, thickness of cell $l = 2$ cm

Absorbance, $\qquad A = \log\dfrac{I_0}{I} = \log\dfrac{100}{20} = \log 5 = 0.6989$

Concentration of solution (c):

$\because$ Absorbance, $A = \varepsilon \cdot l \cdot c \,\therefore\, $ Concentration, $c = A/l \cdot \varepsilon$

$$c = \dfrac{0.6989}{2 \times 3.5 \times 10^3} = \dfrac{6.989 \times 10^{-1}}{7.0 \times 10^4} = \mathbf{8.60 \times 10^{-6} \text{ mol dm}^3}.$$

12. Calculate the force constant of the carbon monooxide (CO) molecule whose fundamental frequency is 2140 cm^{-1}. (Atomic mass of ^{12}C $= 19.9 \times 10^{-27}$ kg and ^{16}O $= 26.6 \times 10^{-27}$ kg).

Solution:

Reduced mass of carbon monooxide, $\mu = \dfrac{M_c \cdot M_o}{M_c + M_o} = \dfrac{19.9 \times 10^{-27} \times 26 \times 10^{-27}}{19.9 \times 10^{-27} + 26 \times 10^{-27}}$

$$= \dfrac{5.2934 \times 10^{-52}}{4.65 \times 10^{-26}} = 1.138^{-26} \text{ kg}$$

We know that the frequency $v = \dfrac{1}{2\pi}\sqrt{\dfrac{k}{\mu}}$

And wavenumber, $\qquad \bar{v} = \dfrac{1}{2\pi c}\sqrt{\dfrac{k}{\mu}} \quad \therefore \quad$ Force constant $k = 4\pi^2 c^2 (\bar{v})^2 \mu$

$k = 4(2.14)^2 \times (3.0 \times 10^{10})^2 \times (2140)^2 \times (1.14 \times 10^{-26})$

$= 1853$ kg s–1 or N m^{-1}.

SUMMARY

Analytical chemistry: "Analytical chemistry may be defined as the science and art of determining the composition of materials in terms of the elements or compounds contained."

Types of chemical analysis: **qualitative analysis** and **quantitative analysis**:

Types of analytical methods: (1) Gravimetric analysis, (2) volumetric analysis, (3) optical methods, (4) electrical method and, (5) separation method.

Thermogravimetric analysis (TGA): "Thermogravimetric analysis (TGA) is an analytical technique used to determine a material's thermal stability and its fraction of volatile components by monitoring the weight change that occurs as a specimen is heated." TGA is of two types—(1) dynamic TGA and (2) static or isothermal TGA.

Differential thermal analysis (DTA): It may be defined formally as a technique of recording the temperature difference between a substance and a reference material against either temperature or time as the two specimens are subjected to identical temperature command in an environment heated or cooled at a controlled rate.

Differential scanning calorimetry (DSC): This technique is very closely related to DTA. In this method, the sample and reference material also subjected to a closely controlled programmed temperature. The heat energy is added to or subtracted from the sample or reference containers in order to maintain both reference and sample at same temperatures. The heat energy is recorded as a function of temperature or time.

Spectroscopy: "Spectroscopy is the branch of science which deals with the study of interaction of electromagnetic radiations with matter and its properties by investigating light, sound or particles that are emitted, absorbed or scattered by the matter under investigation."

Electromagnetic radiations (EMR): The electromagnetic radiations are the radiations in which the electric field E (vector) and the magnetic field B (vector) oscillate perpendicular to each other and both are perpendicular to direction of propagation of wave.

Selection rule: The molecular spectra are governed by this rule which specify the changes in the quantum number accompanying the particular electronic transition. For a diatomic molecule such as NO_2, CO_2, NO_2, etc. the selection rule can be written as:

Pure rotational transition: $\Delta J = \pm 1$, where J is the rotational quantum number (i)

Pure vibrational transition: $\Delta v = \pm 1$, where v is the vibrational quantum number (ii)

Types of molecular spectra: (1) Rotational spectra (microwave region), (2) vibrational and vibrational–rotational spectra, (3) raman spectra, (4) electronic spectra, (5) nuclear magnetic resonance (NMR) spectra, (6) electron spin resonance (ESR) electron paramagnetic resonance (ESR) spectra, (7) mossbauer spectra or nuclear gamma resonance fluorescence (NRF) spectra.

Atomic spectroscopy: This spectroscopy deals with the study of interaction of electromagnetic radiations with atoms which are usually in their lower energy state, i.e. ground state.

Molecular spectroscopy: This spectroscopy deals with the study of interaction of electromagnetic radiations with molecules. This interaction results into the transition between rotational and vibrational energy levels in addition to the electronic transitions which are usually in their lower energy state, i.e. ground state.

UV–visible spectroscopy: "The spectroscopy which deals with the study of interaction of electromagnetic radiations with matter in the ultra violet (200–400 nm) region and visible (400–800 nm) region."

Vibrational rotational–infrared spectroscopy: "This spectroscopy deals with the study of interaction of EMR with material in the infrared region. This spectrum is given by heteronuclear diatomic molecules possessing permanent dipole moment and polyatomic molecules with and without permanent dipole moment".

Electronic spectroscopy: The electronic band spectra of molecules are generally very complex and are investigated with considerable difficulty both in emission and in absorption. They are observed in the ultraviolet and visible regions of the electromagnetic spectrum. Their complexity arises from the fact that a transition between two electronic states is almost invariably accompanied by simultaneous transitions between the vibratinal and rotational energy levels as well.

Spectrophotometry: "The spectrophotometry is the technique which involves the measurement of the extent to which the radiant energy is absorbed by a chemical system as a function of the wavelength (or frequency) of the radiation as well as absorption measurements at a pre-determined wavelength."

Flame photometry (flame emission spectroscopy): "Flame photometery is the technique or instrumental approach which deals with the examination of radiations (energy) emitted from a substance when suitably exited by introducing it into the flame with the help of atomizer. The intensity of emitted radiations depends on the concentration of the solution of the substance."

Conductometric titration: "The titration in which the end point is detected by conductivity measurement is called *conductometric titration*."

Advantages of conductometric titration: (1) Conductometric titrations are more accurate than usual titration, (2) These can be used even for titration of vary dilute solutions, (3) These can be used for titration of coloured solution, (4) These can be used for titration of even weak acid or weak basis, (5) In this type of titrations no any indicator is required as the end point is determined graphically.

EXERCISES

1. What do you understand by analytical methods?
2. Explain the different types of analytical methods.
3. What do you mean by thermal analysis?
4. What is thermogravimetric analysis (TGA)? Explain its principal and applications.
5. What do you understand by differential thermal analysis (DTA)? Explain its principal by taking the example of Holloycite. Give its applications in different fields.
6. Distinguish between TGA and DTA.
7. Describe the principal and applications of differential scanning calorimetry.
8. What is spectroscopy?
9. What are the different regions of the electromagnetic spectrum and for what are they used?
10. What is meant by electromagnetic radiation? Describe various regions of the electromagnetic radiation in terms of wavelength and energy?
11. Discuss the important characteristics of electromagnetic radiation?
12. Discuss the difference between absorption and emission of radiation?
13. How do molecular and atomic spectra differ?
14. Define the following terms:
    (a) Wavelength (b) Frequency
    (c) Wavenumber (d) Amplitude.
15. Discuss important features of various electromagnetic radiation regions?
16. The wavelength of an electromagnetic radiation is 500 nm. Calculate:
    (a) Frequency in cycle s^{-1}.
    (b) Wavenumber in cm^{-1}
    [**Ans.** (a) 6×10^{14} cycle s^{-1} and (b) $2 \times 10^{14} cm^{-1}$]
17. Explain the principles of spectroscopy?

18. Calculate the wavelength, wavenumber and frequency of radiation having energy of 1.5 eV?
19. Why the value of energy of radiation increases or decreases with an increase in wavelength?
20. Give the relationship between wavelength and energy of electromagnetic radiation?
21. What are the essential requirements for sample containers?
    (**Hint:** They must be made of such substances which are transparent in the wavelength region to be studied and they must be reproducible in path length or be designed in such a manner that their path length may be easily determined.)
22. Calculate the absorbance of 1.03×10^{-3} M solution, if $\varepsilon = 720$ and path length is 1.0 cm.
    [**Ans.** 0.4]
23. Calculate the molar absorptivity of the solution having concentrated 1×10^{-4}. Given $A = 0.71$ and path length $= 1.0$ cm?
    [**Ans.** 7.1×10^3 l mol^{-1} cm^{-1}]
24. Explain the laws of absorption.
25. Describe Beer's law and define all the terms used in it.
26. What are the relationships between P/P_0, T, $\%T$ and A?
27. What are the reasons for deviation of Beer's law?
28. Describe the components of a spectrophotometer and discuss the use of each component?
29. Explain and give examples of the types of transitions which occur in organic compounds?
30. What are chromophores? Describe the spectral effects of adding chromophores to a molecule.
31. There is no absorption due to $n \to \sigma^*$ transition in the spectrum of trimethyl amine in acidic solution. Why?
32. Name a compound which contains sigma, pi and n electrons?
    (**Hint:** formaldehyde)
33. What are the types of spectral transitions which are associated with complexes?
34. What types of transitions are thought to be the origin of charge transfer bands?
35. Arrange the following transitions in order of their decreasing energy.
    $n \to \pi^*$, $\pi \to \pi^*$, $\sigma \to \sigma^*$.
    [**Ans.** $\sigma \to \sigma^*$, $\pi \to \pi^*$, $n \to \pi^*$]
36. Why is ethanol a good solvent in UV spectroscopy?
    (**Hint:** 95% ethanol is cheap and good solvent for UV, because it is transparent down to about 210 nm)
37. What is the essential requirement for a solvent to be used in UV spectroscopy?
    (**Hint:** A suitable is one which is transparent within the region under examination or do not absorb in the region of interest.)
38. Discuss the relations between absorption spectrum and colour of a compound, giving suitable examples?
39. Discuss the origin of colour in organic compounds. What are chromophores and auxochromes?
40. What are chromophores? Name some chromophores. And explain how can you identify a particular chromophore in a compound?
41. What are auxochromes? Why and how auxochrome increases the colouring power of a chromophore?

42. Which radiation, infrared or ultraviolet has (a) shorter wavelength and (b) lower energy?
    (**Hint:** the wavelength of UV lights shorter than infrared radiation. The energy of infrared radiation is relatively smaller than UV radiation.)
43. At what wavelength the coloured compounds absorb?
    (**Hint:** longer than 400 nm.)
44. Write short notes on:
    (a) Molar extinction coefficient, (b) laws of absorption, (c) photometric titrations and (d) photometric accuracy.
45. Calculate the energy associated with UV radiation having wavelength 280 nm. Give your answer in kcal mol^{-1}.
    [**Ans.** 100 kcal mol^{-1}].
46. Why the absorption bands appear instead of sharp lines in UV spectrum?
47. Give simple idea about the instrument used for ultraviolet absorption measurements. Discuss the various types of transitions and arrange them in order of increasing energy.
48. Discuss briefly the basic principles of UV spectroscopy. Briefly describe the scanning of UV spectrum of organic compound.
49. What is spectrophotometer? How is it used in the study of complex ions?
50. How absorption measurements are used in analysis of multi-component mixtures?
51. A solution containing 3.00 ppm has a transmittance of 65% in 1.0 cm cube.
    (a) Calculate absorbance of the solution.
    (b) Calculate transmittance and absorbance of a solution containing 5.2 ppm of the solute.
    (c) What is the molar absorptivity if its molecular weight is 155?
    [**Ans.** (a) 0.19, (b) 47%, 0.33, (c) 9800 l mol^{-1} cm^{-1}]
52. Why wavenumber in cm^{-1} is mostly used to measure the positions of a given infrared absorption?
53. Why infrared spectroscopy has widely been used for the identification of organic compounds?
54. Which of the following will show pure rotational spectra and why?
55. Write a short note on modes of vibrations of polyatomic molecules. Illustrate your answer with suitable examples?
56. What is the origin of rotational and vibrational–rotational spectra? Explain how the spectral data obtained can be used in determining the interatomic distances in diatomic molecules.
57. Discuss the basic principle of infrared spectroscopy?
58. What are the different regions of infrared radiation? Explain various types of stretching and bending vibrations with suitable examples.
59. What is necessary condition required for a molecule to absorb infrared radiation?
60. Discuss the various types of stretching and bending vibrations which arise in aromatic compounds in their IR spectra?
61. Discuss the difference between atomic absorption and flame emission spectroscopy?
62. Write short notes on: (a) Doppler broadening and (b) pressure broadening.
63. What is Doppler's effect? Discuss the advantages of atomic absorption over flame photometry.
64. Why flame emission is mainly used for quantitative determination of alkali and alkaline earth metals only?
    (**Hint:** Because of relatively low energy of the flame, all elements cannot be excited as in arc excitation.)

65. Discuss the principle of flame photometry? What are the different components of a flame photometer and what are its uses.
66. Give principle, procedure and important applications of flame photometry?
67. Why the analysis of alkali metals is advantageously performed with low temperature flames?
68. Explain the difference between absorption spectra and emission spectra.
69. What is emission spectroscopy? What are the advantages and disadvantages of emission spectroscopy?
70. Discuss the origin of emission spectroscopy?
71. Discuss the mechanism of spectrum produced by electronic excitation?
72. How flame photometry is used in quantitative analysis?
73. What are the functions of flame in flame emission spectrometry?
    (**Hint:** (a) To convert constituents of liquid sample into vapour state, (b) to decompose theses constituents into atoms or simple molecules and (c) to electronically excite a fraction of resulting atomic or molecular species.)
74. What is conductometric titration? Explain the principle and method of conductometric titrations.
75. Why conductomertric titrations are considered as superior to the normal titrations?
76. Define the following terms: (i) resistance, (ii) specific resistance, (iii) conductivity, (iv) equivalent conductivity, (v) molar conductivity and (vi) cell constant.
77. What are the factors which affect the conductivity of the solutions?
78. Explain the effect of temperature on conductivity of the solutions.

CHEMISTRY LABORATORY EXPERIMENTS

LIST OF EXPERIMENTS

1. Determination of Ca^{2+}, Mg^{2+} Hardness of Water Using EDTA Solution
2. Determination of Alkalinity of Water Sample
3. Determination of Dissolved Oxygen (DO) in the Given Water Sample
4. To Find the Melting and Eutectic Point for a Two-component System by Using a Method of Cooling Curve
5. Determination of Viscosity of Lubricant by Redwood Viscometer (No. 1 and No. 2)
6. To Determine Flash Point and Fire Point of an Oil by Pensky–Marten's Flash Point Apparatus
7. To Prepare Phenol Formaldehyde and Urea Formaldehyde Resin
8. To Find Out the Saponification Number of a Given Oil Sample
9. To Determine the TDS of Water Samples of Different Sources
10. Determination of Concentration of $KMnO_4$ Solution Spectrophotometerically
11. Determination of Strength of HCl Solution by Titrating it Against NaOH Solution, Conductometerically
12. To Determine the Amount of Sodium and Potassium in a Given Water Sample by Flamephotometer
13. Estimation of Total Iron in an Iron Ore
14. Estimation of Calcium in Limestone and Dolomite

Experiment 1: Determination of Ca^{2+}, Mg^{2+} Hardness of Water Using EDTA Solution

AIM

To determine the Ca^{2+} and Mg^{2+} hardness of given water sample by EDTA titration method.

Apparatus/reagent required

Burette, beaker, conical flask, measuring cylinder, standard hard water, EDTA solution, Eriochrome Black–T, calcium precipitating buffer solution, ammonium buffer solution (pH ≈ 10).

Theory

Ethylenediaminetetraacetic acid (EDTA) is a very good complexing agent and acts as chelate. Its disodium salt ionizes to give the following ion:

$$\begin{array}{c} ^{-}OOC-CH_2 \\ HOOC-CH_2 \end{array} N-CH_2-CH_2-N \begin{array}{c} CH_2-COOH \\ CH_2-COO^{-} \end{array}$$

which, for the simplicity, can be represented by H_2Y^{2-}. It forms complexes with Ca^{2+}, and Mg^{2+} and other divalent cations represented by the reactions:

$$M^{2+} + H_2Y^{2-} \longrightarrow MY^{2-} + 2H^+$$

$$M^{n+} + H_2Y^{2-} \longrightarrow MY^{(n-4)} + 2H^+$$

The dissociation of these complexes depends on the pH of the solution, and the complexes with hardness causing divalent ions are stable in alkaline medium (pH ≈ 10).

The indicator used is a complex organic compound [sodium-1-(1-hydroxy-2-napthylazo)-6-nitro-2-naphthol-4-sulphonate], which is commonly known as *Eriochrome Black–T*.

It has two phenolic hydrogen atoms which are ionisable. For simplicity it can be represented as $Na^+H_2In^-$. At different pHs, it shows different colours:

$$\underset{\text{Red}}{H_2In^-} \xrightarrow[5.5]{pH = 7.0} \underset{\text{Blue}}{HIn^{2-}} \xrightarrow[11.0]{11.5} \underset{\text{Yellowish orange}}{In^{3-}}$$

In the pH range 7–11, the indicator reacts with the metal ion to form a weak complex with a wine red colour:

$$\underset{\text{Blue}}{HIn^{2-}} + M^{2+} \longrightarrow \underset{\text{Wine red}}{MIn^-} + H^+$$

The optimum pH for the experiment is 10.0 ± 0.1 and is adjusted by ammonia buffer ($NH_4OH + NH_4Cl$). When a small amount of the indicator solution is added to a hard water sample (pH $\approx$ 10) the indicator reacts with Mg^{2+} ion to produce wine red colour.

$$Mg^{2+} + HIn^{2-} \longrightarrow MgIn^- + H^+$$

As EDTA (H_2Y^{2-}) is added, free Ca^{2+} ions present in water are complexed to CaY^{2-} as it is the most stable complex:

$$Ca^{2+} + H_2Y^{2-} \longrightarrow CaY^{2-} + 2H^+$$

Free Mg^{2+} ions then react to give Mg–EDTA complex.

$$Mg^{2+} + H_2Y^{2-} \longrightarrow MgY^{2-} + 2H^+$$

This Mg–EDTA complex is less stable than Ca–EDTA complex, but more than Mg–indicator complex. Therefore, if an extra drop of EDTA is added after complexing all the free Ca^{2+} and Mg^{2+} ions, EDTA takes up Mg^{2+} from the weak Mg–indicator complex to from stable Mg–EDTA complex and simultaneously the indicator is liberated in the free form:

$$\underset{\text{Wine red}}{MgIn^-} + H_2Y^2 \longrightarrow \underset{\text{Pure blue}}{MgY^{2-} + HIn^2} + 2H^+$$

Completion of the above reaction makes the end point of the titration.

Ca^{2+} in hard water sample is precipitated as calcium oxalate by adding

calcium precipitating buffer solution $\begin{bmatrix} COONH_4 \\ | \\ COONH_4 \end{bmatrix} + NH_4Cl + NH_4OH$

$$\begin{matrix} COO^- \\ | \\ COO^- \end{matrix} + Ca^{2+} \longrightarrow \underset{\text{Calcium oxalate (white ppt.)}}{\begin{matrix} COO \\ | \diagdown \\ COO \diagup \end{matrix} Ca}$$

Procedure

Standarization of EDTA solution with standard hard water

Pipette out 50 ml of standard hard water into 250 ml conical flask. Add 2 ml of buffer solution and four drops of Eriochrome Black–T indicator. A wine red colour appears. Titrate it against EDTA solution (taken in burette) till the colour changes from wine red to pure blue. Record the volume of EDTA used as A ml.

Determination of total hardness

Pipette out 50 ml of the hard water sample into a conical flask. Add 2 ml buffer solution and four drops of the indicator. Titrate it against EDTA till the wine red colour changes to pure blue. Record the volume of EDTA used as B ml. This volume corresponds to the total hardness of the water sample.

Determination of Mg^{2+} hardness

Take 100 ml of the hard water sample into a 500 ml dry beaker and add 20 ml of the calcium precipitating buffer solution while constantly stirring the mixture with a glass rod. Allow the precipitate to settle down for about 1 h and filter through a dry funnel fitted with two pieces of Whattman filter paper no. 42 into a dry flask. Measure 50 ml of the filtrate into a conical flask, 5 ml ammonia buffer (pH $\approx$ 10) and four drops of the Eriochrome Black–T indicator and titrate it against EDTA solution. The volume of EDTA used (C ml) corresponds to magnesium hardness.

Determination of calcium hardness

Calcium hardness is obtained by subtracting magnesium hardness from the total hardness.

Observation and calculation

Strength of standard hard water (SHW) = 1000 pm
Volume of SHW = 50 ml
Volume of water sample = 50 ml
Volume of EDTA used in standardization = A ml
Volume of EDTA used for total hardness = B ml
Volume of EDTA used for Mg^{2+} hardness = C ml

$$\text{Strength of EDTA solution} = \frac{\text{Strength of SHW} \times \text{Volume of SHW}}{\text{Volume of EDTA solution used}}$$

$$\text{Strength of EDTA solution} = \frac{1000 \times 50}{A \text{ ml}} = x \text{ ppm}$$

$$\text{Total hardness (as } CaCO_3) = \frac{\text{Strength of EDTA } (x \text{ ppm}) \times \text{Volume of EDTA}}{\text{Volume of water sample}}$$

$$\text{Total hardness} = \frac{x \text{ ppm} \times B}{50} = y \text{ ppm}$$

$$\text{Magnesium hardness (as } CaCO_3) = \frac{\text{Strength of EDTA } (x \text{ ppm}) \times \text{Volume of EDTA}}{\text{Volume of water sample}}$$

$$\text{Magnesium hardness (as } CaCO_3) = \frac{x \text{ ppm} \times B}{50} = z \text{ ppm}$$

$$\text{Calcium hardness} = x - y = n \text{ ppm}$$

Precautions

1. All the solution should be freshly prepared.
2. Distilled water should be checked with care before use.
3. The same amount of the indicator must be added each time.
4. The reaction mixture should be briskly shaken during the titration.
5. The end point should be observed correctly.
6. pH 10 should be maintained during the titration of water sample.

Experiment 2: Determination of Alkalinity of Water Sample

AIM

To determine the alkalinity of a given water sample.

Apparatus/reagents required

Burette, pipette, conical flask (250 ml capacity), measuring flask, N/50 H_2SO_4, phenolphthalein indicator, methyl orange indicator.

Theory

The knowledge of alkalinity in water is important for: (1) Controlling the corrosion, (2) in conditioning the boiler feed water (internally), (3) for calculating the amounts of lime and soda needed for water softening and (4) in neutralizing the acidic solution produced by the hydrolysis of salts. In boilers for steam generation, high alkalinity in water may lead caustic embrittlement and also the precipitation of sludge and deposition of scales.

The alkalinity of water is due to the presence of: (1) hydroxide ion (OH^-), (2) carbonate ion (CO_3^{2-}) and (3) Bicarbonate ion (HCO_3^-). These can be estimated separately by titrating against standard acid, using phenolphthalein and methyl orange as indicators. The chemical reaction involved can be shown by the following equations:

(i) $OH^- + H^+ \longrightarrow H_2O$
(ii) $CO_3^{2-} + H^+ \longrightarrow HCO_3^-$ $\left.\begin{array}{c}\\\\\end{array}\right] P \quad \Bigg\} M$
(iii) $HCO_3^- + H^+ \longrightarrow H_2O + CO_2$

The titration of the water sample against a standard acid up to phenolphthalein end point shows the completion of reactions (i) and (ii) only. This amount of acid used thus corresponds to hydroxide plus one half of the normal carbonate present. The titration of the water sample against a standard acid to methyl orange end point marks the completion of reactions (i), (ii) and (iii). Hence, the amount of acid used after the phenolphthalein end point corresponds to one half of the normal carbonate plus all the bicarbonates while the total amount of acid used represents the total alkalinity (due to hydroxide, bicarbonate and carbonate ions).

The possible combinations of ions causing alkalinity in water are:

(i) OH^- only, CO_3^{2-} only, HCO_3^- only or
(ii) OH^- and CO_3^{2-} together
(iii) CO_3^{2-} and HCO_3^- together

The possibility of OH^- and HCO_3^- ions together is not possible since they combine together to form CO_3^{2-} ions.

$$OH^- + HCO_3^- \rightarrow H_2O + CO_3^{2-}$$

Procedure

Pipette out 100 ml of water sample into a conical flask. Add 2–3 drops of phenolphthalein indicator and titrate this sample against N/50 H_2SO_4 until the pink colour (caused by phenolphthalein) just disappeared. Note down this reading as phenolphthalein end point (V_1 ml). Now add 3–4 drops of methyl orange indicator in the same solution, it gives light yellow colour to the solution. Continue the titration against N/50 H_2SO_4 until the light yellow colour changes to red. Now note down the total volume of acid as V_2 ml. This is the methyl orange end point. The same experiment is repeated for both the sample, till the concordant readings are obtained.

Observation and calculations

Volume of water sample taken for each titration = 100 ml
Volume of N/50 H_2SO_4 used to phenolphthalein end point = V_1 ml
Additional volume of N/50 H_2SO_4 used to methyl orange end point = V_2 ml

$$\text{Phenolphthalein alkalinity} = \frac{\text{Normality of } H_2SO_4 \times \text{Volume of } H_2SO_4}{\text{Volume of water sample}} \times 50 \times 1000$$

$$\text{Phenolphthalein alkalinity} = \frac{A}{50 \times 100} \times 50 \times 1000 = A \times 10 = x \text{ ppm}$$

$$\text{Methyl orange alkalinity or total alkalinity} = \frac{\text{Normality of } H_2SO_4 \times \text{Volume of } H_2SO_4}{\text{Volume of water sample}} \times 50 \times 1000$$

$$\text{Methyl orange alkalinity or total alkalinity} = \frac{(A+B)}{100} \times 50 \times 1000 = (A+B) \times 10 = y \text{ ppm}$$

$$\text{Hydroxide alkalinity} = \frac{(A-B)}{50 \times 100} \times 50 \times 1000 = 10(A-B) = c \text{ ppm}$$

$$\text{Hydroxide ion concentration} = \frac{(A-B)}{50 \times 100} \times 17 \times 1000 = 10 \times (A-B) \times 0.34 \text{ ppm}$$

$$\text{Carbonate alkalinity} = \frac{2B}{50 \times 100} \times 50 \times 1000 = 2B \times 10 \text{ ppm} \quad (\text{when } A > B)$$

and
$$= 10 \times 2A \quad (\text{when } \leq B)$$

$$\text{Carbonate ion concentration} = 2B \times 10 \text{ ppm} \quad (\text{when } A > B)$$

and
$$= 10 \times 2A \quad (\text{when } = B)$$

$$\text{Bicarbonate alkalinity} = \frac{B-A}{50 \times 100} \times 50 \times 1000 = (B-A) \; 10 \text{ ppm}$$

$$\text{Bicarbonate ion concentration} = \frac{(B-A) \times 10 \times 61}{50}$$

$$\text{Bicarbonate ion concentration} = \text{Bicarbonate alkalinity} \times 1.22 \text{ ppm}$$

Precautions

1. The glass apparatus should be perfectly cleaned before the start of the experiment.
2. Since phenolphthalein and methyl orange indicators are used in the determination of alkalinity the end point titration should be observed carefully.
3. The colour change due to indicator should be stable.
4. The end point should be observed correctly.

Experiment 3: Determination of Dissolved Oxygen (DO) in the Given Water Sample

AIM
To determine the dissolved oxygen (DO) in the given sample of water.

Apparatus and chemicals
Managenous sulphate solution (48%), alkaline potassium iodide, conc. H_2SO_4 acid, standard thiosulphate (hypo) solution (N/40), freshly prepared starch solution, potassium oxalate (2%) solution, pipette, burette, glass bottle, glass rod, conical flask, measuring flask and measuring cylinder.

Theory
In presence of good amount of dissolved oxygen, aerobic bacteria lead to oxidation of organic compound present in water, this is called *aerobic oxidation*. If water is polluted with large amount of organic compounds, a large amount of oxygen is rapidly used up in biological aerobic oxidation. This decreases the dissolved oxygen which in turn decreases the pollution of aquatic life. The dissolved oxygen tests is applied mainly for determining the DO of polluted water and industrial effluents, and constitute a method of controlling pollution of water sources.

The solubility of dissolved oxygen decreases with increase in concentration of salt at one atmosphere pressure. The solubility of dissolved oxygen in water at 30°C is about 7–8 ppm. The solubility is less in saline water and at a given temperature decreases with increases in concentration of impurities. Dissolved oxygen is determined by Wrinker's method which is based on the fact that dissolved oxygen oxidizes KI to I_2. The liberated iodine is titrated against standard sodium thiosulphate solution using starch as an indicator. Since dissolved oxygen is present in the molecular state, it as such cannot oxidize KI. So manganese hydroxide generated by the action of KOH on manganese sulphate is used as an oxygen carrier to bring about the reaction between KI and oxygen.

$$MnSO_4 + 2KOH \longrightarrow Mn(OH)_2 + K_2SO_4$$
$$2Mn(OH)_2 + O_2 \longrightarrow 2MnO(OH)_2$$
$$MnO(OH)_2 + H_2SO_4 \longrightarrow MnSO_4 + 2H_2O + [O]$$
$$2KI + H_2SO_4 + [O] \longrightarrow K_2SO_4 + H_2O + I_2$$
$$I_2 + 2Na_2S_2O_3 \longrightarrow Na_2S_4O_6 + 2NaI$$
$$Starch + I_2 \longrightarrow Starch\ iodide\ (Blue\ coloured\ complex)$$

Procedure
1. Take 300 ml of the water sample in a glass stoppered bottle avoiding as far as possible the contact with air and add to it 0.7 ml of concentrated H_2SO_4.

2. Add 0.2 ml (four drops) of N/10 KMnO$_4$ dropwise by constant stirring, stopper the bottle and mix the contents of the bottle by inverting it a few times. If the permanganate colour disappears within 5 min, add additional amount of KMnO$_4$.
3. Add 0.5 ml of potassium oxalate (2%) solution, stopper the bottle and mix it well. Add additional amount of oxalate solution if the permanganate colour is not discharged within 10 min.
4. Add 2 ml of manganous sulphate solution and 3 ml of alkaline KI solution together.
5. Stopper the bottle and mix it by rotating and inverting the bottle 10 – 15 times and allow to stand for 10 min.
6. When the precipitate is settled, add 1 ml of conc. H$_2$SO$_4$. Shake the bottle until the precipitate has completely dissolved.
7. Allow the solution to stand for 5 min.
8. Take 200 ml of this solution, add 2 ml of starch, blue colour appears and titrate it against N/40 hypo solution (Na$_2$S$_2$O$_3$) till blue colour disappears. Note down the volume of hypo solution used during titration.
9. Repeat the titration to get the concordant readings of hypo solution used.

Observation

Normality of hypo solution (N_2) = N/40
Volume of water sample taken (V_1) = 200 ml

Observation table

S.No.	Volume of sample solution taken (ml)	Burette reading		Volume of standard hypo solution used (V_2 ml)
		Initial	Final	
1.				
2.				
3.				
			Concordant volume (V_2) =	

Calculation

Using Normality equation

$$N_1V_1 = N_2V_2$$
$$\text{(Water)} \quad \text{(Hypo)}$$

$$N_1 \times 200 \text{ ml} = 1/40 \times V_2 \text{ ml}$$

$$N_1 = V_2/8000$$

Normality × Equivalent weight = Strength of oxygen = $N_1 \times 8$ g l^{-1}

$$= V_2/8000 \times 8 \text{ g l}^{-1}$$

$$= V_2/8000 \times 8 \times 1000 \text{ mg l}^{-1}$$

$$= V_2 \text{ mg l}^{-1}$$

Result

The amount of dissolved oxygen in the given sample of water is ...... ppm

Precautions
1. Contact of water with oxygen/air should be avoided.
2. No bubble should be present in the bottle as it will interfere with the reaction and will lead to higher concentration of oxygen.
3. End point should be detected carefully.

Experiment 4: To Find the Melting and Eutectic Point for a Two-Component System by Using a Method of Cooling Curve

AIM
To find out the melting and eutectic point of a two-component (urea and phenol) system.

Apparatus and chemicals
Double walled glass tube, thermometer, J-type glass stirrer, stand, clamp, heating bath, wrist watch, urea, phenol and glycerol or paraffin liquid.

Theory
Phase is the homogeneous, physically distinct, mechanically separable portion of the system, e.g. mixture of water and alcohol is a one-phase system, whereas mixture of water and carbon tetrachloride is a two-phase system. Phase diagram is the plot which gives the conditions of temperature, pressure and compositions for various phases of the system present in equilibrium. Melting point is the temperature at which the last trace of solid melts.

Phase change processes, which accompany a cooling process, e.g. formation of a liquid phase from vapour phase, formation of a solid phase from a liquid phase, are exothermic or heat evolving processes.

If a liquid phase is allowed to cool in steady surroundings and the value of the temperature is plotted against the time, the graph so obtained is called *cooling curve*. The rate of cooling of a single homogeneous phase decreases as it comes closer to the temperature of the surroundings.

For a pure substance, the temperature remains constant till whole of the liquid is completely solidified thereafter the fall in temperature will continue.

For the mixture of two substances the temperature does not remain constant as in the case of pure substance, it decreases continuously but at different rate.

Eutectic point
Eutectic term means easy to melt. A binary system consisting of two substances, which are miscible in all proportion in liquid phase but do not react chemically, is known as *eutectic system*. The two

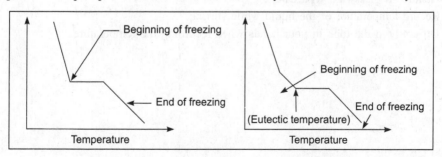

substances involved, have the property of lowering each other's melting point. Solid solution of two or more substances having the lowest freezing point of all the possible mixtures of the components is known as *eutectic mixture*. The minimum freezing point attainable corresponding to the eutectic mixture is termed as the *eutectic point* or it is the maximum temperature up to which a solid phase can exist. Below this point the system completely solidifies or no liquid mixture can exists.

In the case of cooling curve, the longest temperature halt is assumed as the eutectic point of the mixture.

Procedure

1. Weigh 5 g of urea and take it in the inner tube of the eutectic apparatus.
2. Add 10 ml of phenol to it.
3. Fit the tube with a thermometer and J-type stirrer and cover it with a cork.
4. Insert this tube into the glycerol bath and heat the mixture till the entire solid melts completely.
5. Remove the inner tube from the glycerol bath and put it inside the outer tube (air jacket).
6. Fix it in a clamp on an iron stand.
7. Record the temperature every minute with constant stirring till all the content in the system solidified.

Observation

Sl. No.	Time (min)	Temperature (°C)
1	1	
2	2	
3	3	
4	4	
5	5	
6	6	
7	7	
8	8	

Result

The eutectic point of the given mixture is found to be ...... °C

Precautions

1. Handle all glassware with care.
2. Record temperature of the liquid while stirring.
3. Do not hold the tube in your hands when recording the temperature.

Experiment 5: Determination of Viscosity of Lubricant by Redwood Viscometer (No. 1 and No. 2)

AIM

To determine the viscosity of given lubricant (oil) by Redwood viscometer No. 1 or 2

Apparatus

Redwood viscometer No. 1 or 2, beaker, thermometer.

Theory

The viscosity of lubricant oil can be determined with the help of an apparatus called *viscometer*. In a viscometer, a fixed volume of the oil is allowed to flow from a given height, through a standard capillary tube under its own weight and the time of flow (in second) is noted. The time (in seconds) is directly proportional to the true viscosity of lubricating oils. The results are expressed in terms of time taken by oil to flow through the jet of the Redwood viscometer.

The Redwood viscometer No. 1 is commonly used for determining viscosities of this lubricant oil and it has a jet of bore diameter 1.02 mm and length of 10 mm. On the other hand, Redwood viscometer No. 2 is used for measuring viscosities of highly viscous oils like fuel oil. It has a jet of diameter 3.8 mm and length 15 mm. Redwood viscometer consists of the following essential parts (Figure E5.1):

(i) *Oil cup:* It is a brass cylinder of 90 mm height and 46.5 mm diameter.
(ii) *Heating bath:* It is cylindrical copper bath which surrounds the oil cup.
(iii) *Stirrer:* It is placed outside the oil cup carrying four blades.
(iv) *Spirit level:* The lid of the cup is provided with a spirit level for vertical levelling of the jet.
(v) *Levelling screws:* The entire apparatus rests on three legs, provided at their bottom with levelling screws.
(vi) *Kohlrausch flask:* It is a specially shaped flask of receiving the oil from the jet outlet. Its capacity is 50 ml upto the mark of the neck.

Procedure

The levelled oil cup is cleaned thoroughly and ball of rod (valve ball) is placed on the agate jet to close it. Oil under test is filled in the cup up to the pointer level. An empty Kohlrausch flask is kept just below the jet. Water is filled in the bath and heated slowly with constant stirring of the water, when the oil is at the desired temperature heating is stopped and the ball of the valve is lifted. The time taken for the flow of 50 ml of the oil through the jet (collect in the flask) is noted and then the valve is immediately closed to prevent any overflow of the oil. The result is expressed in "Redwood No. 1 seconds" at the particular temperature. Higher the time of flow higher will be the viscosity of the oil.

Observation and result: The viscosity of given lubricant oil is .... Redwood No. 1 seconds.

Experiment 5

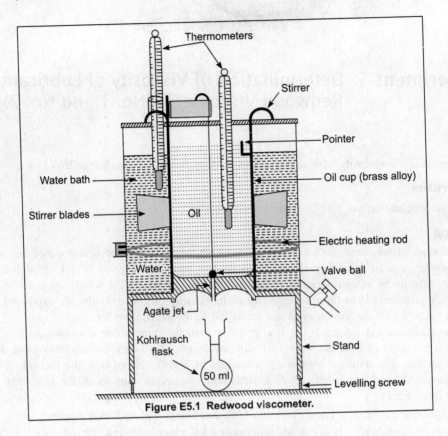

Figure E5.1 Redwood viscometer.

Precautions
1. The viscometer should be in the vertical levelled position.
2. The temperature of the water bath should not exceed the desired temperature.
3. The valve ball should be placed properly on the agate jet to prevent the leakage of oil.
4. The flow time should be noted carefully and accurately.

Experiment 6: To Determine Flash Point and Fire Point of an Oil by Pensky–Marten's Flash Point Apparatus

AIM
To determine the flash and fire point of an oil by Pensky–Marten's flash point apparatus.

Requirement
Pensky–Marten's flash point apparatus, thermometer, lubricating oil.

Theory
The lubricating oil vapours have a tendency to burn at certain temperature which is the indicator of its flash point and fire point: "The flash point of oil is that minimum temperature at which it gives off sufficient vapours that ignite for a moment when a tiny flame is brought near to the surface of the oil under specific conditions, while fire point is that minimum temperature at which the lubricating oil gives off sufficient vapours which burn continuously for five seconds when surface of oil is exposed to tiny flame of specific dimensions under specific conditions." In general, fire point of oil is 5–40°C higher than its flash point.

Flash point and fire point of lubricating oil can be measured by using Penskey–Marten's apparatus (Figure E6.1). It consists of a brass cup, flame exposure device, air bath and stirrer. The oil cup is 5 cm in diameter and 5.5 cm deep. The lid of the cup is provided with four openings of specific sizes. One of the openings in the middle is for the stirrer carrying two brass blades. The second opening is meant for the passage of air and the third one is for introducing the test flame. Second and third opening is controlled by shutter control device. Through fourth opening, the thermometer is inserted into the oil cup to note down the temperature of the sample. The cup is mounted on air bath which is kept on electric heater attached to a controller, or bunsen burner. The flash and fire point determines the applicability of oil at high temperature.

Procedure
The oil is filled in the cup up to the pointer. Then oil cup is placed into the air bath and the apparatus is placed over the heater or burner. It is covered with the lid and thermometer is inserted in the oil. The test flame is lighted and the oil is heated at the heating rate 2–4°C per minute. At every 2°C rise in temperature a small flame is passed over the sample surface. When a flash appear at any point on the surface of oil, the temperature is taken as the flash point. The heating of oil is continued and tested with flame until oil ignites and burns for at least for 5 s. This temperature is recorded as fire point.

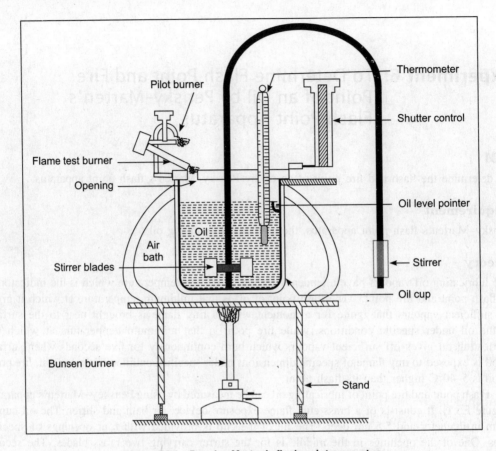

Figure E6.1 Pensky–Marten's flash point apparatus.

Precautions
1. Apparatus should be clean and free from moisture.
2. The flame insertion should be quick process.
3. The oil once heated should not be reused for determination of flash point.
4. The oil should be filled accurately up to the pointer.

Experiment 7: To Prepare Phenol Formaldehyde and Urea Formaldehyde Resin

PART A

AIM
To prepare the crystals of phenol formaldehyde (bakelite) resin (PF).

Apparatus and chemicals
Beaker, measuring cylinder, glass rod, ice bath, chemical balance, phenol, formaldehyde (40%), glacial acetic acid, conc. HCl, etc.

Procedure
Take 5 ml of glacial acetic acid and 2.5 ml of 40% formaldehyde solution (formalin) in a 250 ml beaker and add to it 2 g of phenol. Add small quantity (4–6 drops) of conc. HCl and hexamethylene tetramine. Wrap a cloth loosely round the beaker carefully. Keep it undisturbed in ice bath for a few minutes. Pink crystals of PF resin are formed. Dry the crystal and note down the yield.

Reaction involved

Phenol + HCHO ⟶ o-Hydroxy methyl phenol + p-Hydroxy methyl phenol

⟶ Novolac (a linear polymer) $\xrightarrow{\text{Hexamethylene tetramine}}$

(Contd.)

[Structure diagram of cross-linked phenol-formaldehyde polymer with OH-substituted benzene rings connected by CH₂ bridges]

Cross-linked polymer (bakelite)

Result
A pink, rigid, hard, infusible, insoluble solid is obtained. Yield = ... g

Precautions
1. Phenol should be weight carefully and accurately.
2. Acetic acid and formaldehyde should be added accurately.
3. Conc. HCl should be added carefully and dropwise with the help of glass rod along the walls of the beaker.

PART B

AIM
To prepare the crystals of urea formaldehyde (UF) resin.

Apparatus and chemicals
Beaker, measuring cylinder, glass rod, hot plate, balance, weight box, urea, 40% formaldehyde solution, conc. HCl.

Procedure
Take 20 ml of 40% formaldehyde solution (formaline) in a beaker by measuring cylinder. Add 10 g urea with constant stirring until a saturated solution is obtained. Add few drops of conc. H_2SO_4. Stirr continuously during the addition. Heat the contents at about 50°C (if required). Clear water white solid mass appears. Now wash the residue with water and dry it, weigh the product so formed and note down the yield.

Reactions involved

$$O=C{\overset{NH_2}{\underset{NH_2}{}}} \xrightarrow{HCHO} O=C{\overset{NHCH_2OH}{\underset{NH_2}{}}} \text{ Monomethylol urea}$$

$$\downarrow HCHO$$

$$O=C{\overset{NHCH_2OH}{\underset{NHCH_2OH}{}}} \xrightarrow[\text{polymerization}]{\text{Condensation}} \begin{array}{c} -N-CH_2-N-CH_2-N-CH_2- \\ | \quad\quad | \quad\quad | \\ CO \quad\quad CO \quad\quad CO \\ | \quad\quad | \quad\quad | \\ -N-CH_2-N-CH-N-CH_2- \end{array}$$

2HCHO, Dimethylol urea → Cross-linked polymer (UF resin)

Result

Clear water white crystals of urea formaldehyde resin are obtained. Yield = …….. g.

Precautions

1. Urea should be weighed carefully and accurately.
2. Concentrated H_2SO_4 should be added dropwise with the help of glass rod along the walls of the beaker.
3. Be little away from the beaker as the reaction is sometimes vigorous.
4. Beaker and glass rod should be clean.
5. Contents should not be heated above 50°C otherwise it will stick with the walls of the beaker.

Experiment 8: To Find Out the Saponification Number of a Given Oil

AIM
To determine the saponification value of a given oil sample.

Apparatus and reagent
Alkali resistance conical flask, air condenser, burette, pipette, dropper, measuring cylinder, water bath, oil sample, ethyl methyl ketone, alcoholic KOH (N/2), HCl solution (N/2), phenolphthalein indicator.

Theory
Saponification is the process of formation of soap by the reaction of fat, oil or fatty acid with alkali.

The saponification number may be defined as the number of milligrams of KOH required to saponify one gram fat or oil.

The mineral oils are the hydrocarbon mixture and do not react with alkali to give soap. Although vegetable oils hydrolyzed with alkali to give soap and glycerol.

The mixture of oil sample, solvent and alkali (KOH) is refluxed to yield the soap.

$$\underset{\text{Fat or oil}}{\begin{array}{l} CH_2OOCR \\ | \\ CHOOCR' \\ | \\ CH_2OOCR'' \end{array}} + \underset{\text{Alkali}}{3KOH} \longrightarrow \underset{\text{Glycerol}}{\begin{array}{l} CH_2-OH \\ | \\ CH-OH \\ | \\ CH_2-OH \end{array}} + \underset{\text{Potassium soap}}{\begin{array}{l} RCOOK \\ R'COOK \\ R''COOK \end{array}}$$

where, R, R' and R'' are alkyl groups.

The unreacted KOH is titrated back with standard acid using phenolphthalein indicator.

$$OH^- + H^+ \longrightarrow H_2O$$

Procedure
Take 2.5 g of *oil sample* in the alkali-resistant conical flask (B–24 mouth). Add 25 ml ethyl methyl ketone (solvent) and 25 ml alcoholic KOH with the help of measuring cylinder. Fix the air condenser and keep the conical flask on water bath to reflux the mixture for about one hour. The reflux of mixture results in the saponification of oil. Remove the flask from water bath and cool the contents below tap up to room temperature. Disconnect the air condenser. Add 4–5 drops of phenolphthalein indicator into the flask and titrate the mixture against standard (N/2) acid. At the end point colour changes from pink to colourless. Note down the volume (A ml) of acid used up to end point.

For the *blank determination* take 25 ml ethyl methyl ketone (solvent) and 25 ml alcoholic KOH in the another conical flask with the help of measuring cylinder. Fix the air condenser and keep the conical flask on water bath to reflux the mixture for about one hour. Titrate the mixture against N/2 acid using phenolphthalein indicator. At the end point colour changes from pink to colourless. Note down the volume (B ml) of acid used up to end point.

Observation and calculation

Weight of oil sample taken = 2.5 g
Volume of solvent added to the flask = 25 ml
Volume of alcoholic KOH added to the flask = 25 ml
Volume of N/2 HCl used for sample determination = A ml
Thus, the volume of N/2 HCl equivalent to KOH used for the saponification of w g sample = $(B - A)$ ml

Thus, saponification value of the sample = $\dfrac{(B-A)}{w \times 2} \times 56 = \dfrac{(B-A)}{w} \times 28 = $ ......

Result

Saponification value of given oil sample = ...

Precautions

1. The solution mixture should not be heated above boiling point.
2. Reflux time should be at least one hour.
3. To fasten the reaction the flask should be shaken during the reflux.

Experiment 9: To Determine the TDS of Water Samples of Different Sources

AIM
To determine the amount of total dissolved solids (TDS) in the given water sample.

Apparatus and reagent
Crucible or glass beaker, water bath, measuring cylinder, electric oven, desiccator and balance.

Theory
When a solution of dissolved solids is heated the volatile components are evaporated leaving behind the solid impurities (residue). The weight of solid residue after complete drying gives the total dissolved solids (TDS) present in water sample.

Procedure
Take a glass crucible or beaker, clean it thoroughly and dry it by placing into the electric oven for half an hour at 150°C. Take out the crucible from oven, cool it up to room temperature in a desiccator and take the weight of this empty crucible. Now take 25 ml of water sample into the dried crucible. Heat it on the water bath just to dryness of the contents and keep the crucible in the electric oven (≈ 1 h) for complete dryness and also to remove the water of crystallization. Cool the crucible up to room temperature. Take the weight of crucible containing dry residue. The weight of dry residue gives the value of TDS.

Observation and calculation
Volume of water sample taken = 25 ml

Weight of empty beaker = x g

Weight of beaker containing dry solid residue = y g

Total dissolved solids (TDS) = $\dfrac{(y-x)}{25} \times 1000 = z$ g/l = $z \times 1000$ ppm

Result
Amount of total dissolved solids (TDS) in the given water sample = ... ppm

Precautions
1. The crucible should be cleaned thoroughly and dried completely.
2. Volume of water sample should be measured accurately.
3. The crucible containing the residue should be heated in the electric oven for at least one hour.

Experiment 10: Determination of Concentration of KMnO$_4$ Solution Spectrophotometerically

AIM
Determination of concentration of KMnO$_4$ solution spectrophotometrically.

Requirements
Spectrophotometer, beaker, distilled water, standard KMnO$_4$ solution, tissue paper.

Theory
Spectroscopy is the branch of science which deals with the study of interaction of electromagnetic radiations with matter. Electromagnetic radiation is a type of energy that is transmitted through space at enormous velocities. When monochromatic visible light falls on the homogeneous medium, the intensity of transmitted light is less than that of incident light. A part of incident light may be absorbed.

$$I_o = I_a + I_t$$

where I_o = intensity of incident light, I_a = intensity of absorbed light and I_t = intensity of transmitted light.

A spectrophotometer is a device which detects the percentage transmittance of light radiation, when light of certain frequency range is passed through the sample. The UV–visible spectrophotometer consists of light source, monochromator, detector, amplifier and recording devices. It is based on caution fundamental law, i.e. Lambert–Beer law which states that "when monochromatic light passes through the sample medium, the intensity of that light decreases exponentially with the increase in concentration and thickness of the absorbing medium".

$$I_t = I_o \, e^{-ct}$$

Absorbance or optical density is defined as:

$$A = \log I_o/I_t = \varepsilon c t$$

where I_o = intensity of incident light, I_t = intensity of transmitted light, A = absorbance, ε' = molar extinction coefficient or molar absorptivity, t = thickness of the medium, c = concentration of the absolving medium.

If the same sample cell (i.e. t is constant) is used for measurement of absorbance of solution having different concentration, then $A \propto C$, i.e. extent of absorbance (A) is directly proportional to the concentration (C) of the absorbing medium.

Thus, if a graph is plotted between A and C, we get a straight line for solution obeying Lambert–Beer law. This is known as *calibration curve* (Figure E10.1).

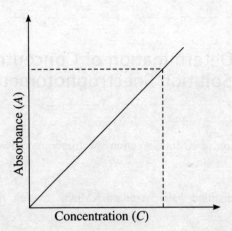

Figure E10.1 Plot of concentration versus absorbance.

For measuring the concentration of given solute, the calibration curve is obtained by measuring the absorbance of standard solutions of different concentration. This calibration curve is then used for measuring the concentration of unknown solution.

Procedure

Initial setting of spectrophotometer

1. Connect the instrument to power supply, switch on the instrument and ensure the glowing of red light.
2. Put instrument on the tungsten filament lamp (T) by rotating the knob marked with meter T and D so that the visible light passes through the shutter.
3. Adjust the wavelength knob to the required wavelength region on the scale. Choose the position of wavelength between 400 and 960 nm wavelengths.
4. Adjust the 'set zero' knob so that monitor shows zero on transmittance (T) scale and 100 on optical density (OD) scale.

Final setting of the spectrophotometer

5. Open the lid of the sample holder and insert a cuvette containing blank solution (distilled water) into the corvette. Close the lid of the sample holder to fix it properly.
6. Adjust the control knob so that monitor shows 100 on transmittance scale and zero on OD.
7. Open the lid and remove the cuvette. Close the lid tightly again.
8. Check zero on the monitor and adjust it, if disturbed.
9. Repeat steps 4–8 till zero OD and 100% transmittance is fixed.

Determination of λ_{max}

10. Insert the cuvette containing standard solution into the sample holder and note down the readings.
11. Change the wavelength by about 20 nm every time and note down the corresponding optical density. Plot the graph between wavelength on X-axis and OD along Y-axis.

Determination of concentration of sample solution

12. Fix the wavelength at λ_{max} position by wavelength controller.
13. Prepare the standard solutions of KMnO$_4$ with concentration 0.2%, 0.5%, 1.0%, 1.5%, 2.0%, 2.5%, 3.0%, etc. (20 ml each).
14. Note down the OD of series of KMnO$_4$ solutions prepared above.
15. Plot the OD against concentration. It should be a straight line.
16. Now take the solution of unknown concentration of KMnO$_4$ and note down optical density.
17. Find out the concentration of the unknown solution from the graph corresponding to the OD of the solution.
18. Report the concentration of unknown solution in mg l^{-1} (ppm).

Precautions

1. Only dilute solutions of known concentrations should be used for the calibration curve.
2. For preparation of standard calibration curve, dilute solutions of known concentrations should be used.
3. λ_{max} should be carefully observed.
4. Cuvette should be properly cleaned.
5. Cuvette should be handled carefully as it is very fragile.

Experiment 11: Determination of Strength of HCl Solution by Titrating it against NaOH Solution Conductometerically

AIM

Determination of strength of HCl solution by titrating against standard NaOH solution, conductometrically.

Requirement

Conductivity meter, conductivity cell, microburette, beaker, pipette, 0.01N KCl solution, 0.1N NaOH and approx. 0.01N HCl solution.

Theory

Conductometeric titrations are carried out to find the strength of unknown solution. In this technique, the end point is detected by the measurement of conductance of solution. This method is based on the principle that conductance of solution depends upon the number and mobility of the ions. In this titration, the conductance values are plotted against the volume of titrant added. The point of intersection of two straight lines gives the end point of the titration.

A conductivity cell dipped in a measuring solution, is placed in the inverting input path of an operational amplifier. AC voltage of constant amplitude and suitable frequency is applied to the system. For a given feed back resistance R_f, the output e_o is linearly proportional to the conductance of the solution (g_i) (Figure E11.1).

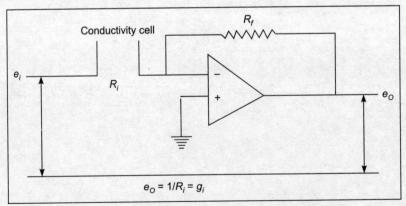

Figure E11.1 Circuit diagram of conductivity meter.

At the beginning of the titration when the NaOH is not added in the solution shows maximum conductivity because H⁺ ions possess greatest mobility of any ion. On adding NaOH the concentration of H⁺ ions decrease because hydrogen ions are replaced by OH⁻ ions. The mobility of OH⁻ is much less, so that the conductivity of the solution decreases rapidly. The solution at neutralization, i.e. at the end point contains only Na^+ and Cl^- ions and will have a minimum conductance.

$$H^+ + Cl^- + Na^+ + OH^- \longrightarrow Na^+ + Cl^- + H_2O$$

After the complete neutralization, further addition of NaOH will increase the conductance due to OH^- ions as they possess second highest mobility. Thus, the conductance will be minimum at equivalence point. The graph obtained in this case is V-shaped as shown in the following figure and point of interaction corresponds to the end point (Figure E11.2).

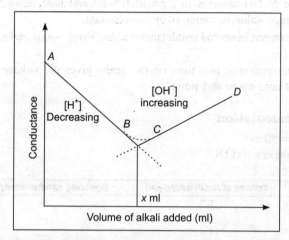

Figure E11.2 Conductometric titration of a strong acid (HCl) against a strong base (NaOH).

Procedure

The conductivity measurements are made by making use of conductivity meter which is based on principle of Wheat stone bridge. Before measuring the conductance of solution the instrument should be calibrated.

Calibration of the instrument

It is carried out in the following steps:

1. Switch on the instrument and wait for 5 min.
2. Connect the conductivity cell to the cell terminals on the instrument.
3. Take 50 ml of 0.01N KCl solutions in a beaker and immerse the cell into the solution so that platinum foils (electrodes) of the cell are completely dipped into solution.
4. Select the proper conductance range by the multiplier switch in the proper range.
5. Set the temperature control at room temperature or other measurement temperature.
6. Set the calibration knob to calebrate position.
7. Set the function knob to conductivity position.
8. Adjust calibration control knob to get the desired value of conductivity of 0.01 KCl solutions on the display.
9. After the calibration of instrument remove the cell from KCl solution and wash thoroughly with distilled water.
10. After the calibration, do not disturb the calibration control switch till the experiment is over.

Titration of HCl versus NaOH solution

1. Pipette out 50 ml of given HCl solution in a 100 ml beaker.
2. Immerse the conductivity cell into the solution so that platinum electrodes are completely dipped into the solution.
3. Select the proper conductance range and put the function knob to conductivity position.
4. Note the conductance of solution.
5. From the burette, add 0.1N NaOH. Mix the solution and note the conductance of solution.
6. Keep adding the NaOH solution in a lot of 0.5 ml and note down the conductance value till the conductance value becomes stable or constant.
7. Plot the graph between observed conductance value along y-axis and volume of NaOH along x-axis.
8. The point of intersection of two lines on the graph gives the volume of alkali used for the neutralization of acid up to end point.

Observation and calculation

Volume of HCl taken = 50 ml
Normality of NaOH solution = 0.1N

S. No.	Volume of NaOH added (ml)	Observed conductance (mho)
1.	0.0	—
2.	0.5	—
3.	1.0	—
4.	1.5	—
5.	2.0	—
6.	2.5	—
7.	—	—

Volume of NaOH for equivalence point = x ml

Applying normality equation, $N_1 V_1 = N_2 V_2$
$\qquad\qquad\qquad\qquad\quad$ (HCl) $\quad$ (NaOH)

$$N_1 \times 50 = 0.1 \times x$$

$$\text{Normality, } N_1 = 0.1 \times \frac{x}{50}$$

Strength = Normality × Equivalent weight
$\qquad\quad = N \times 36.5 = a$ g l^{-1}.

Result

Strength of given HCl solution is g/l.

Precautions

1. Stirring of solution, should be done after each addition of titrant (NaOH).
2. Once instrument is calibrated, calibration knob should not be disturbed.

3. All the precautions regarding the handling of the instrument should be observed as per instrument manual.
4. Conductivity cell should be cleaned by using warm 50% detergent solution but not by strong reagent and should be rinsed with distilled water.
5. Conductivity cell should be used carefully and platinum foil should be dipped into the solution.
6. Cell electrodes should not be touched with finger.

Experiment 12: To Determine Amount of Sodium and Potassium in a Given Water Sample by Flamephotometer

AIM

To determine the amount of Na and K in a given water sample by flame photometer.

Apparatus and chemicals

Flamephotometer, Air compressor, chemicals, three standard solutions, double distilled water, water sample.

Theory

When the metal solution is sprayed into the flame so that the solvent evaporates leaving behind solute elements and undergo dissociation into atoms in vaporized state. Some of the atoms get excited by absorbing thermal energy. The excited atoms radiate energy of different wavelengths which on

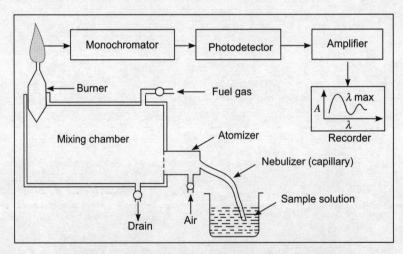

Figure E12.1 Schematic block diagram of flame photometer.

passage through optical filter permits the exciting character wavelength of metal under examination. These are finally recorded by a suitable read out plotter.

Procedure

Air at an appropriate pressure is passed into a nebulizer, the suction produced by it draws the sample under examination into nebulizer from where it emerges as a fine mist and passes into a mixing chamber. In the mixing chamber, this fine mist of sample mixes evenly with the fuel gas stream at a given pressure and the mixture burns at the tip of the burner. Radiation from the resulting flame passes through the lens and an optical filter which permit only the radiation characteristic of the element under investigation in the sample to the photodetector. The output signal from the photodetector is sent to the read out device processed to display concentration of the element in the ppm.

1. **Preparation of stock solution of Na and K:** Stock solution of Na and K is prepared by dissolving the NaCl by dissolving 2.542 g of dry NaCl and 1.909 g KCl (AR grade) in distilled water in 1–litre volumetric flask. The solution is referred as *stock solution* (1000 ppm).
2. **Preparation of standard solutions Na and K:** Dilute the stock solution by adding distilled water to obtain the solution of various concentrations, e.g. 40 ppm, 20 ppm and 10 ppm.

Operation of the Elico flame photometer (CL–361)

3. **Calibration and measurement of Na and K concentration:**
   - Switch on the instrument and air compressor.
   - Set the flame by maintaining the ratio of the air and fuel gas mixture.
   - Calibration: press the <mode> on the screen.
   - Four functions on the screen: (1) Read, (2) Calibrate, (3)View, (4) Print.
   - Press <(2)> on the screen.
   - Select the measurement mode [gen, serum, urin]; press <1> for general mode, Screen displays: (1) Na, (2) K, (3) Li, (4) Ca.
   - Select Na and K by pressing 1 and 2 <enter>
   - Screen display (SD): (1) segmental, (2) quadratic #Press 2 for quadratic
   - SD Enter number of standards (min. 3, max. 20), enter 3 standard, #Press<enter>
   - On screen, feed the value of standard (1) 40 ppm, (2) 20 ppm, (3) 10 ppm
   - After putting the value of standards on SD: Will you verify entered conc. <yes/no> # <press no>
   - SD: Feed the blank press <enter>
   - SD: Please wait ......
   - SD: Feed the distilled water to clean #press<enter>
   - SD: Cleaning the nebulizer
   - SD: Feed the standard solution #press<enter>
   - SD: Feed the distilled water to clean #press<enter>
   - SD: Feed the standard solution (2) #press<enter>
   - SD: Feed the distilled water to clean (1) #press<enter>
   - SD: Feed the standard solution (3) #press<enter>
   - SD: Feed the distilled water #press<enter>
   - SD: Save this curve (yes/no) #press<yes>
   - SD: Enter cal file number * * (1–5) (enter)
   - Enter the value of file number by pressing active keys (1–9) #press<enter>
   - SD: Overwrite the exiting file ? (yes/no) #press<enter>
   - SD: Enter date #press<enter>
   - #SD: Enter time SD: your name #press<enter>
   - SD: General sample name #press<enter> Press <no. 1>

- SD: (1) Sample, (2) Rapid, (3) View, (4) Print, (5) Norm, (6) Save, #press 2 for rapid.
- SD: Feed the smple #press<enter>
- SD: Conc. ppm Na (.......) K (.......) cc [note the reading]
- To take the reading of other sample. Repeat the four steps <Now switch off>

Result

The concentration of Na and K in given water sample is Na = ... ppm and K = ppm.

Precautions

1. The colour of the flame should be maintained as blue by adjusting the air and gas ratio.
2. The flame height should be adequate i.e. it should cover the optical window.
3. The stock solution and standard solutions should be prepared accurately.

Experiment 13: Estimation of Total Iron in an Iron Ore

AIM

To determine the amount of total iron in the iron ore (Fe_2O_3) solution by standard (N/10) potassium dichromate ($K_2Cr_2O_7$) solution.

Apparatus and reagent

Burette, pipette, conical flask, beaker, droper, funnel, hot plate, iron ore solution, standard (N/10) $K_2Cr_2O_7$ solution, stannous chloride solution (5%), mercuric chloride solution (saturated), sodium diphenyl amine sulphonate indicator solution, conc. HCl.

Theory

This titration furnishes an example of redox titration. The iron ore solution contains iron in the Fe^{2+} and Fe^{3+} state. Therefore, to determine the total iron, all the ferric ions (Fe^{3+}) should first be brought to the ferrous state (Fe^{2+}). The Fe^{3+} ions are reduced to Fe^{2+} ions by conc. $SnCl_2$ solution as follows.

To a hot solution of iron ore, conc. $SnCl_2$ solution (3–4N with respect to HCl) is added in slight excess to ensure complete reduction of Fe^{3+} to Fe^{2+}.

$$Sn^{2+} \longrightarrow Sn^{4+} + 2e^-$$
$$[Fe^{3+} + e^- \longrightarrow Fe^{2+}] \times 2$$
$$\overline{Fe^{3+} + Sn^{2+} \longrightarrow 2Fe^{2+} + Sn^{4+}}$$

The solution is then cooled and the slight excess of $SnCl_2$ is removed by adding a saturated solution of mercuric chloride ($HgCl_2$).

$$Sn^{2+} \longrightarrow Sn^{4+} + 2e^-$$
$$2Hg^{2+} + 2Cl^- + 2e^- \longrightarrow Hg_2Cl_2$$
$$\overline{2Hg^{2+} + 2Cl^- + 2e^- \longrightarrow Hg_2Cl_2 \text{ (silky white ppt.)}}$$

Now the total iron is determined by titrating the solution with standard $K_2Cr_2O_7$ using sodium diphenyl amine sulphonate indicator solution. The end point is determined by change of colour from deep green to violet blue.

$$[Fe^{2+} \longrightarrow Fe^{3+} + e^-] \times 6$$
$$Cr_2O_7^{2-} + 14H^+ + 6e^- \longrightarrow 2Cr^{3+} + 7H_2O$$
$$\overline{Cr_2O_7^{2-} + 14H^+ + 6Fe^{2+} \longrightarrow 2Cr^{3+} + 6Fe^{3+} + 7H_2O}$$

Procedure

Take 20 ml of iron ore solution in a titration flask. Add 5 ml of conc. HCl and heat just to boiling. Add dropwise a 5% solution of $SnCl_2$ (in 1:1 HCl) with thorough shaking of the flask until the yellow colour (due to $FeCl_3$) gives a faint green colour. Add 3–4 drops of $SnCl_2$ in excess to ensure complete reduction. Now cool the flask rapidly under tap and add about 5 ml of a saturated solution of $HgCl_2$ in one lot and mix thoroughly. After 2 min, add 6–8 drops of the sodium diphenylamine sulphonate indicator solution and titrate rapidly with standard (N/10) $K_2Cr_2O_7$ solution until there is a change of colour from colourless to green and at the end point from deep green to violet blue. Note down the volume of $K_2Cr_2O_7$ solution used. Take three concordant readings.

Observation

S. No.	Volume of iron ore (ml)	Volume of $K_2Cr_2O_7$		Volume of $K_2Cr_2O_7$ used (ml)
		Initial reading	Final reading	
1				
2				
3				

Calculation

Applying normality equation, $\quad N_1V_1 = N_2V_2$
$\quad\quad\quad\quad\quad\quad\quad\quad\quad\quad\quad\quad$ (HCl) $\quad$ (NaOH)

$$N_1 \times 20 = \frac{x}{10}$$

or $\quad$ Normality $\quad\quad\quad\quad N_1 = \frac{x}{10 \times 20}$

$\quad\quad\quad\quad\quad\quad$ Strength = Normality × Equivalent weight

$\quad\quad\quad\quad\quad\quad\quad\quad\quad\quad = N_1 \times 56 = \ldots\ldots$ g l^{-1}.

Result

Amount of total iron in iron ore solution = ... g l^{-1}

Precautions

1. $SnCl_2$ solution should be added dropwise in a hot solution and should be thoroughly mixed after each addition.
2. $HgCl_2$ solution should be added to the cooled solution in one lot and not drop by drop.
3. The reaction between $SnCl_2$ and $HgCl_2$ being slow, 2–3 minutes should be allowed for its completion.

Experiment 14: Estimation of Calcium in Limestone and Dolomite

AIM
Estimation of calcium in limestone and dolomite.

Requirements
Beaker, measuring flask, conical flask, 0.1N $KMnO_4$, NH_3 (liquor ammonia), NH_4Cl, H_2SO_4, dilute HCl, limestone or dolomite, ammonium oxalate.

Theory
Lime stone consists of largely $CaCO_3$ and a small amount of magnesium carbonate. While dolomite is equimolar mixture of calcium carbonate and magnesium carbonate. Limestone powder is dissolved in HCl and calcium present in solution is precipitated as calcium oxalate using either oxalic acid or ammonium oxalate in the presence of ammonia. The precipitate is filtered washed and dissolved in dilute sulphuric acid and oxalic acid so liberated is titrated with standard $KMnO_4$ solution. From the volume of $KMnO_4$, required for titration, the amount of calcium in the ore can be calculated.

$$CaCO_3 + 2HCl \longrightarrow CaCl_2 + H_2O + CO_2$$

$$CaCl_2 + \begin{array}{c} COONH_4 \\ | \\ COONH_4 \end{array} \longrightarrow \begin{array}{c} COO \\ | \\ COO \end{array}\!\!\!\!> Ca + 2NH_4Cl$$

Ammonium oxalate $\qquad\qquad$ Calcium oxalate (ppt.)

$$\begin{array}{c} COO \\ | \\ COO \end{array}\!\!\!\!> Ca + H_2SO_4 \longrightarrow \begin{array}{c} COOH \\ | \\ COOH \end{array}$$

Oxalic acid

$$KMnO_4 + 3H_2SO_4 \longrightarrow K_2SO_4 + 2MnSO_4 + 5[O] + 3H_2O$$

$$\begin{array}{c} COOH \\ | \\ COOH \end{array} + 5[O] \longrightarrow 5H_2O + 10CO_2$$

Procedure
1. Take one gram of limestone or dolomite into a beaker, dissolve it in dil. HCl, cover the beaker with watch glass so that no CO_2 should leave the beaker.
2. Wash the watch glass with distilled water into the same beaker, make the solution alkaline by adding the ammonia (≈ 2 ml) and 1 g of NH_4Cl and stir it to dissolve. Make up the volume to one litre by adding distilled water.
3. Take 100 ml of this ore, add 20 ml of ammonium oxalate to precipitate the calcium as calcium oxalate.

4. Filter it and wash the precipitate with distilled water to free from chloride and oxalate ions.
5. Transfer the filter paper along with the precipitate into the conical flask.
6. Add 20 ml dilute H_2SO_4 and same volume of distilled water.
7. Dissolve the precipitate by shaking.
8. Take 20 ml of this solution, heat it up to 70°C and titrate against N/10 $KMnO_4$ solution.
9. Note down the volume of $KMnO_4$ solution used up to end point when pink colour just disappears.

Observations

S. No.	Volume of solution taken for each titration	Initial reading of burette	Final reading of burette	Volume of $KMnO_4$ used (ml)
1.	20 ml			
2.	20 ml			
3.	20 ml			

Weight of ore taken $= w$ g

Volume of calcium oxalate solution prepared $= 100$ ml

Volume taken for each titration $= 20$ ml

Let the concordant volume of N/10 $KMnO_4$ used $= x$ ml

$$N_1V_1 = N_2V_2$$
(Calcium solution)   ($KMnO_4$)

$$N_1 \times 20 = \frac{x}{10}$$

or Normality

$$N_1 = \frac{x}{10 \times 20}$$

Strength of calcium ore solution as

$$CaO = \frac{x}{200} \times 28 \text{ g l}^{-1}$$

Weight of calcium (as CaO) in w g sample

$$\frac{x}{100} \times \frac{100}{1000} \times \frac{28}{w} = a \text{ g.}$$

% of calcium (as CaO) in the sample $= \frac{x}{200} \times \frac{100}{1000} \times 28 \times \frac{100}{w}$

$$= \frac{x}{w} \times 1.4 = b\%$$

Precautions

1. The washing liquid should always be taken in the beaker in which the precipitation was carried out and then it should be transferred to the filter paper.

2. A fresh portion of the washing liquid should be used only when the first portion has completely passed through the filter paper.
3. For testing the filtrate the presence of chloride or oxalate, a few drops of the filtrate should be collected directly from the funnel.
4. Any precipitate sticking to the beaker after washing should be dissolved in dilute H_2SO_4 and transferred to the measuring flask.

Sample Paper 1

ENGINEERING CHEMISTRY

Time 3:00 Hrs   Max. Marks 100

Note: Attempt five questions in all, covering at least one from each part. Question Number 1 is compulsory. All questions carry equal marks.

Q.1 Answer the following in brief: [10 × 2 = 20]
   (a) Is it possible to have a quadruple point in a one-component system? Why?
   (b) Define the heterogeneous catalysis giving one example.
   (c) How are exhausted ion-exchange resins regenerated?
   (d) Why can water softened by lime-stone process cause boiler-troubles?
   (e) Why Calgon conditioning is better than phosphate conditioning?
   (f) Iron corrodes faster than aluminium, even though iron is placed below aluminium in the electrochemical series. Why?
   (g) Why the synthetic lubricants have added advantage over natural lubricants?
   (h) Why PVC is tougher and stronger polymer than polyethylene?
   (i) What structural changes occur when natural rubber is subjected to vulcanization?
   (j) Define the absorption spectroscopy.

SECTION 1

Q.2 (a) What is Gibbs' phase rule? Derive the phase rule equation for one-component systems?
Give applications and limitations of phase rule.   [10]
   (b) Calculate the number of components and degree of freedom for the following equilibria: [5 × 2 = 10]
      (i) Dissociation of NH_4Cl in a sealed tube
      (ii) $I_2(s) \Leftrightarrow I_2(H_2O) \Leftrightarrow I_2$ (benzene)
      (iii) Boiling azeotropic mixture
      (iv) $Fe(s) + H_2O(g) \Leftrightarrow FeO(s) + H_2(g)$
      (v) A mixture of $N_2(g)$, $H_2(g)$ and $NH_3(g)$

Q.3 (a) Explain the following with example: [8]
      (i) catalytic promoter   (ii) catalytic poison
      (iii) auto catalyst   (iv) induced catalyst
   (b) Describe the intermediate compound theory of the mechanism of catalytic reactions.
                                                                                                                                                    [8]
   (c) Explain the effect of temperature on enzyme catalysis.   [4]

SECTION 2

Q.4 (a) What do you understand by hard water and soft water? Explain the disadvantages of hard water in various fields. [8]

(b) 100 ml water sample required 20 ml of N/50 H_2SO_4 for neutralization to phenolphthalein end point. After that methyl orange was added to this and further acid required was 15 ml. Calculate the amount and types of alkalinity present in water in terms of $CaCO_3$ equivalent. [6]

(c) What do understand by scale and sludge? What are the disadvantages of scale and sludge formation? [6]

Q.5 (a) Calculate the quantity of lime and soda required for the softening of 50,000 litres of water containing the following impurities: [8]
$Ca(HCO_3)_2$ = 9.7 mg, $Mg(HCO_3)_2$ = 7 mg, $CaSO_4$ = 13.4 mg, $MgSO_4$ = 11 mg, $MgCl_2$ = 2.5 mg, $NaAlO_2$ (used as coagulant) = 4 mg and NaCl = 7.8 mg [Mol. Wt. of $Ca(HCO_3)_2$ = 162, $Mg(HCO_3)_2$ = 146, $CaSO_4$ = 136, $MgSO_4$ = 120, $MgCl_2$ = 95, $NaAlO_2$ = 82 and NaCl = 58]

(b) Write short notes on the following: [12]
  (i) mixed bed deionization,
  (ii) break point chlorination,
  (iii) reverse osmosis for desalination of water.

SECTION 3

Q.6 (a) What do you understand by dry and wet corrosion. Explain the mechanism of rusting of iron. [10]

(b) Describe the cathodic and anodic methods for the protection of metal from corrosion. [10]

Q.7 (a) What are lubricants? Explain the fluid film and boundary line lubrication mechanism of lubrication. [8]

(b) Explain the following properties of lubricating oils: [12]
  (i) viscosity and viscosity index,
  (ii) flash and fire point,
  (iii) aniline point,
  (iv) cloud and pour point.

SECTION 4

Q.8 (a) Describe the effects of structure on properties of polymers. [7]

(b) Give the preparation, properties and uses of PVC and SBR. [8]

(c) Write short note on biodegradable polymers. [5]

Q.9 (a) Explain the principal and applications of thermogravimetric analysis (TGA). [8]

(b) What is spectroscopy? Explain the applications of IR spectroscopy. [6]

(c) Distinguish between absorption and emission spectroscopy. [6]

Sample Paper 2

ENGINEERING CHEMISTRY

Time 3:00 Hrs   Max. Marks: 100

Note: Attempt five questions in all, covering at least one from each part. Question Number 1 is compulsory. All questions carry equal marks.

Q.1 Answer the following in brief: [10 × 2 = 20]
- (a) Define the terms metastable equilibrium and triple point.
- (b) Why the industries prefer to use the catalysts which work at room temperature.
- (c) What will happen if enzymes are not present in the body?
- (d) In the deionization process, water is usually first passed through the cation exchanger and then through the anion exchanger. Why?
- (e) Why do we express hardness of water in terms of $CaCO_3$ equivalent?
- (f) Why impure metals are more susceptible to corrosion than pure metals?
- (g) What is the significance of flash-point?
- (h) Thermosets articles cannot be shaped by application of heat and pressure.
- (i) All simple molecules cannot act as monomers.
- (j) Give the statement of Lambert's and Beer's law.

SECTION 1

Q.2 (a) Explain the following terms: [6]
    (i) phase   (ii) component
    (iii) degree of freedom   (iv) transition point.

(b) Distinguish between water system and CO_2 system.   [4]

(c) Describe the Zn–Mg system or sodium sulphate–water system with the help of T–C phase diagram.   [10]

Q.3 (a) What do you understand by catalyst? Explain the mechanism of homogeneous and heterogeneous catalysis by giving suitable examples.   [7]

(b) Derive the Michaelis–Menten equation for the rate of enzyme catalyzed reaction.   [7]

(c) Describe the adsorption theory to explain the mechanism of heterogeneous catalysis.   [6]

SECTION 2

Q.4 (a) What do you understand by alkalinity of water? How the alkalinity of water is determined? Explain in detail. [7]

(b) 0.28 gm of $CaCO_3$ was dissolved in HCl and the solution made up to 1 litre with distilled water. 100 ml of the above solution required 28 ml of EDTA solution on titration. 100 ml of hard water sample required 33 ml of same EDTA solution on titration. After boiling 100 ml of this water, cooling, filtering and then titration required 10 ml of EDTA solution. Calculate the temporary and permanent hardness of water. [8]

(c) Write short note on caustic embrittlement. [5]

Q.5 (a) Describe the principal and procedure of lime–soda method for the treatment of hard water. [10]

(b) What do you understand by municipal water? Explain various steps used in the treatment of municipal water. [10]

SECTION 3

Q.6 (a) What is corrosion? Explain the mechanisms of differential aeration corrosion, galvanic corrosion and microbiological corrosion. [10]

(b) Explain the factors affecting the rate of corrosion. [7]

(c) How are the metals/metallic articles protected from corrosion by corrosion inhibitors. [3]

Q.7 (a) What do you understand by the additives for lubricants? Explain various additives and their role to improve the quality of lubricants. [8]

(b) Explain the preparation, properties and uses of greases. [6]

(c) What do you understand by biodegradable lubricants? How these lubricants can help to protect the environment from pollution. [6]

SECTION 4

Q.8 (a) What do you understand by polymer and polymerization? Explain the effects of structure on properties of polymers. [8]

(b) Write short notes on the following: [12]
  (i) Silicones,
  (ii) Polymer composites, and
  (iii) Coordination polymerization.

Q.9 (a) Describe the principal and applications of DTA and DSC. [12]

(b) Explain the principle and applications of flame photometry. [8]

Sample Paper 3

ENGINEERING CHEMISTRY

Time 3:00 Hrs Max. Marks: 100

Note: Attempt five questions in all, selecting at least one from each unit. All questions carry equal marks.

SECTION 1

Q.1 (a) Give three statements of second law of thermodynamics. [3]
(b) Derive the Gibbs–Helmholtz equation and give its application. [6]
(c) Prove that $\Delta S = nC_V \ln \dfrac{T_2}{T_1} + nR \ln \dfrac{V_2}{V_1}$, where all symbols have usual meaning. [5]
(d) Prove that entropy is the measure of randomness/disorderness of the system. [2]
(e) The free energy change (ΔG) accompanying a given process is -85.78 kJ at 298 K and -83.67 kJ at 308 K. Calculate the change in free energy (ΔG) for the process at 303 K. [4]

Q.2 (a) What is Gibbs' phase rule? Derive the phase rule equation for one component systems. [6]
(b) Give the applications and limitations of phase rule. [4]
(c) What is eutectic system? Explain the Pb–Ag system with the help of TC diagram. [6]
(d) Define phase and components. [2]
(e) Is it possible to have a quadruple point in one–component system? Explain. [2]

SECTION 2

Q.3 (a) Describe the EDTA titration method for the determination of hardness of water. [6]
(b) What are scale and sludge? What are the disadvantages of scale and sludge formation? [5]
(c) What do you understand by internal conditioning of boiler? [2]
(d) Why Calgon conditioning is better than phosphate conditioning? [2]
(e) 100 ml water sample required 20 ml of N/50 H_2SO_4 for neutralization to phenolphthalein end point. After that methyl orange was added to this and further acid required was 15 ml. Calculate the types of alkalinity present in water in terms of $CaCO_3$ equivalent. [5]

Q.4 (a) What do you understand by brackish water? Explain the electrodilysis method for the desalination of brackish water. [6]

(b) Write short notes on: [8]
  (i) Zeolite method (ii) Break point chlorination.
(c) Describe the principal and procedure of lime–soda method for the treatment of hard water. [6]

SECTION 3

Q.5 (a) What is corrosion? explain the mechanism of electrochemical corrosion. [6]
(b) Define the Pilling–Bedworth's rule. [2]
(c) How is the rate of corrosion affected by the following factors: [6]
Purity of metal, position of metal in the galvanic series, temperature and moisture.
(d) Distinguish between galvanization and electroplating. [3]
(e) Explain the soil corrosion. [3]

Q.6 (a) What do you understand by lubricants? Explain the fluid film mechanism of lubrication. [6]
(b) What are synthetic lubricants? Why they have added advantage over natural lubricants. [4]
(c) Explain the following properties of lubricating oils: [6]
  (i) viscosity and viscosity index (ii) flash and fire point
  (iii) aniline point (iv) cloud and pour point.
(d) Explain the preparation, properties and uses of greases. [4]

SECTION 4

Q.7 (a) Describe the mechanism of step growth and chain growth polymerization. [5]
(b) Explain the effects of structure on properties of polymers. [6]
(c) Give the preparation, properties and uses of PF and SBR. [6]
(d) Why PVC is tougher and stronger polymer than polyethylene? [3]

Q.8 (a) Discuss the principal and working of TGA technique. [6]
(b) What are conductometric titrations? How the conductometric titrations are better than usual titrations? [6]
(c) Define the spectroscopy. Explain the principle and applications of flame photometry. [8]

LOGARITHMS TABLES

	0	1	2	3	4	5	6	7	8	9	Mean Differences								
											1	2	3	4	5	6	7	8	9
10	00000	00432	00860	01284	01703						42	85	127	170	212	254	297	339	381
						02119	02531	02938	03342	03743	40	81	121	162	202	242	283	323	364
11	04139	04532	04922	05308	05690						37	77	116	154	193	232	270	309	348
						06070	06446	06819	07188	07555	37	74	111	148	185	222	259	296	333
12	07918	08279	08636	08991	09342						36	71	106	142	177	213	248	284	319
						09691	10037	10380	10721	11059	34	68	102	136	170	204	238	272	307
13	11394	11727	12057	12385	12710						33	66	98	131	164	197	229	262	295
						13033	13354	13672	13988	14301	32	63	95	126	158	190	221	253	284
14	14613	14922	15229	15534	15836						30	61	91	122	152	183	213	244	274
						16137	16435	16732	17026	17319	29	59	88	118	147	177	206	236	265
15	17609	17898	18184	18469	18752						28	57	85	114	142	171	199	228	256
						19033	19312	19590	19866	20140	28	55	83	110	138	165	193	221	248
16	20412	20683	20951	21219	21484						27	53	80	107	134	160	187	214	240
						21748	22011	22272	22531	22789	26	52	78	104	130	156	182	208	233
17	23045	23300	23553	23805	24055						26	50	76	101	126	151	176	201	227
						24304	24551	24797	25042	25285	25	49	73	98	122	147	171	196	220
18	25527	25768	26007	26245	26482						24	48	71	95	119	143	167	190	214
						26717	26951	27184	27416	27646	23	46	69	93	116	139	162	185	208
19	27875	28103	28330	28556	28780						23	45	68	90	113	135	158	180	203
						29003	29226	29447	29667	29885	22	44	66	88	110	132	154	176	198
20	30103	30320	30535	30750	30963	31175	31387	31597	31806	32015	21	43	64	85	106	127	148	170	190
21	32222	32428	32634	32838	33041	33244	33445	33646	33846	34044	20	41	61	81	101	121	141	162	182
22	34242	34439	34635	34830	35025	35218	35411	35603	35793	35984	20	39	58	77	97	116	135	154	174
23	36173	36361	36549	36736	36922	37107	37291	37475	37658	37840	19	37	56	74	93	111	130	148	167
24	38021	38202	38382	38561	38739	38917	39094	39270	39445	39620	18	35	53	71	89	106	124	142	159
25	39794	39967	40140	40312	40483	40654	40824	40993	41162	40330	17	34	51	68	85	102	119	136	153
26	41497	41664	41830	41996	42160	42325	42488	42651	42813	42975	16	33	49	66	82	98	115	131	148
27	43136	43297	43457	43616	43775	43933	44091	44248	44404	44560	16	32	47	63	79	95	111	126	142
28	44716	44871	45025	45179	45332	45484	45637	45788	45939	46090	15	30	46	61	76	91	107	122	137
29	46240	46389	46538	46687	46835	46982	47129	47276	47422	47567	15	29	44	59	74	88	103	118	132
30	47712	47857	48001	48144	48287	48430	48572	48714	48855	48996	14	29	43	57	72	86	100	114	129
31	49136	49276	49415	49554	49693	49831	49969	50106	50243	50379	14	28	41	55	69	83	97	110	124
32	50515	50650	50786	50920	51054	51188	51322	51455	51587	51720	13	27	40	54	67	80	94	107	121
33	51851	51983	52114	52244	52375	52504	52634	52763	52892	53020	13	26	39	52	65	78	91	104	117
34	53148	53275	53403	53529	53656	53782	53908	54033	54158	54283	13	25	38	50	63	76	88	101	113
35	54407	54531	54654	54777	54900	55023	55145	55267	55388	55509	12	24	37	49	61	73	85	98	110
36	55630	55751	55871	55991	56110	56229	56348	56467	56585	56703	12	24	36	48	60	71	83	95	107
37	56820	56937	57054	57171	57287	57403	57519	57634	57749	57864	12	23	35	46	58	70	81	93	104
38	57978	58092	58206	58320	58433	58546	58659	58771	58883	58995	11	23	34	45	57	68	79	90	102
39	59106	59218	59329	59439	59550	59660	59770	59879	59988	60097	11	22	33	44	55	66	77	88	99
40	60206	60314	60423	60531	60638	60746	60853	60959	61066	61172	11	21	32	43	54	64	75	86	97
41	61278	61384	61490	61595	61700	61805	61909	62014	62118	62221	10	21	31	42	53	63	74	84	95
42	62325	62428	62531	62634	62737	62839	62941	63043	63144	63246	10	20	31	41	51	61	71	82	92
43	63347	63448	63548	63649	63749	63849	63949	64048	64147	64246	10	20	30	40	50	60	70	80	90
44	64345	64444	64542	64640	64738	64836	64933	65031	65128	65225	10	20	29	39	49	59	68	78	88
45	65321	65418	65514	65610	65706	65801	65896	65992	66087	66181	10	19	29	38	48	57	67	76	86
46	66276	66370	66464	66558	66652	66745	66839	66932	67025	67117	9	19	28	37	47	56	65	74	84
47	67210	67302	67394	67486	67578	67669	67761	67852	67943	68034	9	18	27	36	46	55	64	73	82
48	68124	68215	68305	68395	68485	68574	68664	68753	68842	68931	9	18	27	36	45	53	63	72	81
49	69020	69108	69197	69285	69373	69461	69548	69636	69723	69810	9	18	26	35	44	53	62	70	79

(Contd.)

Logarithms Tables (Contd.)

| | 0 | 1 | 2 | 3 | 4 | 5 | 6 | 7 | 8 | 9 | \multicolumn{9}{c}{Mean Differences} |
											1	2	3	4	5	6	7	8	9
50	69897	69984	70070	70157	70243	70329	70415	70501	70586	70672	9	17	26	34	43	52	60	69	77
51	70757	70842	70927	71012	71096	71181	71265	71349	71433	71517	8	17	25	34	42	50	59	67	76
52	71600	71684	71767	71850	71933	72016	72099	72181	72263	72346	8	17	25	33	42	50	58	66	75
53	72428	72509	72591	72673	72754	72835	72916	72997	73078	73159	8	16	24	32	41	49	57	65	73
54	73239	73320	73400	73480	73560	73640	73719	73799	73878	73957	8	16	24	32	40	48	56	64	72
55	74036	74115	74194	74273	74351	74429	74507	74586	74663	74741	8	16	23	31	39	47	55	63	70
56	74819	74896	74974	75051	75128	75205	75282	75358	75435	75511	8	15	23	31	39	46	54	62	69
57	75587	75664	75740	75815	75891	75967	76042	76118	76193	76268	8	15	23	30	38	45	53	60	68
58	76343	76418	76492	76567	76641	76716	76790	76864	76938	77012	7	15	22	30	37	44	52	59	67
59	77085	77159	77232	77305	77379	77452	77525	77597	77670	77743	7	15	22	29	37	44	51	58	66
60	77815	77887	77960	78032	78104	78176	78247	78319	78390	78462	7	14	22	29	36	43	50	58	65
61	78533	78604	78675	78746	78817	78888	78958	79029	79099	79169	7	14	21	28	36	43	50	57	64
62	79239	79309	79379	79449	79518	79588	79657	79727	79796	79565	7	14	21	28	35	41	48	55	62
63	79934	80003	80072	80140	80209	80277	80346	80414	80482	80550	7	14	20	27	34	41	48	54	61
64	80618	80686	80754	80821	80889	80956	81023	81090	81158	81224	7	13	20	27	34	40	47	54	60
65	81291	81358	81425	81491	81558	81624	81690	81757	81823	81889	7	13	20	26	33	40	46	53	59
66	81954	82020	82086	82151	82217	82282	82347	82413	82478	82543	7	13	20	26	33	39	46	52	59
67	82607	82672	82737	82802	82866	82930	82995	83059	83123	83187	6	13	19	26	32	38	45	51	58
68	83251	83315	83378	83442	83506	83569	83632	83696	83759	83822	6	13	19	25	32	38	44	50	57
69	83885	83948	84011	84073	84136	84198	84261	84323	84386	84448	6	12	19	25	31	37	43	50	56
70	84510	84572	84634	84696	84757	84819	84880	84942	85003	85065	6	12	19	25	31	37	43	50	56
71	85126	85187	85248	85309	85370	85431	85491	85552	85612	85673	6	12	18	24	31	37	43	49	55
72	85733	85794	85854	85914	85974	86034	86094	86153	86213	86273	6	12	18	24	30	36	42	48	54
73	86332	86392	86451	86510	86570	86629	86688	86747	86806	86864	6	12	18	24	30	35	41	47	53
74	86923	86982	87040	87099	87157	87216	87274	87332	87390	87448	6	12	17	23	29	35	41	46	52
75	87506	87564	87622	87679	87737	87795	87852	87910	87967	88024	6	12	17	23	29	35	41	46	52
76	88081	88138	88195	88252	88309	88366	88423	88480	88536	88593	6	11	17	23	29	34	40	46	51
77	88649	88705	88762	88818	88874	88930	88986	89042	89098	89154	6	11	17	22	28	34	39	45	50
78	89209	89265	89321	89376	89432	89487	89542	89597	89653	89708	6	11	17	22	28	33	39	44	50
79	89763	89818	89873	89927	89982	90037	90091	90146	90200	90255	6	11	17	22	28	33	39	44	50
80	90309	90363	90417	90472	90526	90580	90634	90687	90741	90795	5	11	16	22	27	32	38	43	49
81	90848	90902	90956	91009	91062	91116	91169	91222	91275	91328	5	11	16	21	27	32	37	42	48
82	91381	91434	91487	91540	91593	91645	91698	91751	91803	91855	5	11	16	21	27	32	37	42	48
83	91908	91960	92012	97064	92117	92169	92221	92273	92324	92376	5	10	16	21	26	31	36	42	47
84	92428	92480	92531	92583	92634	92686	92737	92788	92840	92891	5	10	15	20	26	31	36	41	46
85	92942	92993	93044	93095	93146	93197	93247	93298	93349	93399	5	10	15	20	26	31	36	41	46
86	93450	93500	93551	93601	93651	93702	93752	93802	93852	93902	5	10	15	20	25	30	35	40	45
87	93952	94002	94052	94101	94151	94201	94250	94300	94349	94399	5	10	15	20	25	30	35	40	45
88	94448	94498	94547	94596	94645	94694	94743	94792	94841	94890	5	10	15	20	25	29	34	39	44
89	94939	94988	95036	95085	95134	95182	95231	95279	95328	95376	5	10	15	19	24	29	34	39	44
90	95424	95472	95521	95569	95617	95665	95713	95761	95809	95856	5	10	14	19	24	29	34	38	43
91	95904	95952	95999	96047	96095	96142	96190	96237	96284	96332	5	9	14	19	24	28	33	38	42
92	96379	96426	96473	96520	96567	96614	96661	96708	96755	96802	5	9	14	19	24	28	33	38	42
93	96848	96895	96942	96988	97035	97081	97128	97174	97220	97267	5	9	14	18	23	28	32	38	42
94	97313	97359	97405	97451	97497	97543	97589	97635	97681	97727	5	9	14	18	23	28	32	37	42
95	97772	97818	97864	97909	97955	98000	98046	98091	98137	98182	5	9	14	18	23	27	32	36	41
96	98227	98272	98318	98363	98408	98453	98498	98543	98588	98632	5	9	14	18	23	27	32	36	41
97	98677	98722	98767	98811	98856	98900	98945	98989	99034	99078	4	9	13	18	22	27	31	36	40
98	99123	99167	99211	99255	99300	99344	99388	99432	99476	99520	4	9	13	18	22	26	31	35	40
99	99564	99607	99651	99695	99739	99782	99826	99870	99913	99957	4	9	13	17	22	26	31	35	39

ANTILOGARITHMS

	0	1	2	3	4	5	6	7	8	9	1	2	3	4	5	6	7	8	9
											\multicolumn{9}{c}{Mean Differences}
0.00	10000	10023	10046	10069	10093	10116	10139	10162	10186	10209	2	5	7	9	12	14	16	19	21
0.01	10233	10257	10280	10304	10328	10351	10375	10399	10423	10447	2	5	7	10	12	14	17	19	21
0.02	10471	10495	10520	10544	10568	10593	10617	10641	10666	10691	2	5	7	10	12	15	17	20	22
0.03	10715	10740	10765	10789	10814	10839	10864	10889	10914	10940	3	5	8	10	13	15	18	20	23
0.04	10965	10990	11015	11041	11066	11092	11117	11143	11169	11194	3	5	8	10	13	15	18	20	23
0.05	11220	11246	11272	11298	11324	11350	11376	11402	11429	11455	3	5	8	11	13	16	18	21	24
0.06	11482	11508	11535	11561	11588	11614	11641	11668	11695	11722	3	5	8	11	13	16	19	21	24
0.07	11749	11776	11803	11830	11858	11885	11912	11940	11967	11995	3	5	8	11	14	16	19	22	25
0.08	12023	12050	12078	12106	12134	12162	12190	12218	12246	12274	3	6	8	11	14	17	20	22	25
0.09	12303	12331	12359	12388	12417	12445	12474	12503	12531	12560	3	6	9	11	14	17	20	23	26
0.10	12589	12618	12647	12677	12706	12735	12764	12794	12823	12853	3	6	9	12	15	18	21	24	26
0.11	12882	12912	12942	12972	13002	13032	13062	13092	13122	13152	3	6	9	12	15	18	21	24	27
0.12	13183	13213	13243	13274	13305	13335	13366	13397	13428	13459	3	6	9	12	15	18	21	25	28
0.13	13490	13521	13552	13583	13614	13646	13677	13709	13740	13772	3	6	9	13	16	19	22	25	28
0.14	13804	13836	13868	13900	13932	13964	13996	14028	14060	14093	3	6	10	13	16	19	22	26	29
0.15	14125	14158	14191	14223	14256	14289	14322	14355	14388	14421	3	7	10	13	16	20	23	26	30
0.16	14454	14488	14521	14555	14588	14622	14655	14689	14723	14757	3	7	10	13	17	20	24	27	30
0.17	14791	14825	14859	14894	14928	14962	14997	15031	15066	15101	3	7	10	14	17	21	24	28	31
0.18	15136	15171	15205	15241	15276	15311	15346	15382	15417	15453	4	7	11	14	18	21	25	28	32
0.19	15488	15524	15560	15596	15631	15668	15704	15740	15776	15812	4	7	11	14	18	22	25	29	32
0.20	15849	15885	15922	15959	15996	16032	16069	16106	16144	16181	4	7	11	15	18	22	26	30	33
0.21	16218	16255	16293	16331	16368	16406	16444	16482	16520	16558	4	8	11	15	19	23	26	30	34
0.22	16596	16634	16672	16711	16749	16788	16827	16866	16904	16943	4	8	12	15	19	23	27	31	35
0.23	16982	17022	17061	17100	17140	17179	17219	17258	17298	17338	4	8	12	16	20	24	28	32	36
0.24	17378	17418	17458	17498	17539	17579	17620	17660	17701	17742	4	8	12	16	20	24	28	32	36
0.25	17783	17824	17865	17906	17947	17989	18030	18072	18113	18155	4	8	12	17	21	25	29	33	37
0.26	18197	18239	18281	18323	18365	18408	18450	18493	18535	18578	4	8	13	17	21	25	30	34	38
0.27	18621	18664	18707	18750	18793	18836	18880	18923	18967	19011	4	9	13	17	22	26	30	35	39
0.28	19055	19099	19143	19187	19231	19275	19320	19364	19409	19454	4	9	13	18	22	26	31	35	40
0.29	19498	19543	19588	19634	19679	19724	19770	19815	19861	19907	5	9	14	18	23	27	32	36	41
0.30	19953	19999	20045	20091	20137	20184	20230	20277	20324	20370	5	9	14	19	23	28	32	37	42
0.31	20417	20464	20512	20559	20606	20654	20701	20749	20797	20845	5	10	14	19	24	29	33	38	43
0.32	20893	20941	20989	21038	21086	21135	21184	21232	21281	21330	5	10	15	19	24	29	34	39	44
0.33	21380	21429	21478	21528	21577	21627	21677	21727	21777	21827	5	10	15	20	25	30	35	40	45
0.34	21878	21928	21979	22029	22080	22131	22182	22233	22284	22336	5	10	15	20	25	31	36	41	46
0.35	22387	22439	22491	22542	22594	22646	22699	22751	22803	22856	5	10	16	21	26	31	37	42	47
0.36	22909	22961	23014	23067	23121	23174	23227	23281	23336	23388	5	11	16	21	27	32	37	43	48
0.37	23442	23496	23550	23605	23659	23714	23768	23823	23878	23933	5	11	16	22	27	33	38	44	49
0.38	23988	24044	24099	24155	24210	24266	24322	24378	24434	24491	6	11	17	22	28	34	39	45	50
0.39	24547	24604	24660	24717	24774	24831	24889	24946	25003	25061	6	11	17	23	29	34	40	46	51
0.40	25119	25177	25236	25293	25351	25410	25468	25527	25586	25645	6	12	18	23	29	35	41	47	53
0.41	25704	25763	25823	25882	25942	26002	26062	26122	26182	26242	6	12	18	24	30	36	42	48	54
0.42	26303	26363	26424	26485	26546	26607	26669	26730	26792	26853	6	12	18	24	31	37	43	49	55
0.43	26915	26977	27040	27102	27164	27227	27290	27353	27416	27479	6	13	19	25	31	38	44	50	56
0.44	27542	27606	27669	27733	27797	27861	27925	27990	28054	28119	6	13	19	26	32	39	45	51	58
0.45	28184	28249	28314	28379	28445	28510	28576	28642	28708	28774	7	13	20	26	33	39	46	52	59
0.46	28840	28907	28973	29040	29107	29174	29242	29309	29376	29444	7	13	20	27	34	40	47	54	60
0.47	29512	29580	29648	29717	29785	29854	29923	29992	30061	30130	7	14	21	28	34	41	48	55	62
0.48	30200	30269	30339	30409	30479	30549	30620	30690	30761	30832	7	14	21	28	35	42	49	56	63
0.49	30903	30974	31046	31117	31189	31261	31333	31405	31477	31550	7	14	22	29	36	43	50	58	65

(Contd.)

Antilogarithms Tables (Contd.)

	0	1	2	3	4	5	6	7	8	9	1	2	3	4	5	6	7	8	9
															Mean Differences				
0.50	31623	31696	31769	31842	31916	31989	32063	32137	32211	32285	7	15	22	29	37	44	52	59	66
0.51	32359	32434	32509	32584	32659	32735	32809	32885	32961	33037	8	15	23	30	38	45	53	60	68
0.52	33113	33189	33266	33343	33420	33497	33574	33651	33729	33806	8	15	23	31	39	46	54	62	69
0.53	33884	33963	34041	34119	34198	34277	34356	34435	34514	34594	8	16	24	32	40	47	55	63	71
0.54	34674	34754	34834	34914	34995	35075	35156	35237	35318	35400	8	16	24	32	40	48	56	65	73
0.55	35481	35563	35645	35727	35810	35892	35975	36058	36141	36224	8	16	25	33	41	50	58	66	74
0.56	36308	36392	36475	36559	36644	36728	36813	36898	36983	37068	8	17	25	34	42	51	59	68	76
0.57	37154	37239	37325	37411	37497	37584	37670	37757	37844	37931	9	17	26	35	43	52	61	69	78
0.58	38019	38107	38194	38282	38371	38459	38548	38637	38726	38815	9	18	27	35	44	53	62	71	80
0.59	38905	38994	39084	39174	39264	39355	39446	39537	39628	39719	9	18	27	36	45	54	63	72	82
0.60	39811	39902	39994	40087	40179	40272	40365	40458	40551	40644	9	19	28	37	46	56	65	74	83
0.61	40738	40832	40926	41020	41115	41210	41305	41400	41495	41591	9	19	28	38	47	57	66	76	85
0.62	41687	41783	41879	41976	42073	42170	42267	42364	42462	42560	10	19	29	39	49	58	68	78	87
0.63	42658	42756	42855	42954	43053	43152	43251	43351	43451	43551	10	20	30	40	50	60	70	80	89
0.64	43652	43752	43853	43954	44055	44157	44259	44361	44463	44566	10	20	30	41	51	61	71	81	91
0.65	44668	44771	44875	44978	45082	45186	45290	45394	45499	45604	10	21	31	42	52	62	73	83	94
0.66	45709	45814	45920	46026	46132	46238	46345	46452	46559	46666	11	21	32	43	53	64	75	85	96
0.67	46774	46881	46989	47098	47206	47315	47424	47534	47643	47753	11	22	33	44	54	65	76	87	98
0.68	47863	47973	48084	48195	48306	48417	48529	48641	48753	48865	11	22	33	45	56	67	78	89	100
0.69	48978	49091	49204	49317	49431	49545	49659	49774	49888	50003	11	23	34	46	57	68	80	91	103
0.70	50119	50234	50350	50466	50582	50699	50816	50933	51050	51168	12	23	35	47	58	70	82	93	105
0.71	51286	51404	51523	51642	51761	51880	52000	52119	52240	52360	12	24	36	48	60	72	84	96	108
0.72	52481	52602	52723	52845	52966	53088	53211	53333	53456	53580	12	24	37	49	61	73	85	98	110
0.73	53703	53827	53951	54075	54200	54325	54450	54576	54702	54828	13	25	38	50	63	75	88	100	113
0.74	54954	55081	55208	55336	55463	55590	55719	55847	55976	56105	13	26	38	51	64	77	90	102	115
0.75	56234	56364	56494	56624	56754	56885	57016	57148	57280	57412	13	26	39	52	66	79	92	105	118
0.76	57544	57677	57810	57943	58076	58210	58345	58479	58614	58749	13	27	40	54	67	80	94	107	121
0.77	58884	59020	59156	59293	59429	59566	59704	59841	59979	60117	14	27	41	55	69	82	96	110	123
0.78	60256	60395	60534	60674	60814	60954	61094	61235	61376	61518	14	28	42	56	70	84	98	112	126
0.79	61659	61802	61944	62087	62230	62373	62517	62661	62806	62951	14	29	43	58	72	86	101	115	130
0.80	63096	63241	63387	63533	63680	63826	63973	64121	64269	64417	15	29	44	59	74	88	103	118	132
0.81	64565	64714	64863	65013	65163	65313	65464	65615	65766	65917	15	30	45	60	75	90	105	120	135
0.82	66069	66222	66374	66527	66681	66834	66988	67143	67298	67453	15	31	46	62	77	92	108	123	139
0.83	67608	67764	67920	68077	68234	68391	68549	68707	68865	69024	16	32	47	63	79	95	110	126	142
0.84	69183	69343	69503	69663	69823	69984	70146	70307	70469	70632	16	32	48	64	81	97	113	129	145
0.85	70795	70958	71121	71285	71450	71614	71779	71945	72111	72277	17	33	50	66	83	99	116	132	149
0.86	72444	72611	72778	72946	73114	73282	73451	73621	73790	73961	17	34	51	68	85	101	118	135	152
0.87	74131	74302	74473	74645	74817	74989	75162	75336	75509	75683	17	35	52	69	87	104	121	138	156
0.88	75858	76033	76208	76384	76560	76736	76913	77090	77268	77446	18	35	53	71	89	107	125	142	159
0.89	77625	77804	77983	78163	78343	78524	78705	78886	79068	79250	18	36	54	72	91	109	127	145	163
0.90	79433	79616	79799	79983	80168	80353	80538	80724	80910	81096	19	37	56	74	93	111	130	148	167
0.91	81283	81470	81658	81846	82035	82224	82414	82604	82794	82985	19	38	57	76	95	113	132	151	170
0.92	83176	83368	83560	83753	83946	84140	84333	84528	84723	84918	19	39	58	78	97	116	136	155	175
0.93	85114	85310	85507	85704	85901	86099	86298	86497	86696	86896	20	40	60	79	99	119	139	158	178
0.94	87096	87297	87498	87700	87902	88105	88308	88512	88716	88920	20	41	61	81	102	122	142	162	183
0.95	89125	89331	89536	89743	89950	90157	90365	90573	90782	90991	21	42	62	83	104	125	146	166	187
0.96	91201	91411	91622	91833	92045	92257	92470	92683	92897	93111	21	42	64	85	106	127	149	170	191
0.97	93325	93541	93756	93972	94189	94406	94624	94842	95060	95280	22	43	65	87	109	130	152	174	195
0.98	95499	95719	95940	96161	96383	96605	96828	97051	97275	97499	22	44	67	89	111	133	155	178	200
0.99	97724	97949	98175	98401	98628	98855	99083	99312	99541	99770	23	46	68	91	114	137	160	182	205

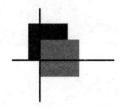

INDEX

Absolute viscosity, 204
Absorbance, 268
Absorption spectroscopy, 267
Absorption spectrum, 262, 275
Acid–base catalysis, 93
Acid-catalyzed reaction, 94
Acid value, 208
Acrylonitrile butadiene styrene (ABS) plastic, 248
Additives, 193, 195
Adsorption theory, 90
    modern, 90
Alkaline hardness, 110
Alkalinity, 305
    bicarbonate, 116
    carbonates, 116
    hydroxide, 116
    types, 116
    water, 115
Allotropy, 55
Amino acids-based biopolymers, 238
Analytical chemistry, 252
Aniline point, 207
Animal oil, 194
Anodic inhibiters, 185
Anodic protection, 184
Atomic spectra, 262
Atomic spectroscopy, 267
Auto catalyst, 88
Auxochrome, 265

Bagging, 120
Bakelite, 232
Barrier protection, 180
Base-catalyzed reaction, 94
Bathochromic groups, 266
Bathochromic shift, 266
Beer's law, 270
Biochemical catalysts, 96
Biodegradation mechanism, 201, 240
Biodegradable lubricants, 200
    advantages of, 202
    disadvantages, 202
    types, 201
Biodegradable polymerization, 240
Biodegradable polymers, 239, 244
    applications, 241
Biodegradation, 200, 240
Biopolymerization, 237
Biopolymers, 237, 238
    cellulose, 238
Blank determination, 321
Bleaching powder, 136
Blended oils, 194
Blue shift, 274
Boiler corrosion, 122
Boiler feed water, 145
Boiler safety, 120
Born–Oppenheimer approximation, 280
Boundary line lubrication, 192

Buna-N, 236
Buna-S, 235

Calcium carbonate equivalent, 111
Calgon conditioning, 121
Calibration curve, 323
Carbohydrates, 237
Carbon dioxide system, 53
Carbon residue, 210
Carbonate conditioning, 121
Carnot cycle, 8
Catalysis, 82, 85
    classification, 85
    theories, 89
Catalyst, 82, 83
    activity, 83
    biochemical, 96
    general characteristics, 84
    induced, 88
    role, 82
    selectivity, 92
    solid, 92
Catalytic poisons, 88
Catalytic promoter or activator, 87
Cathodic inhibitors, 185
Cathodic protection, 183
Caustic embrittlement, 122, 171
Cell constant, 286
Cell potential, 162
Charge-transfer
    absorption, 274
    complexes, 274
Chemical corrosion, 165
Chemical potential, 23
    physical significance, 24
    variation, 24
Chemical reactivity, 229
Chlorination, 136
Chromogen, 265
Chromophore, 265
Clausius–Clapeyron equation, 29
Cloud and pour points, 206
Coagulants, 139
Cold lime–soda process, 139
Colloidal conditioning, 121
Concentration cell, 163
    corrosion, 172

Condensation polymerization, 222
Condensed phase rule equation, 57
Condensed system, 57
Conditioning, 122
    with EDTA, 122
    with sodium aluminate, 122
Conductance, 286
Conductivity meter, 326
Conductometeric titration, 287, 326
    types, 288
Conductometric analysis, 285
Congruent melting point, 60, 62
Congruent system, 60
Cooling curves, 72, 311
Cooling effect on curves, 66
Coordination polymerization, 222, 226
Coordination polymers, 222
Copolymers, 220
Corrosion, 160
    boiler, 122
    control, 179
    differential aeration, 172
    factors affecting, 175
    galvanic, 169
    theory, 167
Crystallinity, 229
Crystallite, 229

Daniel cell, 161
De-chlorination, 138
Degree of freedom, 46
Degree of hardness, 111
Degree of polymerization, 218, 227
Desalination of brackish water, 145
Differential scanning calorimetry (DSC), 258
    curve, 260
    thermobalance, 258
Differential thermal analysis, 256
    curve, 257
Disinfection, 135
    chloramines, 138
    ozone, 138
Dissolved oxygen (DO), 308
Doppler broadening, 264
Double beam spectrophotometer, 282
Drop point, 213
Dry, 165
    ice, 55

Index ♦ 351

EDTA titration method, 112, 301
Elasticity, 229
Elastomers, 221, 235
Electrochemical, 161, 164, 167
    series, 164
Electrodialysis, 146
Electromagnetic radiations, 260
Electron paramagnetic resonance (EPR) spectra, 267
Electronic energy levels, 276
Electronic spectra, 266
Electronic spectroscopy, 279
Electronic transition, 273, 274
Electroplating/electrodeposition, 181
Elico flame photometer, 331
EMF, 162
Emission spectroscopy, 268
Emission spectrum, 262
Emulsions, 198
    types, 198
Enantiotropy, 56
Enthalpy, 7
Entropy, 8
    definition, 10
    physical significance, 16
Entropy change, 11, 14
Entropy of mixing, 16
Entropy of mixture of ideal gases, 15
Enzyme, 96, 100
    catalysis, 96, 101
    characteristics, 100
Enzyme catalysis, 96, 101
    effect of temperature, 101
Enzyme-catalyzed reactions, 96
Equilibrium constant, 83
    theories, 89
Equivalent conductivity, 286
Eriochrome Black-T, 113, 301
ESR, 267
Eutectic mixture, 57, 312
Eutectic point, 57, 311
Eutectic system, 57, 311
External treatment, 121

Fibre reinforced plastics, 247
Fibres, 221
Filtration, 134

First law of thermodynamics, 6
Flame emission spectroscopy, 283
Flamephotometer, 283, 330
Flash and fire point, 205
Flocculants, 122
Fluid film, 192
Foaming, 123
Forbidden transitions, 264
Formalin, 317
Frank–Condon principle states, 280
Free energy, 18
    variation, 20
Free-radical polymerization, 224
Friction, 191

Galvanic cell, 161
Galvanic corrosion, 169
Galvanic series, 176, 177
Galvanization, 182
Gibbs phase rule, 43
Gibbs–Helmholtz equation, 26
Graphite, 197
Gravimetric methods, 252
Grease, 195, 196, 244

Halloysite, 257
Hard water, 114
    disadvantages of domestic use, 114
Hardness, 301
    degree, 111,
    determination, 112
    units, 112
    water, 110
Harmonic oscillator model, 277
Heat, 5
Heterogeneous catalysis, 86
Homogeneous catalysis, 85
Homo-polymers, 219
Hooke's law equation, 278
Hot lime–soda process, 140
Hydrodynamic lubrication, 192
Hydroxide alkalinity, 116
    hyper-filtration, 147
Hypsochromic groups, 266
Hypsochromic shift, 266

Ice skating, 53
Impressed current cathodic protection, 183
Incongruent melting compounds, 64
Incongruent systems, 64
Inhibiters, 185
Intermediate compound formation theory, 89
Internal energy, 5
Internal treatment (sequestration), 121
Intrinsic water, 145
Iodine number, 209
Iodine value, 209
Ion exchange or deionization or demineralization method, 142
Ionic polymerization, 225
IR spectra, 276

Key-lock model, 98
Kinetics
    acid-base catalyzed reactions, 94
    enzyme-catalyzed reaction, 97
Koettsdoerfer number, 208

Lambert's law, 268
Lambert–Bouguer's law, 268
Lead–silver (Pb–Ag) system, 57
Lifetime broadening, 264
Lime-soda method, 138
Limestone, 335
Liquid metal corrosion, 166
Lubricant, 191, 313
    functions, 191
    liquid, 194
    semisolid, 195
    solid, 197
    synthetic, 199
Lubricating oils, 194
Lubrication, 191, 193

Mechanical stability, 211
Mechanism of biodegradation, 201, 240
Melting point, 311
Meritectic temperature, 64
Metal cladding, 182
Metastable curve, 51
Metastable equilibria, 51, 70
Michaelis–Menten equation, 98

Microbiological corrosion, 175
Micro-organism, 135
Mineral oils, 194
Minimum weight loss temperature, 255
Mixed bed deionization, 144
Molar absorptivity or molar extinction coefficient or molar absorbance index, 268
Molar conductivity, 286
Molecular energy levels, 263
Molecular spectra, 262
Molecular spectroscopy, 267
Molecular vibrations, 277
Molybdenum disulphide, 197
Mossbauer spectra, 267
Multiplication factor, 111

Natural rubber, 235
NBR, 236
Negative catalyst, 87
Neutralization number, 208
NMR spectra, 267
Non-alkaline hardness, 110
Nucleic acids, 238

Oiliness, 206
Oils, 194
    mineral or petroleum, 194
One-component system, 50
Optical methods, 253

Partial molar properties, 21
Passivation, 175
Passivity, 175
Pattinson's process, 60
Peak area, 257
Pensky–Marten's apparatus, 205, 316
Peritectic temperature, 64
Permanent or non-carbonate hardness, 110
Permutit method, 141
Petroleum oils, 194
pH dependence of rate constant, 102
Phase, 44
    derivation, 46
    diagram, 49
    gibbs, 43
    rule, 43
    transition, 14

Phase rule equation, 43
Phosphate conditioning, 121
Pilling Bedworth's rule, 166
Pitting corrosion, 170
Polished water, 145
Polycarbonate (PC), 248
Polydimethylphenylene, 248
Polymer alloys, 248
Polymer blends, 248
Polymer composite, 247
Polymerization, 218
    addition or chain, 122
Polymers, 218, 220
    addition, 220
    biodegradable, 239, 241, 244
    branched chain, 219
    condensation, 220
    cross-linked, 219
    inorganic, 247
    molecular mass, 246
    physical state, 228
    strength, 226
    toughness, 226
    types, 219
Polymorphism, 55
Positive catalyst, 87
Potable water, 133
Priming, 123
Proteins, 228
Proton-catalyzed reaction, 93
PVA, 231
PVC, 230

Quantum mechanics, 281

Radioactive conditioning, 122
Raman scattering, 266
Raman spectra, 266
Rapid pressure filter, 135
Rapid-gravity filter, 134
Rate of corrosion, 175
    factors affecting, 175
Red shift, 274
Redox titration, 333
Reduced phase rule equation, 57
Redwood viscometer, 204
Reflux, 320
Reverse osmosis, 147

Rotational energy levels, 276
Rotational spectra, 266
Rusting of iron, 167

Sacrificial anodic protection, 183
Salt hydrate system, 72
Saponification value, 208, 320
SBR, 235
Scales, 119, 120
    removal, 120
Screening, 134
Season cracking, 171
Second law of thermodynamics, 7
Sedimentation, 134
    with coagulation, 134
Selection rule, 263
Separation methods, 253
Shape-selective catalysts, 93
Silicones, 242
    fluid, 244
    gums/greases, 244
    resins, 245
    rubbers, 245
Simple harmonic oscillator, 277, 278
Single beam spectrophotometer, 282
Sliding/flow of glacier, 53
Slow sand filters, 134
Sludge, 118
Sodium aluminate, 122
Sodium sulphate–water system, 67
Sodium–potassium (Na–K) system, 64
Soil corrosion, 174
Specific conductance, 286
Specific gravity, 209
Specific resistance, 286
Spectrometer, 274
Spectrophotometer, 281, 323
Spectrophotometry, 281
Spectroscopy, 260
Spontaneous processes, 8
Starch biopolymers, 238
State functions, 2
Stereo-specific polymerization, 226
Stress corrosion, 171
Sugar-based biopolymers, 237
Super-chlorination, 137
Super-cooled water, 51
Surroundings, 2

Synthetic rubber, 235
Synthetic-based biopolymers, 239
System, 2

Teflon, 232
Temporary or carbonate hardness, 110
Thermodynamic equilibrium, 3
Thermodynamics, 1
    processes, 3
Thermogramme, 255
Thermogravimetric analysis, 253
Thermoplastics, 221
Thermosets or thermosetting polymers, 221
Thick film, 192
Thin film, 192
Third law of thermodynamics, 31
Tinning, 182
Titrations, 253
    types, 253
Total dissolved solids (TDS), 322
Total iron, 333
Transition, 264
    point, 69
    reaction, 64
    rule, 263
    temperature, 56, 64
Transmittance, 268
Triple point, 52, 54
Two-component systems, 57

Unavailable energy, 17
Urea formaldehyde (UF) resin, 234, 318
UV–visible spectroscopy, 272

Vapour phase inhibiters, 185
Vegetable oil, 194

Vibrational frequency, 279
Vibrational spectra, 276
Vibrational-rotational spectra, 266
Viscosity, 202, 313
Viscosity index (VI), 202
Volatility, 211
Volumetric methods, 253
Vulcanization, 235

Water, 115
    alkalinity of, 115
    hardness, 301
    impurities, 109
    polished, 145
    softening, 118
    source, 108
    sterilization, 135
    treatment, 133
    system, 50
Water-line corrosion, 173
Wave number, 261
Wavelength, 261
Wear, 191
Wet corrosion, 167
Work, 3
    function, 18
Wrinker's method, 308

Yield value, 212

Zeolite method, 141
Zero-point energy, 278, 279
Zeroth law of thermodynamics, 31
Ziegler–Nata catalyst, 226
Zinc–magnesium (Zn–Mg) system, 60